# Mouse Genetics

## Concepts and Applications

Lee M. Silver

D0075373

*New York   Oxford*
OXFORD UNIVERSITY PRESS
1995

Oxford University Press

Oxford   New York   Toronto
Delhi   Bombay   Calcutta   Madras   Karachi
Kuala Lumpur   Singapore   Hong Kong   Tokyo
Nairobi   Dar es Salaam   Cape Town
Melbourne   Auckland   Madrid

and associated companies in
Berlin   Ibadan

Published by Oxford University Press, Inc.
198 Madison Avenue, New York, New York 10016-4314

Oxford is a registered trademark of Oxford University Press

Library of Congress Cataloging-in-Publication Data
Silver, Lee M.
Mouse genetics: concepts and applications/Lee M. Silver.
p.   cm.   Includes bibliographical references and index.
ISBN 0-19-507554-4
1. Mice—Genetics.   2. Mice—Breeding.   3. Mice as laboratory
animals.   I. Title.
[DNLM: 1. Mice—genetics.   2. Gene Expression—genetics.   3. Mice,
Transgenic—genetics.   4. Mutation—genetics
QH 470.M52 S587m 1995]
QH432.S56   1995
599.32′33—dc20
94–34127

9 8 7 6 5 4 3 2

Printed in the United States of America
on acid-free paper

To Susan
in celebration of a decade of love


To Rebecca, Ari and Max
for explaining the meaning of life

# Contents

# Preface

My aim in writing this book has been two-fold. First, to provide students, in the broadest sense of the term, with a comprehensive introduction to the mouse as a model system for genetic analysis. Second, to provide practicing scientists with a detailed guide for performing breeding studies and interpreting experimental results. The impetus to write this book arose through a decade of formal and informal teaching at Princeton University. I became increasingly frustrated during this time with the lack of a current book on mouse genetics that could be provided to new students and postdoctoral fellows. I also found myself spending increasing amounts of my own time answering mouse questions not only from Princeton students but colleagues as well. Although some may consider it to be an extreme response, I wrote this book to answer all of the questions that I have been asked in the past and those that I can imagine being asked in the future.

In consideration of the broad range of students and scientists that may find a book of this type useful, I have made few assumptions about background knowledge. I have attempted to develop each topic completely from the kinds of first principles that are taught in a contemporary introduction to biology course for undergraduates. In particular, critical concepts in genetics and molecular biology have been fully explained.

It is my hope that this book will serve the interests of three different types of readers. First is the advanced undergraduate or graduate student who may be taking a course or entering a mouse laboratory for the first time. This reader will find a complete description of the laboratory mouse, the molecular tools used for its analysis, and the procedures used for carrying out genetic studies. Second is the established molecular biologist who intends to incorporate the mouse into his or her future studies. This reader will be able to skip over molecular topics, and focus on background material and chapters devoted to genetics. Finally, there is the practicing mouse geneticist. My inclusion of graphs, charts, and statistical formulations for the interpretation of genetic data is meant to provide this reader with a single toolbox for day-to-day use in data analysis. I should like to highlight the fact that, although this book is directed toward the genetics of the mouse, many of the genetic and molecular areas covered in the second half of the book will apply equally well to the genetic analysis of other mammalian species.

It is important for the reader to know what not to expect here. The mouse is used in many areas of biological research today, but I have focused this book entirely on its genetics with a particular emphasis on the application of molecular techniques to traditional breeding studies. In particular, I have not covered one very hot area of research—developmental biology—partly because this field is so well covered in a number of other recently published books. These are listed in the appendix along with other classic and contemporary books on other aspects of mouse biology.

In 1977, I began my career in mouse genetics as a postdoctoral fellow with Karen Artzt in Dorothea Bennett's laboratory at the Sloan-Kettering Institute. Dorothea had a rule (handed down from her mentors, L. C. Dunn and Salome Waelsch) that each scientist in the lab, herself included, had to spend at least two mornings each week in the "mouse room" examining animals, recording litters, and setting up new matings. Mouse room time was meant to serve two purposes. The first was to maintain and track a large breeding colony with hundreds of genetic variants in various experimental crosses. The second purpose was to provide each student and postdoc passing through the lab with an intimate look at the creature that is the mouse. Although my initial choice of the mouse as an experimental system was based on the many purely rational reasons presented in the first chapter of this book, my experience in the Bennett mouse room radically changed my outlook on science. As Dorothea intended, I acquired what the brilliant corn geneticist Barbara McClintock referred to as "a feeling for the organism" (Keller, 1983). Seventeen years later, the mouse—with its fascinating habits, amazing variation, elaborate social structures, and rich history—continues to amuse me even as it provides a tool for nearly all of my scientific investigations. I always find myself in special company when talking to other self-described "mousers," many of whom continue to work in their own mouse rooms long after their colleagues have retired to their offices.

I have presented this one aspect of my personal history as an explanation for the scattered sections throughout this book where I provide details about the mouse that may seem extraneous to some readers. My intent, and my hope, is that in some small way, these details will provide readers with some points of departure on their own paths to a feeling for the organism that is the mouse. Of course, it should go without saying that such a feeling can only be acquired through a "hands-on" approach to the animal itself. The student who successfully follows such a path will develop an intangible, yet powerful, tool for enlightened interpretation of experimental data.

I am indebted to those who guided me in the development of my own feeling for the mouse including Karen Artzt, Dorothea Bennett, Mary Lyon, Salome Waelsch, Jan Klein, and Vicki Sato. I am also indebted to other teachers who helped me mature as a scientist including Sandy Schwartz, Sally Elgin, and Joe Sambrook, and a long-lost college friend named Barry Gertz who explained the wonders of modern biology to an ignorant physics major, and in so doing, caused that physics major to alter the course of his career in a direction that led ultimately to the publication of this book. There are many other colleagues and students who have taught me as much as I have taught them. I would like to especially thank those who provided critical commentary on particular chapters in this book or insight into particular topics that I covered. These include Jiri Forejt, Kristen Ardlie, François Bonhomme, Bill Paven, Muriel Davisson,

Tom Vogt, Shirley Tilghman, Ken Manly, Eva Eicher, Joe Nadeau, Jean-Louis Guénet, Ken Paigen, members of the Princeton Play Group, and a cyberspace pen pal (whose voice I have never heard) named Karen Rader. I am extremely delighted that both Earl and Margaret Green found the time to read the entire manuscript—from start to finish—with insightful comments and corrections of various errors and misconceptions. In the end, though, any mistakes that remain are of my own making. Finally, this book would never have been written without the wholehearted support of my family. I want to thank my parents for their support and encouragement throughout my life, and my wife and children for their patience and understanding over the three years that I spent  hovering in front of a computer screen; now Becka and Ari, its your turn to play.

Lee M. Silver
*Princeton, New Jersey*
*June 1994*

# Mouse Genetics

# An Introduction to Mice

## 1.1 OF MICE, MEN, AND A WOMAN

### 1.1.1 The origin of the house mouse

What is a mouse? Ask any small child and you will hear the answer—a small furry creature with big ears and a big smile. Today, in Japan, Europe, America, and elsewhere, this question evokes images of that quintessential mouse icon— Mickey; but even before the age of cinema, television, and theme parks, mice had entered the cultures of most people. In the English-speaking world, they have come down through the ages in the form of nursery rhymes sung to young children including "Three blind mice" and "Hickory, dickory dock, the mouse ran up the clock. . . ." Artistic renditions of mice in the form of trinkets, such as those shown in Fig. 1.1, are sold in shops throughout the world. Why has the

**Figure 1.1** Some examples of mouse trinkets. The materials used to make these mouse figurines vary from plastic, wood, felt, and fur to wax. The images presented vary from anthropomorphic to realistic to abstract.

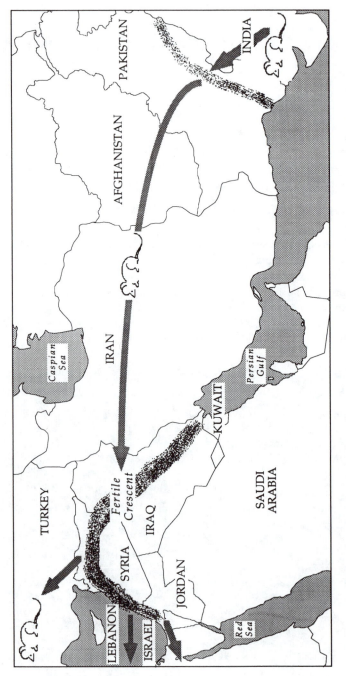

**Figure 1.2** The origin of the house mouse in the context of early human civilizations. This map of the Near East and Southwest Asia illustrates the movement of ancestors of the contemporary house mouse out of the Indian subcontinent into the steppes of Pakistan and from there into the fertile crescent.

mouse been in the minds of people for so long? The most obvious reason is that one particular type of mouse—the so-called house mouse—has lived in close association with most, if not all, human populations since the dawn of civilization.

This dawn occurred at the end of the last ice age, some 10,000 years ago, across a region retrospectively called the Fertile Crescent that extends from modern-day Israel up through Lebanon and Syria and curves back down through Iraq toward the Persian Gulf (Fig. 1.2). It was in this region at this time—known as the neolithic transition—that tribes of nomadic hunters and gatherers began to cultivate plants and domesticate animals as a means for sustenance (Ammerman and Cavalli-Sforza, 1984). Farming eliminated the need for constant migration and brought about the formation of villages and the construction of permanent shelters for both people and their livestock. With the seasonal planting of crops, families needed to store dry food, in the form of grain, for both themselves and their animals. With food reserves in granaries and cupboards, the house mouse began its long interwoven history with humankind.

The ancestors of the house mouse, who were concentrated in the steppes of present-day Pakistan at that time (Fig. 1.2), had been living happily oblivious to people for eons, but suddenly (in terms of evolutionary time), migrants to the new neolithic villages found mouse paradise in the form of a secure shelter with unlimited food (Auffray et al., 1990) . With their ability to squeeze through the tiniest of holes—adults can pass through apertures as small as a single centimeter in width (Rowe, 1981)—our furry friends were clearly pre-adapted to take advantage of these neolithic edifices, and with their agility and speed, they were able to stay one step ahead of the cleaver wielded by the farmer's wife. This pre-adaptation, and the opportunistic ability to eat almost anything, has allowed the house mouse to become the second most successful mammalian species living on earth today (Berry, 1981; Sage, 1981).[1]

When people wandered out from the Middle East in search of new lands to cultivate, mice followed as stowaways within the vehicles used to carry household belongings. Later, they would travel with ship-borne merchants going to and from distant lands. In this millennium, it is not too farfetched to imagine mice traveling on the *Santa Maria* with Columbus to the New World, and on horse-drawn buggies with families emigrating from the original American colonies to the western part of the continent. As people overcame harsh environments through the construction of artificial habitats, these became the natural environment for the house mouse. Freeloading on people has allowed the house mouse to enjoy an incredibly wide range. Today, house mice can be found wherever there are permanent populations of people (as well as many places where there are none), in both urban and rural areas, on all of the continents, at altitudes as high as 15,600 feet (4,750 m), as far north as the Bering Sea and as far south as sub-Antarctic islands (Berry and Peters, 1975; Sage, 1981).

### 1.1.2 Domestication and the fancy mouse

The fact that many "grown-up" humans and mice have had an adversarial relationship through most of history is evident in the derivation of the name that

English speakers use to describe these creatures. *Mouse* can be traced back through the Latin *mus* and the Greek *mys* to the ancient Sanskrit *mush*, meaning "to steal." There was little that adults in the ancient world could do to prevent mice from overrunning granaries until the discovery of the natural predilection of cats for rooting out and destroying small rodents. In fact, Keeler (1931) has suggested that the deification of the cat by the ancient Egyptians was due mostly to the role that it played in reducing house mouse populations. And an ancient Persian legend, from the millennium before Christ, says that "the moon chases the clouds as a cat chases mice" (Keeler, 1931) . In somewhat later times (AD 900), the Welsh fixed the price of cats based on their mouse-catching experience (Sage, 1981) . This image of the cat as a veritable biological pesticide is prevalent in many early cultures, and could explain the original rationale for its domestication.

Although mice and farmers may not have seen eye to eye, one can imagine the potential for a very different type of relationship between mice and people not directly affected by their dastardly deeds. This is because mice are often viewed in a very different light than other animals as best summed up in the words of a contemporary artist:

> The mouse is a great friend to artists, then, because we like him. He doesn't seem to have any specially bad characteristics—at worst, his life is a little drab, but we all suspect our lives of being just that. . . Not enough like us to unnerve us, he is a tiny creature (therefore clearly inferior) who looks up to us and fears us (therefore reassuring), who is not directly useful to us (therefore not a menial), and can be a pleasant furry companion without making extensive demands on us (therefore a true friend). No wonder artists appreciate the mouse; put him in a work and you win your human audience instantly. . . (Feingold, 1980)

The house mouse was highly visible to children growing up on farms as well as in towns, and legend has it that the tame animals wandering in and out of Walt Disney's original cartoon studio in Kansas provided the inspiration for the creation of Mickey Mouse (Updike, 1991). House mice can express a high level of interesting activity in a small amount of space when presented with various playthings. They can breed easily in captivity, their diets are simply satisfied, they can be housed in small spaces, and one can select artificially for increased docility in each generation. With continuous human contact from birth, mice acclimate to touch and can be handled quite readily.

Early instances of mouse domestication, and even worship, by the ancient Greeks and Romans is described in detail by Keeler (1931). From the classical period onward, the domesticated mouse has appeared in various Eurasian cultures. Of particular importance to the history of the laboratory mouse was the fondness held for unusual-looking mice by the Chinese and Japanese. This fondness led Asian breeders to select and develop a variety of mutant lines with strikingly different coat colors, some of which can be seen in detailed paintings from the 18th and 19th centuries. During the 19th century, the house mouse became "an object of fancy" in Europe as well (Sage, 1981), and British, Chinese, and Japanese traders brought animals back and forth to develop new breeds. By the beginning of the 20th century, European and American fanciers were familiar with lines of mice having fanciful names like white English sable, creamy buff, red cream, and ruby-eyed yellow (Sage, 1981).

A critical link between the mouse fanciers and early American mouse geneticists was Miss Abbie Lathrop, a retired school teacher who began, around 1900, to breed mice for sale as pets from her home in Granby, Massachusetts (Morse, 1978). Conveniently, Lathrop's home and farm were located near to the Bussey Institute directed by William Castle of Harvard University (see Section 1.2.2). Not only did Lathrop provide early mouse geneticists—including Castle and his colleagues at Harvard, and Leo Loeb at the University of Pennsylvania—with a constant source of different fancy mice for their experiments, but she conducted her own experimental program as well with as many as 11,000 animals breeding on her farm at any one time between 1910 and her death in 1918 (Morse, 1978). Many of the common inbred lines so important to mouse geneticists today—including C57BL/6 and C57BL/10 (commonly abbreviated as B6 and B10)—are derived entirely from animals provided by Abbie Lathrop. A more detailed account of her contributions along with photographs of her breeding records and her farm can be found in a historical review by Morse (1978).

## 1.2 THE ORIGIN OF MICE IN GENETIC RESEARCH

### 1.2.1 The mouse and Mendel

The mouse played a major role in early genetic studies begun immediately after the rediscovery of Mendel's laws in 1900. All of the initial findings were based on work carried out entirely with plants and there was much skepticism in the scientific community as to how general Mendel's laws would be (Dunn, 1965, p.86). Did the laws explain all aspects of inheritance from individuals? Were there some species groups, such as ourselves and other mammals, where the laws did not apply at all? In particular, the competing theory of *blending inheritance* was defended by Galton during the latter part of the 19th century. The main tenet of this theory was that a *blending* of the traits expressed by each of the parents occurred within each offspring. Blending inheritance and Mendelism have strikingly different predictions for the future descendants of a cross that brings a new "character" into a pure-bred race. According to the blending theory, the new character would remain in all of the descendants from the original "contaminating" cross: even upon sequential backcrosses to the pure-bred parental strain, the contaminating character would only slowly be *diluted out*. Of course, the Mendelian prediction is that a contaminating allele (to use current language) can be eliminated completely within a single generation.

The main support for blending inheritance came through a cursory observation of common forms of variation that exist in animal as well as human populations. It can certainly appear to be the case that human skin color and height do *blend* together and *dilute* from one generation to the next. However, skin color, height, and nearly all other common forms of natural variation are determined not by alternative alleles at a single loci, but instead by interactions of multiple genes, each having multiple alleles leading to what appear to be continua of phenotypes.[2] Mendel's leap in understanding occurred because he chose to ignore such complicated forms of inheritance and instead focused his efforts on traits that came in only two alternative "either/or" forms. Of equal importance was his decision to begin his crosses with pairs of inbred lines that differed by only a single trait, rather than many. It was only in this manner that Mendel was able to see through the noise of commonplace multifactorial traits to derive his principles of segregation, independent assortment, and dominant–recessive relationships between alleles at single loci.

How could one investigate the applicability of Mendel's laws to mammals with the use of natural variants alone? The answer was with great difficulty—not only does natural variation tend to be multifactorial, there is just not very much of it that is *visible* in wild animals, and without visible variation, there could be no formal genetics in 1900. The obvious alternative was to use a species in which numerous variants had been derived and were readily available within pure-breeding lines, and thus begun the marriage between the fancy mice and experimental genetics.

Evidence for the applicability of Mendel's laws to mammals—and by implication, to humans—came quickly, with a series of papers published by the French geneticist Cuénot on the inheritance of the various coat color phenotypes (Cuénot, 1902, 1903, 1905). Not only did these studies confirm the simple dominant and recessive inheritance patterns expected from "Mendelism," they also brought to light additional phenomena such as the existence of more than two alleles at a locus, recessive lethal alleles, and epistatic interactions among unlinked genes.

### 1.2.2 Castle, Little, and the founders of mouse genetics

The most significant force in early genetic work on the mouse was William Ernest Castle, who directed the Bussey Institute at Harvard University until his retirement in 1936 (Morse, 1985).[3] Castle brought the fancy mouse into his laboratory in 1902, and with his numerous students began a systematic analysis of inheritance and genetic variation in this species as well as in other mammals (Castle, 1903; Morse, 1978, 1981; Snell and Reed, 1993). The influence of Castle on the field of mammalian genetics as a whole was enormous. Over a period of 28 years, the Bussey Institute trained 49 students, including L.C. Dunn, Clarence Little, Sewall Wright, and George Snell; thirteen were elected to the National Academy of Sciences in the US (Morse, 1985), and many students of mouse genetics today can trace their scientific heritage back to Castle in one way or another.[4]

A major contribution of the Castle group, and Clarence Little in particular, was the realization of the need for, and development of, inbred genetically homogeneous lines of mice (discussed fully in Section 3.2). The first mating to produce an inbred line was begun by Little in 1909, and resulted in the DBA strain, so-called because it carries mutant alleles at three coat color loci—dilute (*d*), brown (*b*), and non-agouti (*a*). In 1918, Little accepted a position at the Cold Spring Harbor Laboratory, and with colleagues that followed—including Leonell

Strong, L. and E. C. MacDowell—developed the most famous early inbred lines including B6, B10, C3H, CBA, and BALB/c. Although an original rationale for their development was to demonstrate the genetic basis for various forms of cancer,[5] these inbred lines have played a crucial role in all areas of mouse genetics by allowing independent researchers to perform experiments on the same genetic material, which in turn allows results obtained in Japan to be compared directly with those obtained halfway around the world in Italy. A second, and equally important, contribution of Little to mouse genetics was the role that he played in founding the Jackson Laboratory in Bar Harbor, Maine, and acting as its first director (Russell, 1978). The laboratory was inaugurated in 1929—as "the natural heir to the Bussey" (Snell and Reed, 1993)—with eight researchers and numerous boxes of the original inbred strains.

### 1.2.3 The mouse as a model prior to the recombinant DNA revolution

With the demonstration in the mouse of genetic factors that impact upon cancer, millions upon millions of animals were used to elucidate the roles of these factors in more detail. However, for the most part, these biomedical researchers did not breed their own animals. Rather, they bought ready-made, *off-the-shelf*, specialized strains from suppliers like Taconic Farms, Charles River Laboratories, and the Jackson Laboratory (addresses of these and other suppliers are provided in Appendix A). The strong focus of mouse research in the direction of cancer can be seen clearly in the table of contents from the first edition of the landmark book *Biology of the Laboratory Mouse* published in 1941: five of 13 chapters are devoted to cancer biology, with only two chapters devoted to other aspects of genetic analysis (Snell, 1941).

Until the last decade, the community of geneticists that actually performed its own in-house breeding studies on the mouse was rather small. For the most part, individual mouse geneticists worked in isolation at various institutions around the world. Typically, each of these researchers focused on a single locus or well-defined experimental problem that was amenable to analysis within a small breeding colony. Members of the mouse community kept track of each other's comings and goings through a publication called *The Mouse Newsletter*. In its heyday during the 1960s, more than 60 institutions would routinely contribute "a note" to this effect. These contributed notes served the additional purpose of providing researchers with a means for announcing and reading about the various strains and mutations that were being bred around the world. A characteristic of the genetics community, during this period, was the openness with which researchers freely traded specialized mouse stocks—not available from suppliers—back and forth to each other.

Apart from this cottage industry style of conducting mouse genetics, there were three institutions where major commitments to the field had been made in terms of personnel and breeding facilities. These three institutions were the Oak Ridge National Laboratory in Oak Ridge, Tennessee, the Atomic Energy Research Establishment in Harwell, England, and the Jackson Laboratory (JAX) in Bar Harbor, Maine. The genetics programs at both Harwell and Oak Ridge were initiated at the end of the Second World War with the task of defining the effects of radiation on mice as a model for understanding the consequences of nuclear fallout on human beings. Luckily, researchers at both of these institutions—prominently including Bill and Lee Russell at Oak Ridge, and T. C. Carter, Mary Lyon, and Bruce Cattanach at Harwell—appreciated the incredible usefulness of the animals produced as byproducts of these large-scale mutagenesis studies in providing tools to investigate fundamental problems in mammalian genetics (see Section 6.1).

The third major center of mouse genetics—the Jackson Laboratory—has always had, and continues to maintain, a unique place in this field. It is the only non-profit institution ever set up with a dedication to basic research on the genetics of mammals as a primary objective. Although the JAX originally bred many different species (including dogs, rabbits, guinea pigs and others), it has evolved into an institution that is almost entirely directed toward the mouse. Genetic mapping and descriptions of newly uncovered mutations and variants have been a focus of research at the laboratory since its inception in 1929. But in addition to its own in-house research, the JAX serves the worldwide community of mouse geneticists in three other capacities. The first is in the maintenance and distribution of hundreds of special strains and mutant stocks. The second is as a central database resource. The third is in the realm of education in mouse genetics and related fields with various programs for non-scientists, and high school and college students, as well as summer courses and conferences for established investigators.

Even with three major centers of mouse research and the cottage industry described above, genetic investigations of the mouse were greatly overshadowed during the first 80 years of the 20th century by studies in other species, most prominently, the fruit fly, *Drosophila melanogaster*. The reasons for this are readily apparent. Individual flies are exceedingly small, they reproduce rapidly with large numbers of offspring, and they are highly amenable to mutagenesis studies. In comparison to the mouse, the fruit fly can be bred more quickly and more cheaply, both by many orders of magnitude. Until the 1970s, *Drosophila* provided the most tractable system for analysis of the genetic control of development and differentiation. In the 1970s, a competitor to *Drosophila* appeared in the form of the nematode, *Caenorhabditis elegans*, which is even more tractable to the genetic analysis of development as well as neurobiology. So why study the mouse at all?

The answer is that a significant portion of biological research is aimed at understanding ourselves as human beings. Although many features of human biology at the cell and molecular levels are shared across the spectrum of life on earth, our more advanced organismal-based characteristics are shared in a more limited fashion with other species. At one extreme are a small number of human characteristics—mostly concerned with brain function and behavior—that are shared by no other species or only by primates, but at a step below are a whole host of characteristics that are shared in common only with mammals. In this vein, the importance of mice in genetic studies was first recognized in the intertwined biomedical fields of immunology and cancer research, for which a mammalian model was essential. Although it has long been obvious that many other aspects of human biology and development should be amenable to mouse models, until recently, the tools just did not exist to allow for a genetic dissection of these systems.

## 1.3 THE NEW BIOLOGY AND THE MOUSE MODEL

### 1.3.1 All mammals have closely related genomes

The movement of mouse genetics from a backwater field of study to the forefront of modern biomedical research was catalyzed by the recombinant DNA revolution, which began 20 years ago and has been accelerating in pace ever since. With the ability to isolate cloned copies of genes and to compare DNA sequences from different organisms came the realization that mice and humans (as well as all other placental mammals) are even more similar genetically than they were thought to be previously. An astounding finding has been that nearly all human genes have counterparts in the mouse genome, which can almost always be recognized by cross-species hybridization. Thus, the cloning of a human gene usually leads directly to the cloning of a mouse homolog, which can be used for genetic, molecular, and biochemical studies that can then be extrapolated back to an understanding of the function of the human gene. In only a subset of cases are mammalian genes conserved within the genomes of *Drosophila* or *C. elegans*.

This result should not be surprising in light of current estimates for the time of divergence of mice, flies and nematodes from the evolutionary line leading to humans. In general, three types of information have been used to build phylogenetic trees for distantly related members of the animal kingdom— paleontological data based on radiodated fossil remains, sequence comparisons of highly conserved proteins, and direct comparisons of the most highly conserved genomic sequences, namely the ribosomal genes. The most parsimonious model is one in which flies (*Drosophila*) and nematodes (*C. elegans*) diverged apart from the line leading to mammals just prior to the time of the earliest fossil records in the pre-Cambrian period which occurred 570 million years ago. The divergence of mice and people occurred relatively recently at 60 million years before present (see Section 2.2.1). These numbers are presented graphically in Fig. 1.3, where a quick glance serves to drive home the fact that humans and mice are ten times more closely related to each other than either is to flies or nematodes.

Although the haploid chromosome number associated with different mammalian species varies tremendously, the haploid content of mammalian DNA remains constant at approximately three billion basepairs. It is not only the size of the genome that has remained constant among mammals; the underlying genomic organization (discussed in Chapter 5) has also remained the same as well. Large genomic segments—on average, 10–20 million basepairs—have been conserved virtually intact between mice, humans, and other mammals as well. In fact, the available data suggest that a rough replica of the human genome could be built by simply breaking the mouse genome into 130–170 pieces and pasting them back together again in a new order (Nadeau, 1984; Copeland et al.,

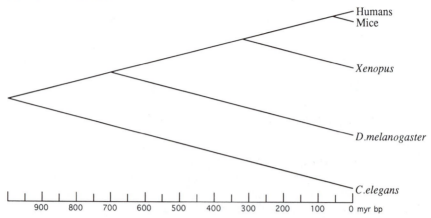

**Figure 1.3** Evolutionary relationships among commonly used model organisms. The approximate times of divergence of humans, mice, frogs (*Xenopus*), flies (*D. melanogaster*), and nematodes (*C. elegans*) from common ancestors is indicated along the time scale in millions of years before present.

1993). Although all mammals are remarkably similar in their overall body plan, there are some differences in the details of both development and metabolism, and occasionally these differences can prevent the extrapolation of mouse data to humans and vice versa (Erickson, 1989) . Nevertheless, the mouse has proven itself over and over again as being the model experimental animal *par excellence* for studies of nearly all aspects of human genetics.

### 1.3.2 The mouse is an ideal model organism

Among mammals, the mouse is ideally suited for genetic analysis. First, it is among the smallest mammals known with adult weights in the range of 25—40 g, 2,000–3,000-fold lighter than the average human adult. Second, it has a short generation time—in the order of 10 weeks from being born to giving birth. Third, females breed prolifically in the laboratory with an average of 5—10 pups per litter and an immediate postpartum estrus. Fourth, an often forgotten advantage is the fact that fathers do not harm their young, and thus breeding pairs can be maintained together after litters are born. Fifth, for developmental studies, the deposition of a vaginal plug allows an investigator to time all pregnancies without actually witnessing the act of copulation and, once again, without removing males from the breeding cage. Finally, most laboratory-bred strains are relatively docile and easy to handle.

The high resolution genetic studies to be discussed later in this book require the analysis of large numbers of offspring from each of the crosses under analysis. Thus, a critical quotient in choosing an organism can be expressed as the number of animals bred per square meter of animal facility space per year. For mice, this number can be as high as 3,000 pups/m$^2$ including the actual space for racks (five shelves high) as well as the inter-rack space as well. All of the reasons listed here make the mouse an excellent species for genetic analysis and have helped to make it the major model for the study of human disease and normative biology.

### 1.3.3 Manipulation of the mouse genome and micro-analysis

The close correspondence discovered between the genomes of mice and humans would not, in and of itself, have been sufficient to drive workers into mouse genetics without the simultaneous development, during the last decade, of increasingly more sophisticated tools to study and manipulate the embryonic genome. Today, genetic material from any source (natural, synthetic or a combination of the two) can be injected directly into the nuclei of fertilized eggs; two or more cleavage-stage embryos can be teased apart into component cells and put back together again in new "chimeric" combinations; nuclei can be switched back and forth among different embryonic cytoplasma; embryonic cells can be placed into tissue culture, where targeted manipulation of individual genes can be accomplished before these cells are returned to the embryo proper. Genetically altered live animals can be obtained subsequent to all of these procedures and these animals can transmit their altered genetic material to their offspring. The protocols involved in all of these manipulations of embryos and genomes have become well-established and *cookbook* manuals (Joyner, 1993; Wassarman and DePamphilis, 1993; Hogan et al., 1994) as well as a video guide to the protocols involved (Pedersen et al., 1993) have been published.

While it is likely that none of these manipulations has yet been applied to human embryos and genomes, it is ethical, rather than technical, roadblocks that impede progress in this direction. The mental image invoked is of a far more sophisticated technology than the so-called futuristic scenario of embryo farms described in Huxley's *Brave New World* (1932).

Progress has also been made at the level of molecular analysis within the developing embryo. With the polymerase chain reaction (PCR) protocol, DNA and RNA sequences from single cells can be characterized, and enhanced versions of the somewhat older techniques of *in situ* hybridization and immuno-staining allow investigators to follow the patterns of individual gene expression through the four dimensions of space and time (Wassarman and DePamphilis, 1993; Hogan et al., 1994). In addition, with the omnipresent micro-techniques developed across the field of biochemistry, the traditional requirement for large research animals like the rat, rabbit, or guinea pig has all but evaporated.

### 1.3.4 High resolution genetics

Finally, with the automation and simplification of molecular assays that has occurred over the last several years, it has become possible to determine chromosomal map positions to a very high degree of resolution. Genetic studies of this type are relying increasingly on extremely polymorphic microsatellite loci (Section 8.3) to produce anchored linkage maps (Chapter 9), and large insert cloning vectors, such as yeast artificial chromosomes (YACs), to move from the observation of a phenotype, to a map of the loci that cause the phenotype, to clones of the loci themselves (Section 10.3). Thus, many of the advantages that

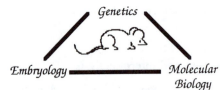

**Figure 1.4** The three fields that come together in modern studies of the mouse.

were once uniquely available to investigators studying lower organisms, such as flies and worms, can now be applied to the mouse through the three-way marriage of genetics, molecular biology, and embryology represented in Fig. 1.4. It is the intention of this book to provide the conceptual framework and practical basis for the new mouse genetics.

# 2

# Town Mouse, Country Mouse

## 2.1 WHAT ARE MICE?

To most people, all small rodents are virtually indistinguishable from each other, and as such, they are lumped together and considered to be mice of one kind or another. In *Webster's Third New International Dictionary*, one finds the following definition for a mouse: "any of numerous small rodents typically resembling diminutive rats with pointed snout, rather small ears, elongated body and slender hairless or sparsely haired tail, including all the small members of the genus *Mus* and many members of other rodent genera and families having little more in common than their relatively small size".[6] In fact, the order *Rodentia* (in the kingdom Animalia, phylum Chordata, and subphylum Vertebrata) is very old and highly differentiated with 28 separate families, numerous genera, and over 1,500 well-defined species accounting for 40% of all mammalian species known to be in existence today (Corbet and Hill, 1991). All families, subfamilies and genera in this order that contain animals commonly referred to as mice are listed in Table 2.1. The family Muridae encompasses over 1,000 species by itself including mice, rats, voles, gerbils, and hamsters. Within this family is the subfamily Murinae, which contains over 300 species of Old World mice and rats, and within this subfamily is the genus *Mus*. The *Mus* genus has been divided into four subgenera, of which one is also called *Mus*. This subgenus contains all of the "true Old World mice" including the house mouse *M. musculus*, the main focus of this book. A humorous view of mouse evolution is reproduced in Fig. 2.1, and a more serious phylogenetic tree with all extant members of the *Mus* subgenus is presented in Fig. 2.2.

There is still a great deal of confusion in the field of rodent systematics, and the proper classification of species into and among genera is now undergoing serious revision with the results of new molecular analyses. Just recently, it was suggested that the guinea pig is not a rodent at all, contrary to long-held beliefs (Graur et al., 1991). Also in other studies (based on DNA-DNA hybridization and quantitative immunological cross-reactivity), a series of African species known as "spiny mice" were found to be more closely related to gerbils than to true Old World mice (Wilson et al., 1987; Chevret et al., 1993).

**Table 2.1** Systematics of rodents that are commonly referred to as mice[a]

| Family name | Genus (and species) | Common name and (range) |
|---|---|---|
| Heteromyidae | *Perognathus* | Pocket mice (southwestern USA) |
| | *Microdipodops* | Kangaroo mice (Nevada, USA) |
| | *Liomys* | Spiny pocket mice (Central America) |
| | *Heteromys* | Forest spiny pocket mice (Central and South America) |
| Gliridae | *Glis, Muscardinus, Eliomys, Dryomys, Glirurus, Myomimus, Graphiurus* | Dormice (Africa, Europe, Asia Minor) |
| Seleviniidae | *Selevinia betpakdalensis* | Desert dormouse (Kazakhstan) |
| Zapodidae | *Zapus* | American jumping mice (western USA, Canada) |
| | *Sicista* | Birch mice (northern Eurasia) |
| | *Eozapus setchuanus* | Szechuan jumping mouse (western China) |
| | *Napaeozapus insignis* | Woodland jumping mouse (West USA, Canada) |
| Muridae | | |
| (Subfamily names) | | |
| Gerbillinae (muridae?) | *Acomys* | African spiny mice |
| Dendromurinae | 10 genera | African climbing mice |
| Hydomyinae | *Neohydromys, Pseudohydromys, Microhydromys, Mayermys* | Shrew mice (New Guinea) |
| Microtinae | *Microtus* ( > 40 species) | Voles (America, Eurasia) |
| Platacanthomyinae | *Platacanthomys, Typhlomys* | Dormice (India, China) |
| Hesperomyinae | *Akodon* | South American field mice |
| | *Andinomys edax* | Andean mouse (Bolivia) |
| | *Baiomys* | American pygmy mice (USA) |
| | *Blarinomys breviceps* | Brasilian shrew mouse |
| | *Chilomys instans* | Colombian forest mouse |
| | *Chinchillula sahamae* | Chinchilla mouse (Peru) |
| | *Eligmodontia* | Highland desert mice (Chile) |
| | *Neacomys* | Bristly mice (northern South America) |
| | *Neotomodon* | Volcano mouse (Mexico) |
| | *Neusticomys monticolus* | Fish-eating mouse (Ecuador) |
| | *Notiomys* | Mole mice (southern South America) |
| | *Ochrotomys nuttalli* | Golden mouse (Southeastern USA) |
| | *Onychomys* | Grasshopper mice (western North America) |
| | *Peromyscus* (58 species) | Deer mice, field mice (North and Central America) |
| | *Punomys lemminus* | Puna mouse (Peru) |
| | *Reithrodontomys* | American harvest mice |
| | *Rheomys* | Water mice (Central America) |
| | *Rhipidomys* | Climbing mice (South America) |
| | *Scolomys melanops* | Ecuador spiny mouse |
| | *Scotinomys* | Brown mice (Central America) |
| | *Wiedomys pyrrhorhinos* | Red-nosed mouse (eastern Brazil) |
| | *Zygodontomys* | Cane mice (Central and South America) |

16

**Table 2.1**   continued

| Family name | Genus (and species) | Common name and (range) |
| --- | --- | --- |
| Murinae | *Apodemus sylvaticus* (and others) | Wood mice, long-tailed field mice (Eurasia) |
| | *Chiropodomys* | Pencil-tailed tree mice (Southeast Asia) |
| | *Haeromys* | Pygmy tree mice (Borneo) |
| | *Micromys minutus* | Harvest mouse (Northern Eurasia) |
| | *Notomys* | Australian hopping mice |
| | *Pseudomys* | Australian mice (South Pacific) |
| | *Rattus norvegicus* | Norway or common rat |
| | *Vandeleuria* | Palm mouse (Southeast Asia) |
| | *Vernaya fulva* | Verney's climbing mouse (China) |
| | ***Mus*** | |
| | (subgenera in the genus *Mus*) | |
| | *Coelomys pahari, mayori* | Gairdner shrew mouse (Indochina, Sri Lanka) |
| | *Nannomys minutoides, setulosus* | African pygmy mice |
| | *Pyromys platythrix, shortridgei, saxicola* | Spiny mice (India, Southeast Asia) |
| | *Mus* | (See text and Fig. 2.2) |

[a] The information in this table is abstracted from Corbet and Hill (1991) with revisions suggested by Bonhomme (1986).

The major reason for the confusion is that classical systematics has always been dependent on taxonomy, and taxonomy has always been dependent on the demonstration of distinct morphological differences—measurable on a macroscopic scale—that can be used to distinguish different species. Unfortunately, many small rodent species have developed gross morphological characteristics that are convergent with those present in other relatively distant species. Thus, traditional taxonomy can fail to provide an accurate systematic description of mice. (An illustration of the close similarity of *Mus* species can be seen in Fig. 3.3.) Fortunately, the tools of molecular phylogenetics—and, in particular, DNA sequence comparisons—have proven highly effective at sorting out the evolutionary relationships that exist among different mouse groups. With continued molecular analysis, it may be possible to clear up all of the confusion that now exists in the field.

Excellent sources of information concerning the systematics and phylogeny of *Mus* and related species are in the proceedings from two conferences, *The Wild Mouse in Immunology* (Potter et al., 1986) and *The Fifth International Theriological Congress* (Berry and Corti, 1990) as well as a review by the Montpellier group (Boursot et al., 1993). A concise description of *Mus* systematics is provided by Bonhomme and Guénet (1989).

**Figure 2.1**   "Evolution of a mouse" by William Stout and Jim Steinmeyer. Pen and Ink, watercolor. From *The Art of Mickey Mouse* compiled by Craig Yoe and Janet Morra-Yoe. ©1991 The Walt Disney Company (Yoe and Morra-Yoe, 1991). Reprinted by permission of Hyperion.

## 2.2  WHERE DO MICE COME FROM?

### 2.2.1  Mice, people, and dinosaurs

The common ancestor to mice and humans was an inconspicuous rodent-like mammal that scurried along the surface of the earth some 65 million years (myr) before present (BP). It needed to be inconspicuous because the earth was

# The *Mus* species group

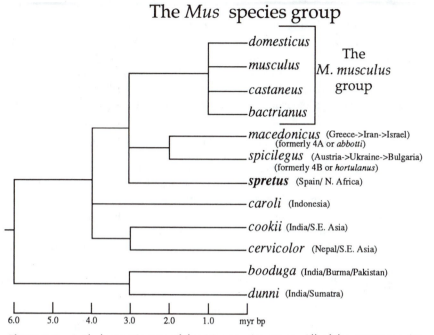

**Figure 2.2** A phylogenetic tree of the *Mus* species group. All of the extant species within the *Mus* subgenus of the *Mus* genus are shown. Table 2.1 lists the other subgenera in the *Mus* genus. *Domesticus, musculus, castaneus,* and *bactrianus* are all subspecies within the *M. musculus* group. All other lines represent well-defined species. The geographical distributions of *M. musculus* subspecies are illustrated in Fig. 2.3. The general distributions of all other species are indicated next to their names here. The indicated times of divergence between evolutionary lines in millions of years before present (myr BP) are consensus estimates based on data from various sources (Ferris et al., 1983b; Bonhomme, 1986; Berry and Corti, 1990; Boursot et al., 1993; Hammer and Silver, 1993).

ruled by enormous dinosaurs, many of whom would have eaten any small mammal that could be caught. The glorious age of the dinosaurs came to an abrupt end with the collision of one or a few large extraterrestrial objects— perhaps asteroids or comets—into the earth's surface over a relatively short period of time (Alvarez and Asaro, 1990; Sheehan et al., 1991). Possible sites at which these impacts may have occurred have been identified in the Yucatan peninsula of Mexico and the state of Iowa (Kerr, 1991, 1992, 1993). It has been hypothesized that the impact resulted in the formation of a thick cloud of dust that dispersed and shrouded the earth for a period of years, leading to a scenario like a nuclear winter with the demise of all green life, and with that, all large animals that depended either directly on plants for survival or indirectly on the animals that ate the plants. At least a small number of our rodent-like ancestors were presumably able to survive this long sunless winter as a consequence of their small size which allowed them to "get by" eating seeds alone. When the sun finally returned, the seeds lying dormant on the ground sprung to life and the world became an extremely fertile place. In the absence of competition from the dinosaurs, mammals were able to become the dominant large animal group, and they radiated out into numerous species that could take advantage of all the newly unoccupied ecological niches. It was in this context that the demise of the dinosaurs brought forth both humans and mice as well as most other mammalian species on earth today.

### 2.2.2 From Asia to Europe and from Europe to the New World

The Muridae family of rodents, which includes both "true" mice and rats, originated in the area across present-day India and Southeast Asia. Phylogenetic and paleontological data suggest that mice and rats diverged apart from a common ancestor 10–15 myr BP (Jaeger et al., 1986), and by 6 myr BP, the genus *Mus* was established. The *Mus* genus has since diverged into a variety of species (listed in Fig. 2.2) across the Indian subcontinent and neighboring lands.

At the beginning of the Neolithic transition some 10,000 years ago, the progenitors to the house mouse (collectively known as *Mus musculus*, as discussed later in this chapter) had already undergone divergence into four separate populations that must have occupied non-overlapping ranges in and around the Indian subcontinent. Present speculation is that the *domesticus* group was focused along the steppes of present-day Pakistan to the west of India (Auffray et al., 1990); the *musculus* group may have been in Northern India (Horiuchi et al., 1992; Boursot et al., 1993); the *castaneus* group was in the area of Bangladesh, and the founder population—*bactrianus*—remained in India proper.

The house mouse could only begin its commensal association with humans after agricultural communities had formed. Once this leap in civilization had occurred, mice from the *domesticus* group in Pakistan spread into the villages and farms of the fertile crescent as illustrated in Fig. 1.2 (Auffray et al., 1990); mice from the *musculus* group may have spread to a second center of civilization in China (Horiuchi et al., 1992); and finally, *bactrianus* and *castaneus* animals went from the fields to nearby communities established in India and Southeast Asia, respectively.

Much later ( ~ 4,000 yrs BP), the *domesticus* and *musculus* forms of the house mouse made their way to Europe. The *domesticus* animals moved with migrating agriculturalists from the Middle East across Southwestern Europe (Sokal et al., 1991) and the development of sea transport hastened the sweep of both mice and people through the Mediterranean basin and North Africa. Invasion of Europe by *musculus* animals occurred by a separate route from the East. Chinese voyagers brought these mice along in their carts and wagons, and they migrated along with their hosts across Russia and further west to present-day Germany where their spread was stopped by the boundary of the *domesticus* range (Fig. 2.3). Finally, it is only within the last millennium that mice have spread to all inhabited parts of the world including sub-Saharan Africa, the Americas, Australia, and the many islands in-between.

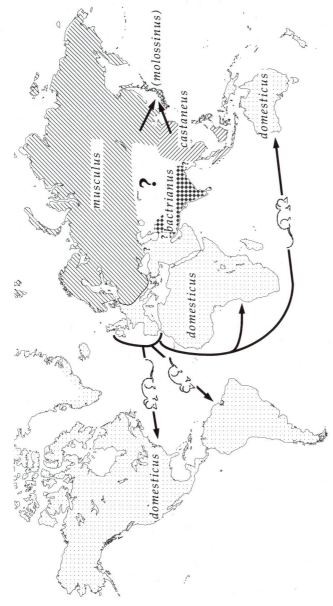

**Figure 2.3** Geographical ranges of subspecies in the *M. musculus* group. The range of each subspecies is indicated with a different pattern. The presumed paths of dispersion of the *M. m. domesticus* group during the last 600 years are indicated with mouse icons and arrows. The movement of the two Asian subspecies onto the Japanese islands (also indicated with arrows) to form the hybrid *faux*-species *M. m. molossinus* is likely to have occurred much earlier. The information presented here is based on figures shown in several publications authored by François Bonhomme and his colleagues (Bonhomme, 1986; Bonhomme and Guénet, 1989).

### 2.2.3 Tracing the movement of humankind with mice as markers

One interesting sidelight of the stowaway tendency of mice is that it is sometimes possible to observe the origin of human populations within the context of the mice that have come along with them. A clear example of this concordance is seen in the *domesticus* mice that have colonized all of North America, South America, Australia, and sub-Saharan Africa in conjunction with their Western European human partners (Fig. 2.3). A more complex example is observed in the Japanese islands where the native mice were long thought to be a separate subspecies or species group referred to in the literature as *Mus molossinus*. In fact, molecular phylogenetic studies have demonstrated that Japanese mice do not represent a distinct evolutionary line at all.

Instead, they appear to have been derived by hybridization of two other house mouse groups on the mainland nearby—*musculus* in China and *castaneus* in Southeast Asia (Yonekawa et al., 1988, Fig. 2.3). The hybrid character of the mice parallels the hybrid origin of the Japanese people themselves.

Finally, there is the interesting observation of a pocket of mice from the *castaneus* group that has recently been uncovered in Southern California (Gardner et al., 1991). This is the only documented example of an established natural house mouse population in the Americas that is *not* derived from the Western European *domesticus* group. This finding is a testament to the strong wave of 20th century Asian migration to the West Coast of the United States.

## 2.3 SYSTEMATICS OF THE *MUS* SPECIES GROUP, THE HOUSE MOUSE, AND THE CLASSICAL INBRED STRAINS

### 2.3.1 Commensal, feral, and aboriginal animals

Animals that are members of the genus *Mus* can be further classified according to their relationship to humankind. The house mouse represents one group within this genus that is characterized by its ability to live in close association with people. Animals dependent on human shelter and/or activity for their survival are referred to as *commensal* animals.[7] As discussed later in this chapter, all commensal mice appear to be members of a single species—*Mus musculus*— that can be subdivided into four distinct subspecies groups with different geographical ranges.

Although the success of *M. musculus* throughout the world is dependent on its status as a commensal species, in some regions with appropriate environmental conditions, animals have reverted back to a non-commensal state, severing their dependence on humankind. Such mice are referred to as *feral*. The return to the wild can occur most readily with a mild climate, sufficient vegetation or other food source, and weak competition from other species. Feral mice have successfully colonized small islands off Great Britain and in the South Atlantic (Berry et al., 1987), and in Australia, *M. musculus* has replaced some indigenous species. Although feral populations exist in North America and Europe as well, here they seem to be at a disadvantage relative to other small indigenous rodents such as *Apodemus* (field mice in Europe), *Peromyscus* (American deer mice), and *Microtus* (American voles). In some geographical areas, individual house mice will switch back and forth from a feral to a commensal state according to the season —in mid-latitude temperate zones, human shelters are much more essential in the winter than in the summertime.

None of the remaining species in the genus *Mus* (indicated in Fig. 2.2) have the ability to live commensally. These animals are not, and their ancestors never have been, dependent on humans for survival. Such animals are referred to as *aboriginal*.

## 2.3.2 Systematics of the house mouse

Although the average person cannot distinguish a field mouse from a house mouse, taxonomists have gone in the opposite direction describing numerous *types* of house mouse species. In the book *The Genetics of the Mouse* published in 1943, Grünberg wrote "The taxonomy of the *musculus* group of mice is in urgent need of revision. About fifty names of reputed 'good species', sub-species, local varieties and synonyms occur in the literature, all of which refer to members of this group". (Grüneberg, 1943). *M. brevirostris, M. poschiavinus, M. praetextus1, and M. wagneri* are among the 114 species names for various house mice present in the literature by 1981 (Marshall, 1981). One reason for this early confusion was the high level of variation in coat color and tail length that exists among house mice from different geographical regions. In particular, the belly can vary in color from nearly white to dark gray (Sage, 1981). A second reason for more recent taxonomic subdivisions was the discovery of a large variation in chromosome number among different European populations (discussed in Chapter 5). These differences and others led traditional taxonomists to conclude the existence of numerous house mouse species.

Over the last decade, the power of molecular biology has been combined with a more detailed investigation of breeding complementarity to sort out the true systematics of the house mouse group [see review by Boursot et al. (1993)]. Much of the credit for this comprehensive analysis goes to two groups of researchers—one at Berkeley including Sage, Wilson, and their colleagues (Sage, 1981; Ferris et al., 1983b) , and the second in Montpellier including Thaler, Bonhomme, and their colleagues (Bonhomme et al., 1978a, 1978b, 1984; Britton and Thaler, 1978; Bonhomme and Guénet, 1989; Auffray et al., 1991). Moriwaki and his colleagues have also contributed to this analysis (Yonekawa et al., 1981, 1988). The accumulated data clearly demonstrate the existence of four primary forms of the house mouse—*domesticus, musculus, castaneus*, and *bactrianus* (Fig. 2.2). Two of these four groups—*domesticus* and *musculus*—are each relatively homogenous at the genetic level whereas the other two are not (Boursot et al., 1993). In particular, mice from the *bactrianus* group show a high level of genetic heterogeneity. The Montpellier team has interpreted these findings as strong supporting evidence for the hypothesis that the Indian subcontinent represents the ancestral home of all house mice and that *bactrianus* animals are descendants of this very old founder population. In contrast, the *musculus* and *domesticus* groups have more recent founders that derive from the ancestral *bactrianus* population (Boursot et al., 1993).[8]

Although the four groups can be distinguished morphologically and molecularly, and have different non-overlapping ranges around the world (Fig. 2.3), it is clear at the DNA level that individuals within all these groups are descendants of a common ancestor that lived between 800,000 and 1 myr ago. Individuals representing pure samples from each of the four groups can interbreed readily in the laboratory to produce fertile male and female offspring. The high level of morphologic and karyotypic variation that has been observed among house mice from different regions must be a consequence of rapid adaptation to aspects of the many varied environments in which the house mouse can survive and thrive. The previously identified "false species" *M. brevirostris, M. poschiavinus*, and *M. praetextus* are not distinguishable genetically and are all members of the *domesticus* group.

### 2.3.3 Hybrid zones and the species debate

Although mouse systematicists have reached a consensus on the structure of the *Mus musculus* group—with the existence of only four well-defined subgroups—there is still a question as to whether each of these subgroups represents a separate species, or whether each is simply a subspecies, or race, within a single all-encompassing house mouse species. The very fact that this question is not simply answered attests to the clash that exists between: (1) those who would define two populations as separate species only if they could not produce fully viable and fertile hybrid offspring, whether in a laboratory or natural setting: and (2) those who believe that species should be defined strictly in geographical and population terms, based on the existence of a natural barrier (of any kind) to gene flow between the two populations (Barton and Hewitt, 1989).

The first question to be asked is whether this is simply a semantic argument between investigators without any bearing on biology. At what point in the divergence of two populations from each other is the magic line crossed when they become distinct species? Obviously, the line must be fuzzy. Perhaps, the house mouse groups are simply in this fuzzy area at this moment in evolutionary time, so why argue about their classification? The answer is that an understanding of the evolution of the *Mus* group in particular, and the entire definition of species in general, is best served by pushing this debate as far as it will go, which is the purpose of what follows.

Each of the four primary house mouse groups occupies a distinct geographical range as shown in Fig. 2.3. Together, these ranges have expanded out to cover nearly the entire land mass on the globe. In theory, it might be possible to solve the species versus subspecies debate by examining the interactions that occur between different house mouse groups whose ranges have bumped up against each other. If all house mice were members of the same species, barriers to interbreeding might not exist, and as such, one might expect boundaries between ranges to be extremely diffuse with broad gradients of mixed genotypes. This would be the prediction of laboratory observations, where members of both sexes from each house mouse group can interbreed readily with individuals from all other groups to produce viable and fertile offspring of both sexes that *appear* to be just as fit in all respects as offspring derived from matings within a group.

However, just because productive interbreeding occurs in the laboratory does not mean that it will occur in the wild where selective processes act in full force. It could be argued that two populations should be defined as separate species if the offspring that result from interbreeding are less fit *in the real world* than offspring obtained through matings within either group. It is known that subtle effects on fitness can have dramatic effects in nature and yet go totally unrecognized in captivity. If this were the case with hybrids formed between different house mouse groups, the dynamics of interactions between different populations would be quite different from the melting-pot prediction described above. In particular, since interspecific crosses would be "non-productive," genotypes from the two populations would remain distinct. Nevertheless, if the two populations favored different ecological niches, their ranges could actually overlap even as each group (species) maintained its genetic identity; such species are considered to be *sympatric*. Examples of sympatric species within the context of the broader *Mus* genus are described in Section 2.3.5.

Species that have only recently become distinct from each other would be more likely to demand the same ecological niches. In this case, ranges would not overlap, since all of the niches in each range would already be occupied by the species members that were there first. Instead, the barrier to gene flow would result in the formation of a distinct boundary between the two ranges. Boundary regions of this type are called hybrid zones because along these narrow geographical lines, members of each population can interact and mate to form viable hybrids, even though gene flow across the entire width of the hybrid zone is generally blocked (Barton and Hewitt, 1989).

The best-characterized house mouse hybrid zone runs through the center of Europe and separates the *domesticus* group to the West from the *musculus* group to the East (Fig. 2.3). If, as the one-species protagonists claim, *musculus* and *domesticus* mice simply arrived in Europe and spread toward the center by different routes—*domesticus* from the southwest and *musculus* from the east—then upon meeting in the middle, the expectation would be that they would readily mix together. This should lead to a hybrid zone which broadens with time until eventually it disappears. In its place initially, one would expect a continuous gradient of the characteristics present in the original two groups.

In contrast to this expectation, the European hybrid zone does not appear to be widening. Rather, it appears to be stably maintained at a width of less than 20 kilometers (Sage et al., 1986). Since hybridization between the two groups of mice does occur in this zone, what prevents the spreading of most genes beyond it? The answer seems to be that hybrid animals in this zone are less fit than those with pure genotypes on either side. One manner in which this reduced fitness is expressed is through the inability of the hybrids to protect themselves against intestinal parasites. Sage (1986) has shown through direct studies of captured animals that hybrid zone mice with mixed genotypes carry a much larger parasitic load, in the form of intestinal worms. This finding has been independently confirmed (Moulia et al., 1991). Superficially, these "wormy mice" do not appear to be less healthy than normal; however, one can easily imagine a negative effect on reproductive fitness through a reduced life span and other changes in overall vitality.

Nevertheless, for a subset of genes and gene complexes, the hybrid zone does *not* act as a barrier to transmission across group lines. In particular, there is evidence for the flow of mitochondrial genes from *domesticus* animals in Germany to *musculus* animals in Scandinavia (Ferris et al., 1983a; Gyllensten and Wilson, 1987) with the reverse flow observed in Bulgaria and Greece (Boursot et al., 1984; Vanlerberghe et al., 1988; Bonhomme and Guénet, 1989). An even more dramatic example of gene flow can be seen with a variant form of chromosome 17—called a *t* haplotype—that has passed freely across the complete ranges of all four house mouse groups (Silver et al., 1987; Hammer et al., 1991).

In contrast to the stable hybrid zone in Europe, other boundaries between different house mouse ranges are much more diffuse. The extreme form of this situation is the complete mixing of two house mouse groups—*castaneus* and *musculus*—that has taken place on the Japanese islands (Yonekawa et al., 1988; Fig. 2.3). So thorough has this mixing been that the hybrid group obtained was considered to be a separate group unto itself—with the name *Mus molossinus*—until DNA analysis showed otherwise.

In the end, there is no clear solution to the one species versus multiple species debate and it comes down to a matter of taste. However, the consensus has been aptly summarized by Bonhomme: "None of the four main units is completely genetically isolated from the other three, none is able to live sympatrically with any other. In those locations where they meet, there is evidence of exchange ranging from differential introgression . . . to a complete blending. It is therefore necessary to keep all these taxonomical units, whose evolutionary fate is unpredictable, within a species framework" (Bonhomme and Guénet, 1989). Thus, in line with this consensus, I will describe the four house mouse groups by their subspecies names *M. m. musculus, M. m. domesticus, M. m. castaneus*, and *M. m. bactrianus*. I will use *M. musculus* as a generic term in general discussions of house mice, where the specific subspecies is unimportant or unknown.

### 2.3.4 Origin of the classical inbred strains

As presented in Chapter 1, the original inbred strains were derived almost exclusively from the fancy mice purchased by geneticists from pet mouse breeders like Abbie Lathrop and others at the beginning of the 20th century. Mouse geneticists have always been aware of the multi-faceted derivation of the fancy mice from native animals captured in Japan, China, and Europe. Thus, it is not surprising that none of the original inbred strains are truly representative of any one house mouse group, but rather each is a mosaic of *M. m. domesticus, M. m. musculus, M. m. castaneus*, and perhaps *M. m. bactrianus* as well (Bonhomme et al., 1987). Nevertheless, the accumulated data suggest that the most prominent component of this mosaic is *M. m. domesticus*.

In early comparative DNA studies carried out with the use of restriction enzymes, the classical inbred lines were analyzed to determine the derivation of two particular genomic components—the mitochondrial chromosome and the Y chromosome. The findings were surprising. First, all of the classical inbred strains were found to carry mitochondria derived exclusively from *domesticus* (Yonekawa et al., 1980; Ferris et al., 1982). Even more surprising was the fact that the mitochondrial genomes present in all of the inbred strains were identical, implying a common descent along the maternal line back to a female who could have lived as recently as 1920.

The Y chromosome results also showed a limited ancestry but, in contrast to the mitochondrial results, the great majority of the classical inbred strains have a common paternal-line ancestor that came from *musculus* (Bishop et al., 1985; Tucker et al., 1992). Again, a large number of what are thought to be independent inbred strains (including B6, BALB/c, LP, LT, SEA, 129, and others) carry indistinguishable Y chromosomes (Tucker et al., 1992). Ferris and colleagues (1982) suggest that, contrary to the published records, early interstrain contaminations may have been responsible for a much closer relationship among many of the inbred lines than had been previously assumed. It was, in fact, the absence of sufficient inter-strain variation that served as the impetus to use more novel approaches to linkage analysis in the mouse such as the interspecific crosses described in the next section and in more detail in Chapter 9. Atchley and Fitch (1991) have constructed a phylogenetic tree that shows the relative overall genetic relatedness among 24 common inbred strains.

For many biological studies, use of the classical inbred strains is perfectly acceptable even though they are not actually representative of any race found in nature. However, in some cases, especially in studies that impact on aspects of evolution or population biology, it obviously does make a difference to use animals with genomes representative of naturally occurring populations. It is only in the last decade that a major effort has been devoted to the generation of new inbred lines directly from wild mice certified to represent particular *M. musculus* subgroups. It is now possible to purchase inbred lines representative of *M. m. domesticus*, *M. m. musculus*, and *M. m. castaneus* (as well as the *M. m. molossinus* hybrid race) from the Jackson Laboratory. Many other inbred lines have been derived from mice captured in particular localities and a list of investigators that maintain such lines has been published (Potter et al., 1986).

### 2.3.5 Close relatives of *Mus musculus* and interpopulation hybrids

A phylogenetic tree showing the relationships that exist among close relatives of the house mouse *M. musculus* is presented in Fig. 2.2. All *Mus* species have the same basic karyotype of 40 acrocentric chromosomes.[9] The three closest known relatives of *Mus musculus* are aboriginal species with restricted ranges within and near Europe. All three species—*M. spretus*, *M. spicilegus*, and *M. macedonicus*—are sympatric with *M. musculus* but interspecific hybrids are not produced in nature. Thus, there is a complete barrier to gene flow between the house mice and each of these aboriginal species. The ability of two animal populations to live sympatrically—with overlapping ranges—in the absence of gene flow is the clearest indication that the two populations represent different species. Nevertheless, in the forced, confined environment of a laboratory cage, Bonhomme and colleagues were able to demonstrate the production of interspecific $F_1$ hybrids between each of these aboriginal species and *M. musculus* (Bonhomme et al., 1978b, 1984).

The best characterized of the aboriginal species is *Mus spretus*, a western mediterranean short-tailed mouse with a range across the most southwestern portion of France, through most of Spain and Portugal, and across the North African coast above the Sahara in Morocco, Algeria, and Tunisia (Bonhomme and Guénet, 1989). *M. spretus* is sympatric with the *M. m. domesticus* group across its entire range. In 1978, Bonhomme and his colleagues reported the landmark finding that *M. spretus* males and laboratory strain females could be bred to produce viable offspring of both sexes. Although all male hybrids are sterile, the female hybrid is fully fertile and can be backcrossed to either *M. musculus* or *M. spretus* males to obtain fully viable second-generation offspring.[10]

In a series of subsequent papers, Bonhomme and colleagues demonstrated the power of the interspecific cross for performing multilocus linkage analysis with molecular and biochemical makers (Bonhomme et al., 1979, 1982; Avner et al., 1988; Guenet et al., 1990). With the large evolutionary distance that separates the two parental species, it is possible to readily find alternative DNA and biochemical alleles at nearly every locus in the genome. This finding stands in stark contrast to the high level of non-polymorphism observed at the majority of loci examined within the classical inbred lines. The significance of the interspecific cross for mouse genetics cannot be understated: it was the single most important factor in the development of a whole genome linkage map based on molecular markers during the last half of the 1980s. A detailed discussion of the actual protocols involved in such a linkage analysis will be presented in Chapter 9.

Two other well-defined aboriginal species have non-overlapping ranges in Eastern Europe. *Mus spicilegus* (previously known as *M. hortulanus* or species 4B) is commonly referred to as the mound-building mouse. Its range is restricted to the steppe grassland regions north and west of the Black Sea in current-day Bulgaria, Romania, and Ukraine (Bonhomme et al., 1978, 1983; Sage, 1981). *Mus macedonicus* (previously known as *M. abbotti, M. spretoides*, or species 4A) is restricted in range to the eastern Mediterranean across Greece and Turkey; this very short-tailed species is the Eastern European equivalent of *M. spretus* in terms of ecological niches. *M. spicilegus* and *M. macedonicus* are an interesting pair of species in that they are barely distinguishable from each other morphologically, and yet they fail to interbreed in the wild, and successful attempts at interbreeding in the laboratory have yet to be published.[11] Nevertheless, males from both species can be bred with *M. musculus* to give an outcome identical to that obtained with the *M. spretus-M. musculus* cross—both male and female hybrid offspring are fully viable, however, only the females are fertile (Bonhomme et al., 1984).

Presumably, both of these interspecific hybrid types could be used for linkage analysis in the same manner as that described above. However, in general, these crosses would not provide any obvious advantage over the *spretus-musculus* cross. The one exception to this statement would be in chromosomal regions where *spretus* and *musculus* were distinguished by an inversion polymorphism that did not distinguish *musculus* from either *macedonicus* or *spiciligus*. The presence of an inversion polymorphism will prevent recombination in $F_1$ hybrids and can lead to false estimates of gene distances. In only one instance to date has such a polymorphism between *spretus* and *musculus* been demonstrated—in the proximal region of chromosome 17 (Hammer et al., 1989). This inversion can cause a suppression of recombination over a chromosomal region that extends far beyond the inverted region itself (Himmelbauer and Silver, 1993). In the case of this particular chromosomal region, *musculus* and *macedonicus* have been shown to share the same gene order leading to the occurrence of normal recombination in the *macedonicus-musculus* $F_1$ hybrid (Hammer et al., 1989).

The failure to find other inversion polymorphisms does not mean that they do not exist. Inversions can only be demonstrated formally by creating a linkage map for *M. spretus* by itself and comparing the gene order on this map to the gene order on a *M. musculus* map. This has not been done for any chromosome other than the seventeenth. Nevertheless, a recent comparison of linkage maps constructed from the *spretus-musculus* cross and an intersubspecific *domesticus-castaneus* cross points to several additional regions where inversion polymorphisms are implicated based on the observation of localized recombination suppression in the interspecific cross only (Copeland et al., 1993). Cryptic inversions could have serious consequences for those using linkage map distances as means for estimating the physical length of DNA that must be walked from a cloned marker to a gene of interest as discussed more fully in chapter 10.

Other more distant members of the genus *Mus* have evolved in and around India. These include *M. caroli, M. cooki, M. cervicolor, M. booduga,* and *M. dunni*. None of these species can produce interspecific hybrids with any representatives of the *M. musculus* complex under normal laboratory conditions. However, with artificial insemination, Chapman and colleagues were able to demonstrate fertilized embryos representing $F_1$ hybrids between *M. musculus* and *M. caroli, M. cervicolor,* or *M. dunni* (West et al., 1977). However, the embryos formed with *M. musculus,* and either *M. dunni* or *M. cervicolor* never gave rise to live-born animals, with most *cervicolor–musculus* hybrids failing to undergo even the first cleavage division, and most *dunni–musculus* hybrids failing at the blastocyst stage. Although most *caroli–musculus* embryos also died prenatally, a small number actually made it through to a live birth. These interspecific hybrids were all delivered by Caesarian section; they were usually small and only a few survived after fostering to nursing mothers. None were shown to be fertile, although the sample size was exceedingly small.

Inbred lines developed from a number of different *Mus* species, including *M. spretus* and *M. spicilegus* (*M. hortulanus*) are available for purchase from the Jackson Laboratory. In addition, outbred stocks representing most of the other Mus species are maintained by individual investigators [listed in Potter et al. (1986)].

## 2.4 LIFESTYLES AND ADAPTABILITY OF WILD HOUSE MICE

### 2.4.1 Shelter, food, and water

*Mus musculus* can live in an incredible variety of different habitats. Commensal animals can live in all types of human-made structures including houses, barns, other buildings, haystacks, ruins, and in coal mines, 1,800 feet below the ground. The possibilities are virtually unlimited—animals have been found in climates as different as frozen-food lockers and central heating ducts (Bronson, 1984). Feral animals can live in agricultural fields, meadows, and scrublands (Sage, 1981). They do not normally live in woodlands or forest, but even this is possible in some areas, such as islands, where natural predators do not exist (Berry and Jakobson, 1974; Berry et al., 1987). The survival of feral mice is often dependent on the production of nests and burrow systems, which act to ameliorate the prevailing air temperatures (Sage, 1981). Both sexes construct nests that can range from very simple to highly complex enclosed structures used for food storage as well as nesting. Feral animals can display a highly developed homing behavior and are capable of returning to their nests after long distance (250 m) displacement (Sage, 1981). Mice can eat almost anything—cereals, grass, seeds, roots and stems of various plants, adult insects, and even larvae (Rowe, 1981). Animals can also subsist with very little water, especially if their food is high in moisture content (Grüneberg, 1943). In many locations, the morning dew can probably provide much of the daily water requirement (Rowe, 1981). These traits provide the house mouse with great adaptability and have played an important role in its dispersion among many different habitats, both commensal and feral.

### 2.4.2 Population structures and reproduction

The paradigm population structure for animals living under commensal conditions is that of independent, relatively stable *demes*, or families. The classic deme will have a single dominant male who patrols a well-defined home range and sires most of the young; up to ten breeding female members of the deme will confine their own ranges to that of the single dominant male.

Different dominant males will have mutually exclusive territories. Males will tolerate their own offspring, but will kill offspring born to females that belong to other demes. In highly structured populations of this type, the level of interdemic migration is very low, even between nesting sites located within a few meters of each other (Sage, 1981).

In reality, the picture of demes is an idealized situation that may actually define the structure of some populations but not others, and at some points in time, but not others. In the presence of an ample food supply and in the absence of predators or competitors, populations appear to retain a higher degree of deme structure. However, demes can vary in size from two animals to at least 100; the amount of interdemic migration can vary between none and continuous; and the detailed structure of a population can change drastically in response to changes in the environment.

Under optimal environmental conditions with plenty of food and nesting material, commensal mice living inside temperature-controlled buildings can breed throughout the year (Rowe, 1981; Sage, 1981). In strictly feral populations in temperate climates, breeding activity tends to be seasonal, from spring to early autumn (Rowe, 1981). The average litter size has been found to vary from as few as three pups to as many as nine. Although mice from some laboratory lines can survive as long as 3 years, free-living wild animals are likely to die much earlier from disease, competition, or predators.

The usual structure of feral populations may be very different from that of commensal populations in that animals living outdoors appear to move over much larger distances, and deme structures appear to be much less stable (Berry and Jakobson, 1974). However, the ability and desire of mice to migrate over long distances is complex and highly variable. Many animals appear to live their entire lives in very small and well-defined home ranges (defined as the area in which an animal spends the vast majority of its time) of less than 10 m across. Others will move constantly over much larger distances, traveling kilometers daily, and some will migrate long distances between home ranges that are very small. All possible permutations are possible, and the distribution of animals in each class varies greatly among different populations. The lifestyle of the house mouse has been described aptly by Berry: "The house mouse is a weed: quick to exploit opportunity, and able to withstand local adversity. . . . A consequence of the repeated formation of new populations by small numbers of founders is that every population is likely to be unique" (Berry, 1981).

### 2.4.3 Adaptability and success

The incredible adaptability of *M. musculus* to new environments can be accounted for almost entirely by the enormous plasticity that exists in its behavioral traits (Bronson, 1984). In the case of nearly all other species, specific behaviors are highly defined by genomes, and adaptability to new environments can occur only slowly with changes in behavior as well as physiology and/or morphology driven by natural selection. In contrast, the house mice can disembark from ships in sub-antarctic islands or in equatorial Africa, and adapt immediately to survive and prosper. "To be introduced into a radically new environment is one thing; to be able to reproduce there and so to establish a new population is quite another. The planet-wide spread of the house mouse in both manmade and natural habitats suggests an extreme reproductive adaptability, probably the most extreme among the mammals" (Bronson, 1984). Only humans are as adaptable (some would say less so).

Thus, the defining characteristic of the species *M. musculus* is the decoupling of genetics and behavior. At some point during evolution, the ancestral house mouse population broke away from its previous behavioral constraints and, once this occurred, the success of the species was assured. With men and women as chauffeurs and guides, the global conquest of the house mouse began.

# 3

# Laboratory Mice

## 3.1 SOURCES OF LABORATORY MICE

One of the unique advantages of working with mice, rather than other experimental organisms, is the availability of standard strains such as C57BL/6 (abbreviated B6), BALB/c, and many others that are used in thousands of laboratories around the world each year. With the use of the same standard inbred strain, it is possible to eliminate genetic variability as a complicating factor in comparing results obtained from experiments performed in Japan, Canada, Germany, or any other country in the world. Furthermore, for the most part, results obtained in 1992 can be directly compared to results obtained in 1962 or any other year. But, where do these standard strains come from and how can one be sure that a mouse advertised as BALB/c is actually a BALB/c mouse?

When two animals have the same strain name, such as BALB/c, it means that they can both trace their lineage back through a series of brother–sister matings to the very same mating pair of inbred animals. The breeding protocol through which these original progenitors became inbred is discussed later in this chapter. However, the important point is that, unlike the world of computers, where there can be many independent imitation models of a standard such as the IBM PC, there is no such thing as an imitation BALB/c mouse. Two animals either have a common heritage or they do not. If not, they cannot share the same name. Thus, a strain name implies a history, and the histories of the traditional inbred strains are well documented (see Table 3.1).

A handful of US suppliers provide various strains of mice to researchers. Addresses and phone numbers for each are provided in the appendix; all will provide free catalogs upon request. The Jackson Laboratory (or the JAX as it is commonly abbreviated) maintains an extensive mouse breeding facility with a very large collection of commonly used (and not so commonly used) strains for sale to other scientists.[12] Their 1991 catalog lists hundreds of different inbred strains and substrains of many different types including all of the "standards" as well as newly developed strains and mice that carry various mutant alleles or chromosomal aberrations. Other US suppliers have a more limited selection, but the largest of these—Charles River Laboratory, Taconic Farms, and Harlan Sprague Dawley—may actually sell more mice than the JAX. Each of these three companies stocks a set of common inbred strains—including BALB/c, C57BL/6,

**Table 3.1** Some important inbred strains

| Strain | Color | Major use | Other characteristics | Origin | Generation[a] |
|---|---|---|---|---|---|
| 129/Sv-$Sl/+$ ,$c^+$ $p^+$ (129) | Agouti | Source of most ES cell lines and genetic material used for homologous recombination. Used for studies of embryology and reproduction. | Relatively high testicular teratoma incidence. Relatively small size. Resistant to radiation. | ~1930 (Dunn) | F79 (JAX) |
| BALB/c | Albino | Used in immunological studies and for the production of hybridomas. A new congenic strain BALB/cByJ-Rb(8.12)5Bnr available from JAX is most efficient for hybridoma production. | Docile females. Males of the J substrain only are extremely aggressive. Relatively poor breeders, but variation among sublines. Sensitive to radiation. | 1913 (Bagg) | F105 (CR) F180 (JAX) F195 (Taconic) |
| C57BL/6 (B6) | Black | Standard strain for genetic studies; common backcross partner for congenic construction and mapping panels. | Relatively long lived and hearty. Excellent breeders. Resistant to radiation. | 1921 (Little) | F160 (CR) F187 (JAX) F155 (Taconic) |
| C57BL/10 (B10) | Black | Commonly used in genetic studies performed outside of the US and for the construction of congenics at the $H$-$2$ complex. | Common ancestry with B6. | 1921 (Little) | F192 (JAX) |
| CAST/Ei | Agouti | Used in matings with traditional inbred strains to create $F_1$ hybrids with high levels of heterozygosity for linkage studies. Better intercross reproductive performance than M. spretus strains. | Derived from wild animals of the subspecies M. m. castaneus. Male $F_1$ hybrids formed with lab strains are fertile. | 1971 (Marshall to Eicher) | F53 (JAX) |
| C3H | Agouti | Used commonly in genetic studies. | High mammary tumor incidence. Large adults. | 1920 (Strong) | F167 (CR) F139 (JAX) F160 (Taconic) |
| DBA/2 | Dilute brown | Used in crosses with B6 to produce the best-characterized set of RI strains, and for the production of standard $F_1$ hybrid animals. | | 1909 (Little) | F164 (CR) F183 (JAX) F165 (Taconic) |
| FVB/N | Albino | Recently derived strain with special characteristics that are ideal for the production of transgenic mice. An agouti congenic is currently under construction. | Very large litters, easy to handle, large and prominent pronuclei in zygotes. | 1975 (NIH) | >F30 (CR) F57 (JAX) F40 (Taconic) |
| MOLF/Ei | Agouti | Used in matings with traditional inbred strains to create $F_1$ hybrids with high levels of heterozygosity for linkage studies. Better intercross reproductive performance than M. spretus strains. | Derived from wild animals of the faux subspecies M. m. molossinus (Section 2.3). | ~1975 (Potter to Eicher) | F50 (JAX) |
| SPRET/Ei | Agouti | Used in matings with traditional inbred strains to create highly heterozygous backcross panels for linkage analysis. | Derived from the species M. spretus. Small animals. Difficult to handle. | 1988 (Eicher) | F37 (JAX) |

[a] As of 1993.

C3H, DBA/2 and several others—as well as hybrids and non-inbred strains. Two other US companies—Hilltop Lab Animals and Life Sciences, Inc.—have more focused lists of strains with the latter devoted essentially to the sale of special athymic strains used in various immunological studies. All of these suppliers provide high-quality, disease-free animals that are monitored constantly for genetic purity.

Once a supplier has been chosen for a particular set of experiments, it is best to stay with that supplier for all future orders of mice. The reason for this is that, even though all suppliers propagate their stocks with constant brother–sister matings, and B6 mice sold by JAX, Charles River, or Taconic can all trace their pedigree back to a common pair of founder animals, it is still the case that each independently maintained line will slowly drift apart genetically from its ancestors and distant cousins. Most of the standard inbred strains sold by companies are derived from a genetic resource maintained by the National Institutes of Health (NIH). The NIH inbred lines have been maintained separately from corresponding Jackson Laboratory lines since at least the early 1950s. The B6 strain was only at generation F32 when this separation occurred; by the beginning of 1994, the JAX strain had reached generation F187 and NIH-derived strains sold by Taconic and Charles River had reached generations 155 and 160, respectively (Table 3.1). The small number of differences that may have accumulated over this period will usually not have an impact upon the particular genetic characteristics of importance to any particular experiment, but it is critical to be aware of this possibility. To foster this awareness, independently maintained inbred strains are given different "substrain" designations, which follow the standard name and provide an account of past history. For example, the full name for the standard B6 mouse sold by the Jackson Laboratory is C57BL/6J, where J is the symbol for JAX. The B6 mice sold by both Charles River and Taconic have substrain symbols with multiple parts including N as an indication of their NIH derivation and BR to indicate that they are maintained in barrier facilities. Each supplier has also incorporated a sub-symbol that uniquely identifies the animals that they sell—the full names are C57BL/6NCrlBR for B6 mice supplied by Charles River Laboratory and C57BL/6NTacfBR for B6 supplied by Taconic Farms.

## 3.2 MOUSE CROSSES AND STANDARD STRAINS

### 3.2.1 Outcrosses, backcrosses, intercrosses, and incrosses

A formal classification system has been developed to describe the various types of crosses that can be set up between mice having defined genetic relationships relative to each other at one or more loci. For the sake of simplicity in describing these crosses, I will arbitrarily use a single locus (the *A* locus) with two alleles (*A* and *a*) to represent the situation encountered for the whole genome. With a simple two-allele system, there are only four generalized classes of crosses that can be carried out: each of these is defined in Table 3.2 and described in more detail in the following discussion.

**Table 3.2** Categories of genetic crosses

| Designation | Types of matings | Offspring genotypes | Major uses |
|---|---|---|---|
| Backcross | (1) $A/a \times A/A$<br>(2) $A/a \times a/a$ | (1) $A/a, A/A$<br>(2) $A/a, a/a$ | Linkage analysis; production of congenic strains |
| Incross | (1) $A/A \times A/A$<br>(2) $a/a \times a/a$ | (1) $A/A$<br>(2) $a/a$ | Maintenance of an inbred strain |
| Intercross | $A/a \times A/a$ | $A/a, A/A, a/a$ | Linkage analysis |
| Outcross | (1) $A/A \times a/a$<br>(2) $a^1/a^2 \times a^3/a^4$ | (1) $A/a$<br>(2) $[a^1/a^3, a^1/a^4,$<br>$a^2/a^3, a^2/a^4]$ | Initial step in strain production and linkage analysis; production of $F_1$ hybrids |

At the start of most breeding experiments, there is usually an *outcross*, which is defined as a mating between two animals or strains considered unrelated to each other. In many experiments, the starting material for this outcross is two inbred strains. As described in the next section, all members of an inbred strain are, for all practical purposes, homozygous across their entire genome and genetically identical to each other. Thus, an outcross between two inbred strains can be symbolized as $A/A \times a/a$, and the offspring resulting from such a cross are called the *first filial* generation, symbolized by $F_1$. All $F_1$ animals that derive from an outcross between the same pair of inbred strains are identical to each other with a heterozygous genome symbolized as $A/a$. However, when either or both parents are *not* inbred, as indicated in the second more generalized outcross mating shown in Table 3.2, $F_1$ siblings will *not* be identical to each other.

An outcross between two inbred strains or between one inbred strain and a non-inbred animal that contains a genetic variant of interest is almost always the first breeding step performed in a linkage analysis. The $F_1$ animals obtained from this outcross can be used in two types of crosses commonly performed by mouse geneticists—*backcrosses and intercrosses*. A mating between a heterozygous $F_1$ animal (with an $A/a$ genotype) and one that is homozygous for either the $A$ or $a$ allele is called a *backcross*. This term is derived from the vision of an $F_1$ animal being mated "back" to one of its parents. In actuality, a backcross is usually accomplished by mating $F_1$ animals with other members of a parental strain rather than a parent itself. The two-generation outcross–backcross combination is one of the major breeding protocols used in linkage analysis as described in detail in Chapter 9. From Mendel's first law of segregation, we know that the offspring from a backcross to the $a/a$ parent will be distributed in roughly equal proportions between two genotypes at any single locus—approximately 50% will be heterozygous $A/a$, and approximately 50% will be homozygous $a/a$.

A mating set up between brothers and sisters from the $F_1$ generation, or between any other two animals that are identically *heterozygous* at a particular locus under investigation, is called an *intercross*. The two-generation outcross–intercross series was the classic breeding scheme used by Mendel in the formulation of his laws of heredity, and it is the second major breeding protocol used today for linkage analysis in mice. Again, according to Mendel's first law, the offspring from an intercross will be distributed among three genotypes at any single locus—50% will be heterozygous *A/a*, 25% will be homozygous *A/A*, and 25% will be homozygous *a/a*. The particular uses of each of the two major protocols for linkage analysis—outcross–backcross and outcross–intercross— are discussed in Chapter 9.

A mating between two members of the same inbred strain, or between any two animals having the same homozygous genotype is called an *incross*. The incross serves primarily as a means for maintaining strains of animals that are inbred or carry particular alleles of interest to the investigator. All offspring from an incross will have the same homozygous genotype that is identical to that present in both parents.

## 3.2.2 The generation of inbred strains

The offspring that result from a mating between two $F_1$ siblings are referred to as members of the "second filial generation" or $F_2$ animals, and a mating between two $F_2$ siblings will produce $F_3$ animals, and so on. An important point to remember is that the filial (F) generation designation is only valid in those cases where a protocol of brother–sister matings has been strictly adhered to at each generation subsequent to the initial outcross. Although all $F_1$ offspring generated from an outcross between the same pair of inbred strains will be identical to each other, this does not hold true in the $F_2$ generation, which results from an intercross where three different genotypes are possible at every locus. However, at each subsequent filial generation, genetic homogeneity among siblings is slowly recovered in a process referred to as *inbreeding*. Eventually, this process will lead to the production of *inbred* mice that are genetically homogeneous and homozygous at all loci. The *International Committee on Standardized Nomenclature for Mice* has ruled that a strain of mice can be considered "inbred" at generation $F_{20}$ (Committee on Standardized Genetic Nomenclature for Mice, 1989).[13]

The process of inbreeding becomes understandable when one realizes that at each generation beyond $F_1$, there is a finite probability that the two siblings chosen to produce the subsequent generation will be homozygous for the same allele at any particular locus in the genome. If, for example, the original outcross was set up between animals with genotypes *A/A* and *a/a* at the *A* locus, then at the $F_2$ generation, there would be animals with three genotypes—*A/A, A/a*, and *a/a* present at a ratio of 0.25:0.50:0.25. When two $F_2$ siblings are chosen randomly to become the parents for the next generation, there is a defined probability that these two animals will be identically homozygous at this locus as shown in Fig. 3.1. Since the genotypes of the two randomly chosen animals are independent events, one can derive the probability of both events occurring simultaneously by multiplying the individual probabilities together according to the "law of the product." Since the probability that one animal will be *A/A* is 0.25, the probability that both animals will be *A/A* is $0.25 \times 0.25 = 0.0625$ (Fig. 3.1). Similarly, the probability that both animals will be *a/a* is also 0.0625. The

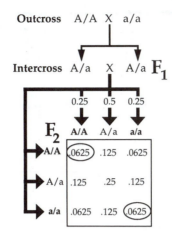

**Figure 3.1** Consequences of inbreeding at the F$_2$ generation. The first cross illustrated is an outcross between animals homozygous for alternative alleles at the *A* locus. The F$_1$ offspring are all identically heterozygous *A/a*. An intercross between two F$_1$ animals will produce an F$_2$ generation with three possible genotypes having the probabilities shown. The final box indicates the different combinations of F$_2$ by F$_2$ matings that are possible and the probabilities associated with each one. The two mating combinations that cause fixation for one allele are circled. This illustration of the process of inbreeding is based on two simplifying assumptions: (1) genetic homogeneity in each of the original parents used for the outcross, and (2) contrasting alleles at every locus in these original parents. In actuality, inbreeding is often begun with parents that, on the one hand, are not homogeneous but, on the other hand, do share many alleles with each other.

probability that either of these two mutually exclusive events will occur is derived by simply adding the individual probabilities together according to the "law of the sum" to obtain 0.0625 + 0.0625 = 0.125.

If there is a 12.5% chance that both F$_2$ progenitors are identically homozygous at any one locus, then approximately 12.5% of all loci in the genome will fall into this state at random. The consequence for these loci is dramatic: all offspring in the following F$_3$ generation, and all offspring in all subsequent filial generations will also be homozygous for the same alleles *at these particular loci*. Another way of looking at this process is to consider the fact that once a starting allele at any locus has been lost from a strain of mice, it can never come back, so long as only brother–sister matings are performed to maintain the strain.

At each filial generation subsequent to F$_3$, the class of loci *fixed* for one parental allele will continue to expand beyond 12.5%.[14] This is because all fixed loci will remain unchanged through the process of incrossing, while all unfixed loci will have a certain chance of reaching fixation at each generation. At each locus that has not been fixed, matings can be viewed as backcrosses, outcrosses, or intercrosses, which are all inherently unstable since they can all yield offspring with heterozygous genotypes as shown in Table 3.2.

Figure 3.2 shows the level of homozygosity reached by individual mice at each generation of inbreeding along with the percentage of the genome that is fixed

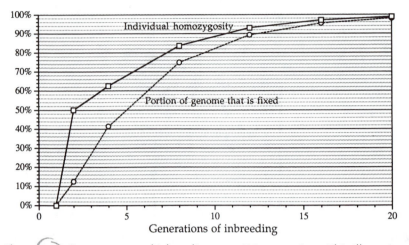

**Figure 3.2** Consequences of inbreeding up to 20 generations. This illustration of inbreeding is based on the same simplifying assumptions described in the legend to Fig. 3.1. Points on the solid line indicate the portion of the genome that will be homozygous in any individual animal at each generation. Points on the dashed line indicate the portion of the genome that will be fixed identically in the two animals chosen to generate the following generation of animals.

identically in both animals chosen to produce the next filial generation according to the formulas given by Green (1981). After 20 generations of inbreeding, 98.7% of the loci in the genome of each animal should be homozygous (Green, 1981). This is the operational definition of *inbred*. At each subsequent generation, the level of heterozygosity will fall off by 19.1%, so that at 30 generations, 99.8% of the genome will be homozygous and at 40 generations, 99.98% will be homozygous.

These calculations are based on the simplifying assumption of a genome that is infinitely divisible with all loci assorting independently. In reality, the size of the genome is finite and, more importantly, linked loci do not assort independently. Instead, large chromosomal chunks are inherited as units, although the boundaries of each chunk will vary in a random fashion from one generation to the next. As a consequence, there is an ever-increasing chance of complete homozygosity as mice pass from the 30th to 60th generation of inbreeding (Bailey, 1978). In fact, by 60 generations, one would be virtually assured of a homogeneous homozygous genome if it were not for the continual appearance of new spontaneous mutations (most of which will have no visible effect on phenotype). However, every new mutation that occurs will soon be fixed or eliminated from the strain through further rounds of inbreeding. Thus, for all practical purposes, mice at the $F_{60}$ generation or higher can be considered 100% homozygous and genetically indistinguishable from all siblings and close relatives (Bailey, 1978) . All of the classical inbred strains (including those in Table 3.1 and many others) have been inbred for at least 60 generations.

### 3.2.3 The classical inbred strains

During the first three decades of the 20th century, a series of inbred strains were developed from mice obtained through the fancy trade (see Chapter 1). A small number of these "classical strains" have, through the years, become the standards for research in most areas of mouse biology. The most important of these strains are listed in Table 3.1 along with their uses, other characteristics, and the number of generations of sequential brother–sister matings that had been accomplished, as of 1993, in the colonies of the major suppliers. Other characteristics relevant to the reproductive performance of many of the classical inbred strains are given in Chapter 4 (see Table 4.1). Pictures of several classical and newly derived mouse strains are presented in Fig. 3.3.

### 3.2.4 Segregating inbred strains

A special class of inbred strains is produced and maintained by brother–sister mating in the same manner as just described with one major exception. Instead of selecting animals randomly at each generation for further matings to maintain the strain, an investigator purposefully selects individuals heterozygous for a mutant allele at a particular locus of interest. This "forced heterozygosity" at each generation results in the development of a "segregating inbred strain" with the same properties as all other inbred strains in regions of the genome not linked to the "segregating locus". In almost all cases, segregating inbred strains are developed around mutant loci that cause lethality, severely reduced viability, or sterility in the homozygous state. Some mutant genes—including Steel (*Sl*), Yellow (*Ay*), Brachyury (*T*), and Disorganization (*Ds*)—can be recognized through the expression of a dominant phenotype that allows direct selection of heterozygotes at each generation. With other mutant genes, heterozygotes cannot be recognized directly and must be identified by progeny testing or through closely linked marker alleles that are recognizable in the heterozygous state.

At each generation of breeding, a segregating inbred strain will produce two classes of animals: those that carry the mutant allele and those that do not. Thus, it is possible to use sibling animals as "experimental" and "control" groups to investigate the phenotypic effects of the mutation in a relatively uniform genetic background. Segregating inbred strains are conceptually similar to congenic strains and the reader should read Section 3.3.3 for more information on the advantages and limitations of this approach to genetic analysis.

### 3.2.5 Newly derived inbred strains

When the genomes of the traditional inbred strains were first analyzed with molecular probes during the 1980s, it became clear that their common origin from the fancy mouse trade had led to a great reduction in interstrain polymorphism at many loci (as discussed in Section 2.3.4). Since polymorphisms are essential for formal linkage analysis, crosses between the traditional inbred strains were less than ideal for this purpose. This problem could be overcome with the development of new inbred strains that were genetically distinct from

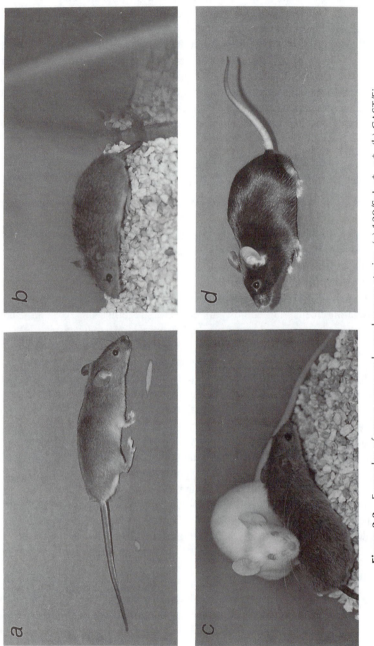

**Figure 3.3** Examples of some commonly used mouse strains. (a) 129/Svj-$c^+$, $p^+$, (b) CAST/Ei (*M. castaneus*); (c) CD-1 (albino, Charles River Breeding Laboratories) with SPRET/Ei; and (d) C57BL/6J.

the traditional ones. Another driving force in the development of new strains *from scratch* was the realization that none of the traditional strains were derived from a single subspecies or population; instead, they were all undefined genomic mixtures from two or more subspecies. Thus, the classical laboratory mice do not actually represent any animal that exists in nature. Although for many investigators, this would not appear to be an important problem, it is likely to become more relevant in future studies that are focused on the interactions among multiple genes rather than single genes in isolation. Within the traditional strains, unnatural combinations of alleles could have subtle unnatural effects on the operation of polygenic traits. To overcome this problem, new inbred strains are routinely derived from a pair of animals captured from a single well-defined wild population. Over the last several decades, inbred strains have been developed from animals representing each of the major subspecies in the house mouse group as well as somewhat more distant species that still form fertile hybrid females with *M. musculus*. Inbred individuals from *M. m. musculus* (CZECH II/Ei), *M. m. domesticus* (WSB/Ei, ZALENDE/Ei), *M. m. castaneus* (CAST/Ei), *M. spiciligus* (previously *M. hortulanus*; PANCEVO/Ei), *M. spretus* (SPRET/Ei), and the *faux* subspecies *M. molossinus* (MOLF/Ei) can all be purchased from the Jackson Laboratory (see Fig. 2.2 for the phylogenetic relationships that exist among these various species and subspecies).

The major hurdle that must be overcome in the development of new inbred strains from wild populations is *inbreeding depression*, which occurs most strongly between the $F_2$ and $F_8$ generations. The cause of this depression is the load of deleterious recessive alleles that are present in the genomes of wild mice as well as all other animal species. These deleterious alleles are constantly generated at a low rate by spontaneous mutation but their number is normally held in check by the force of negative selection acting upon homozygotes. With constant replenishment and constant elimination, the load of deleterious alleles present in any individual mammal reaches an equilibrium level of approximately ten. Different unrelated individuals are unlikely to carry the same mutations, and as a consequence, the effects of these mutations are almost never observed in large randomly mating populations.[15]

Thus, it not surprising that during the early stages of mouse inbreeding, many of the animals will be sickly or infertile. At the $F_2$ to $F_8$ generations, the proportion of sterile mice is often so great that the earliest mouse geneticists thought that inbreeding was a theoretical impossibility (Strong, 1978). Obviously they were wrong. But, to succeed, one must begin the production of a new strain with a very large number of independent $F_1 \times F_1$ lines followed by multiple branches at each following generation. Most of these lines will fail to breed in a productive manner. However, an investigator can continue to breed the few most productive lines at each generation—these are likely to have segregated away most of the deleterious alleles. The depression in breeding will begin to fade away by the $F_8$ generation with the elimination of all of the deleterious alleles. Inbreeding depression should not normally occur when a new inbred strain is begun with two parents who are themselves already inbred because no deleterious genes are present at the outset in this special case.

### 3.2.6 F₁ hybrids

The most obvious advantage of working with inbred strains is genetic uniformity over time and space. Researchers can be confident that the B6 mice used in experiments today are essentially the genetic equivalent of B6 mice used ten years ago. Furthermore, one can be confident that there will always be B6 mice around on which to conduct experiments. Thus, the existence of inbred strains serves to eliminate the contribution of genetic variability to the interpretation of experimental results. However, there is a serious disadvantage to working with inbred mice in that a completely inbred genome is an abnormal condition with detrimental phenotypic consequences. The lack of genomic heterozygosity is responsible for a generalized decrease in a number of fitness characteristics including body weight, life span, fecundity, litter size, and resistance to disease and experimental manipulations.

It is possible to generate mice that are genetically uniform without suffering the consequences of whole genome homozygosity. This is accomplished by simply crossing two inbred strains. The resulting $F_1$ hybrid animals express *hybrid vigor* in all of the fitness characteristics just listed with an overall life span that will exceed that of both inbred parents (Green and Witham, 1991). Furthermore, as long as there are both B6 mice and DBA mice, for example, it will be possible to produce $F_1$ hybrids between the two, and all $F_1$ hybrids obtained from a cross between a B6 female and a DBA male will be genetically identical to each other over time and space. This particular $F_1$ hybrid is the most common of those used and is available directly from most suppliers. All $F_1$ hybrid animals are named with an abbreviated form of the female progenitor first, followed by the male progenitor and the "F1" symbol. The $F_1$ hybrid generated from a cross between B6 females and DBA/2 males is named B6D2F1. Of course, uniformity will not be preserved in the offspring that result from an "intercross" between two $F_1$ hybrids; instead random segregation and independent assortment will lead to $F_2$ animals that are all genotypically distinct.

### 3.2.7 Outbred stocks

A large number of the laboratory mice sold and used by investigators around the world are considered to be *outbred* or *random bred*. Popular stocks of such mice in the US include CD-1 (Charles River Breeding Laboratories), Swiss Webster (Taconic Farms), and ICR and NIH Swiss (both from Harlan Sprague Dawley). Outbred mice are used for the same reasons as $F_1$ hybrids—they exhibit hybrid vigor with long life spans, high disease resistance, early fertility, large and frequent litters, low neonatal mortality, rapid growth, and large size. However, unlike $F_1$ hybrids, outbred mice are genetically undefined. Nevertheless, outbred mice are bought and used in large numbers simply because they are less expensive than any of the genetically defined strains.

Outbred mice are useful in experiments where the precise genotype of animals is not important and when they will not contribute their genome toward the establishment of new strains. They are often ideal as a source of material for biochemical purification and as stud males for the stimulation of pseudopregnancy in females to be used as foster mothers for transgenic or chimeric embryos. It is unwise to use outbred males as progenitors for any strain of mice that will be maintained and studied over multiple generations; the random-bred parent will contribute genetic uncertainty, which could result in unexpected results down-the-road.

If a stock of mice were truly random bred, it would be maintained through matings that were set up randomly among the breeding-age members of the population. Accordingly, matings would sometimes occur between individuals as closely related as siblings. In fact, most commercial suppliers follow breeding schemes that avoid crosses between closely related individuals in order to maintain the maximal level of heterozygosity in all offspring. Thus, random bred is a misnomer; stocks of this type should always be called non-inbred or outbred.

## 3.3 COISOGENICS, CONGENICS, AND OTHER SPECIALIZED STRAINS

### 3.3.1 The need to control genetic background

With the many new tools of molecular genetics described throughout this book, it has become easier and easier to clone genes defined by mutant phenotypes. Often, mutant phenotypes involve alterations in the process of development or physiology. In these cases, simply having a cloned copy of a gene is often not enough to examine critically the full range of effects exerted by that gene on the developmental or physiological process. In particular, normal development and physiology can vary significantly from one strain of mice to the next, and in the analysis of mutants, it is often not possible to distinguish subtle effects due to the mutation itself from effects due to other genes within the background of the mutant strain. To make this distinction, it is essential to be able to compare animals in which differences in the genetic background have been eliminated as a variable in the experiment. This is accomplished through the placement of the mutation into a genome derived from one of the standard inbred strains. It is then possible to perform a direct comparison between mutant and wild-type strains that differ only at the mutant locus. Phenotypic differences that persist between these strains must be a consequence of the mutant allele.

### 3.3.2 Coisogenic strains

In the best of all possible worlds, the mutation of interest will have occurred spontaneously within a strain of mice that is already inbred. In this case, one can be reasonably confident that the mutant animal differs at only a single locus from non-mutant animals of the same strain. If the mutation allows homozygous viability and fertility, it can be propagated as a strain unto itself by inbreeding offspring from the original mutant animal.[16] If the mutation cannot be propagated in the homozygous state, it will be maintained by continuous backcrossing of heterozygous animals to the original inbred strain. In both cases, the new mutant strain is considered *coisogenic* because its genome is identical (isogenic) to that of its sister strain except at the mutant locus. In the past, coisogenic strains could only be obtained by luck—when a spontaneous mutation happened to occur within an inbred strain. Today, one can initiate the production of coisogenic strains at any cloned locus through the use of the gene targeting technology described in Section 6.4.

Coisogenic strains are named with a compound symbol consisting of two parts separated by a hyphen: the first part is the full or abbreviated symbol for the original inbred strain; the second part is the symbol for the mutation or variant allele. If the mutation is maintained in a homozygous state within the coisogenic strain, the mutant symbol is used alone; if the mutation is maintained in a heterozygous state, the $+/m$ genotype symbol is used (where $m$ is the mutation). For example, if the mutation nude (*nu*) appeared in the BALB/cJ strain and the new coisogenic strain was homozygous for this mutation, its complete symbol would be [BALB/cJ-*nu*]; if the semidominant lethal mutation $T$ appeared in the C57BL/6J strain, and the new coisogenic strain was maintained by backcrossing to the parental strain, its symbol would be [B6-$T$/+].

### 3.3.3 Congenic and related strains

#### 3.3.3.1 Historical perspective: the major histocompatibility complex

A large number of mouse mutations and variants with interesting phenotypic effects have been identified and characterized over the last 90 years. Most of these mutations were not found within strains that were already inbred and, to date, most of the genes that underlie these mutations remain uncloned. Thus, in all of these cases, coisogenicity is not a possibility. However, even when a gene has been cloned and the generation of a coisogenic mutant through the gene targeting technology is a possibility, this approach is still extremely tedious and, at the time of this writing, there is no guarantee of a successful outcome. There are other reasons why spontaneous mouse mutations are often important even when the gene underlying the mutation has been cloned. The spontaneous mutation may not be a "knockout" but instead may exert a more subtle effect on gene function that could provide special insight into the action of the wild-type allele. Furthermore, the phenotypic effects of many older mutations have been studied in tedious detail by classical embryologists and other scientists, and it can be advantageous to a contemporary scientist to build upon these classical studies.

The "low-tech" solution to the elimination of genetic background effects in the analysis of an established mutation, or any other genetic variant, is to use breeding protocols, rather than molecular biology, to generate strains of mice that approximate coisogenics to the greatest extent possible. Mice that have been bred to be essentially isogenic with an inbred strain except for a selected *differential chromosomal segment* are called *congenic* strains. The conceptual basis for the development of congenic mice was formulated by George Snell at the Jackson Laboratory during the 1940s, and it led to the first and only Nobel prize for work strictly in the field of mouse genetics.

Snell was interested in the problem of tissue transplantation. Long before 1944, it was known that tissues could be readily transplanted between individuals of the same inbred strain without immunological rejection, but that mice of different strains would reject tissue transplants from each other. Although these observations were a clear indication of the fact that genetic differences were responsible for tissue rejection, the number and types of genes involved remained entirely unknown. In absentia, these genes were named *histocompatibility* (or *H*) loci. The assumption was that the histocompatibility genes were responsible— directly or indirectly—for the production of tissue (or "histological") markers that could be distinguished as "self" or "non-self" by an animal's immune system. If transplanted tissue and a host recipient carried identical genotypes at *all H* loci, there would be no immunological response and the transplant would "take." However, if a single foreign allele at *any H* locus was present in the tissue, it would be recognized as foreign and attacked.

Although the number of histocompatibility loci was unknown, it was assumed to be large because of the rarity with which unrelated individuals—both mice and humans—accept each other's tissues. The logic behind this assumption was the empirical finding that polymorphic loci are most often di-allelic and not usually associated with more than three common alleles. If *H* loci showed a similar level of polymorphism, a large number would be required to ensure that there would almost always be at least one allelic difference between any two unrelated individuals. The experimental problem was to identify and characterize each of the histocompatibility loci in isolation from all of the others.

Snell's approach to this problem was to use a novel multigeneration breeding protocol based on repeated backcrossing to trap a single *H* locus from one mouse strain (the *donor*) in the genetic background of another (the *inbred partner*). The basic approach (developed mathematically in the following section) caused the newly forming congenic strain to become increasingly similar to the inbred partner at each generation, but only those offspring who remained *histo- incompatible* with the inbred partner were selected to participate in the next round of backcrossing. It was assumed that a difference at any one *H* locus would be sufficient to allow full histo-incompatibility. Thus, at the end of the process, Snell expected to find that each independently derived congenic line would have trapped the donor strain allele at a single random *H* locus. With random selection, all *H* loci could be isolated in different congenic strains so long as a large enough number were generated.

With this outcome in mind, Snell began the production of histo-incompatible congenic strains (originally called "congenic resistant" strains) with 125 independent lines of matings (Snell, 1978) . Of these, 27 were carried through to the point at which it was possible to determine which *H* locus had been trapped. Surprisingly, 22 of the 27 lines had trapped the same locus, which was given the name *H*-2 (by chance, it was the second one identified). Contrary to expectations, the *H*-2 locus (now called the *H2* complex since it is known to be a tightly linked complex of genes) acts, for all effective purposes, as the only strong determinant of histocompatibility. Snell and his predecessors were misled by the false assumption that only a limited number of alleles are possible at any one locus. Instead, a subset of genes within the *H2* complex—known as the class I genes— are the most polymorphic in the genome with hundreds of alleles at each individual locus. The generic term "major histocompatibility complex" (*MHC*) is now used to designate this complex locus in mice as well as its homolog in all other mammalian species including humans, where it was historically called *HLA* (for human leukocyte antigen).

### 3.3.3.2 Creation of a congenic strain

In the past, there were several different breeding schemes used to produce congenic mice depending on whether animals heterozygous for the donor allele at the differential locus were phenotypically distinguishable—through a dominant form of expression—from those not carrying the donor allele. It was often the case that the heterozygote could not be distinguished and, as a consequence, congenic strains had to be created through complex breeding schemes that allowed the generation of homozygotes for the variant allele in alternating generations. Today, identifying the heterozygote is almost never a problem, since one will almost certainly map the locus of interest before undertaking the production of a congenic strain, and with a map position will come closely linked DNA markers. Therefore, the following discussion will be limited to the most direct, simple, and efficient method of congenic construction known as the backcross or NX system, which is illustrated in Fig. 3.4 (Flaherty, 1981).[17]

The backcross system of congenic strain creation is straightforward in both concept and calculation. The first cross is always an outcross between the *recipient inbred partner* and an animal that carries the *donor allele*. The donor animals need not be inbred or homozygous at the locus of interest, but the other partner must be both. The second generation cross and all those that follow to complete the protocol are backcrosses to the recipient inbred strain. At each generation, only those offspring who have received the donor allele at the *differential locus* are selected for the next round of backcrossing.

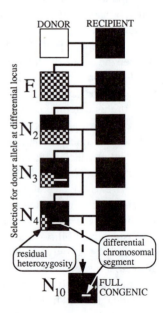

**Figure 3.4** Creation of a congenic strain. A highly schematic representation of the relative contributions of donor and recipient alleles at sequential generations of backcrossing. The donor contribution is indicated in white and the recipient contribution is indicated in black with the checkerboard pattern indicative of heterozygous loci. By the tenth generation of backcrossing, the differential segment around the selected locus (Fig. 3.6) will represent the major contribution from the donor genome.

The genetic consequences of this breeding protocol are easy to calculate. First, one can start with the conservative assumption that the donor (D) and recipient (R) strains are completely distinct with different alleles at every locus in the genome.

Then, all $F_1$ animals will be 100% heterozygous $D/R$ at every locus. According to Mendel's laws, equal segregation and independent assortment will act to produce gametes from these $F_1$ animals that carry $R$ alleles at a random 50% of their loci and $D$ alleles at the remaining 50%. When these gametes combine with gametes produced by the recipient inbred partner (which, by definition, will have only $R$ alleles at all loci), they will produce $N_2$ progeny having genomes in which approximately 50% of all loci will be homozygous $R/R$ and the remaining loci will be heterozygous $D/R$ as illustrated in Fig. 3.4. Thus, in a single generation, the level of heterozygosity is reduced by about 50%. Furthermore, it is easy to see that at every subsequent generation, random segregation from the remaining heterozygous alleles will cause a further ~50% overall reduction in heterozygosity.

In mathematical terms, the fraction of loci that are still heterozygous at the Nth generation can be calculated as $[(1/2)^{N-1}]$, with the remaining fraction $[1 - (1/2)^{N-1}]$ homozygous for the inbred strain allele. These functions are represented graphically in Fig. 3.5. At the fifth generation, after only four backcrosses, the developing congenic line will be identical to the inbred partner across ~94% of the genome. By the tenth generation, identity will increase to ~99.8%. It is at this stage that the new strain is considered to be a certified congenic. As one can see by comparing Figs. 3.2 and 3.5, the development of a congenic line will take approximately half the time that it takes to develop a simple inbred line from scratch. The reason for this more rapid pace is based on the fact that one of the two mates involved at every generation of congenic development is already inbred.

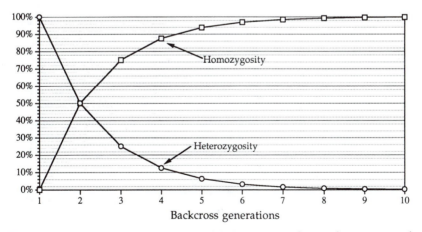

**Figure 3.5** Genomic homogeneity and heterogeneity during the creation of a congenic strain. The reciprocal points on each line indicate the percentage of loci in the genome of each individual animal that will be homozygous for the inbred partner allele or heterozygous with the donor allele at each generation of backcrossing.

Backcrossing can continue indefinitely after the tenth generation, but if the donor allele does not express a dominant effect that is visible in heterozygous animals, it will be easier to maintain it in a homozygous state. To achieve this state, two tenth generation or higher carriers of the selected donor allele are intercrossed, and homozygous donor offspring are selected to continue the line through brother–sister matings in all following generations. The new congenic strain is now effectively inbred, and in conjunction with the original inbred partner, the two strains are considered a "congenic pair."

In some cases, it will be possible to distinguish animals heterozygous for the donor allele from siblings that do not carry it. In a subset of these cases, as well as others, a donor allele may have recessive deleterious effects on viability or fertility. In all such instances, it is advisable to maintain the congenic strain by a continuous process of backcrossing and selection for the donor allele at every generation. Congenic strains that are maintained in this manner are considered to be in a state of "forced heterozygosity." There are two major advantages to pursuing this strategy whenever possible. First, the level of background heterozygosity will continue to be reduced by ~50% through each round of breeding. Second, the use of littermates with and without the donor allele as representatives of the two parts of the congenic pair will serve to reduce the effects of extraneous variables on the analysis of the specific phenotypic consequences of the donor allele.

The rapid elimination of heterozygosity occurs only in regions of the genome that are *not* linked to the donor allele which, of course, is maintained by selection in a state of heterozygosity throughout the breeding protocol. Unfortunately, linkage will also cause the retention of a significant length of chromosome flanking the differential locus, which is called the *differential chromosomal*

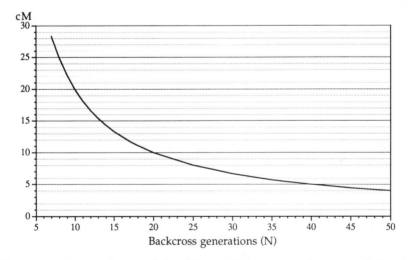

**Figure 3.6** Average length of the differential chromosomal segment during the creation of a congenic strain. The average differential segment length at each generation is shown in centimorgans. The average length of the differential segment on either side of the selected locus will be half the value shown on the graph.

*segment*. Even for congenic lines at the same backcross generation, the length of this segment can vary greatly because of the inherently random distribution of crossover sites. Nevertheless, the expected average length of the differential chromosomal segment in centimorgans can be calculated as $[200\,(1-2^{-N})/N]$ where $N$ is the generation number. For all values of $N$ greater than five, this equation can be simplified to $[200/N]$. This function is represented graphically in Fig. 3.6. As one can see, the average size of the differential segment decreases very slowly. At the tenth generation, there will still be, on average, a 20 cM region of chromosome encompassing the differential locus derived from the donor strain.

It is possible to reduce the length of the differential chromosomal segment more rapidly by screening backcross offspring for the occurrence of crossovers between the differential locus of interest and nearby DNA markers. As an example of this strategy, one could recover 50 congenic offspring from the tenth backcross generation and test each for recipient allele homozygosity at DNA markers known to map at distances of 1–5 cM on both sides of the locus of interest. It is very likely that at least one member of this backcross generation will show recombination between the differential locus and a nearby marker. The animal with the closest recombination event can be backcrossed again to the recipient strain to produce congenic mice of the 11th backcross generation. By screening a sufficient number of these $N_{11}$ animals, it should be possible to identify one or more that show recombination on the opposite side of the differential locus. In this manner, an investigator should be able to obtain a founder for a congenic strain with a relatively well-defined differential chromosomal segment of 5 cM or less after just 11 generations of breeding.

As the preceding discussion indicates, congenic strains differ from the previously described coisogenic strains in two important respects, which must always be considered in the interpretation of unexpected data. First, congenic strains, especially those that have undergone only a minimum number of backcrosses, will have small random remnants of the donor strain—so-called *passenger loci*—scattered throughout the genome. In congenic strains maintained by inbreeding, the same passenger loci will be present in all members of the strain. In rare instances, traits attributed to the selected donor allele may actually result from one of these cryptic passenger loci. Such effects can be sorted out by breeding the congenic strain back to its original inbred partner. If a trait is due to a passenger locus, it will assort independently of the donor locus in subsequent backcrosses.

The second difference between a congenic strain and a coisogenic strain is in the chromosomal vicinity of the differential locus. Congenic strains will always differ from their inbred partner along a significant length of chromosome flanking the differential locus; coisogenic strains will only differ at the differential locus itself and nowhere else. Thus, there is always the possibility that phenotypic differences between the two members of a congenic pair are actually caused by a closely linked gene rather than the selected differential locus. This potential problem is much more difficult to resolve by simple breeding protocols.

### 3.3.3.3 Nomenclature

The nomenclature used for congenic strains is so similar to that used for coisogenic strains that it is sometimes not possible to distinguish between the two by name alone. In such cases, it is necessary to go back to the original source publication for clarification. There are, however, two nomenclature components that are unique to congenic strains. The first is used in those cases where a mutant or variant allele is transferred from one defined genetic background onto another. For example, one might wish to transfer the albino (c) mutation from the BALB/c strain onto a B6 background. In cases of this type, the strain that "donates" the variant allele is symbolized after the recipient strain with the two strain symbols separated by a period. This is followed by a hyphen and the symbol for the variant allele. Thus, in the example just described, the congenic strain would be named B6.BALB-c.

The final nomenclature component is an indication of the number and type of crosses that have occurred subsequent to the original mating between the recipient and donor animals. In the derivation of any new congenic strain, the first cross is always an outcross, and the offspring are considered members of the $F_1$ generation. The second cross is always a backcross, and the offspring are considered members of the $N_2$ generation. (Note that there is no such thing as an $N_1$ generation). The letter 'N' is always used, followed by a subscripted number ($N_i$), to describe a series of backcross events leading to a particular generation of animals. However, remember that $N_{10}$ generation offspring are the result of one outcross followed by an uninterrupted sequence of nine backcrosses to the same parental strain. Once a congenic strain is established, backcrossing to the parental stain is often stopped, and future generations are propagated by a simple inbreeding protocol. The number of generations of inbreeding is indicated, as always, with the filial generation symbol 'F'. For example, suppose that the albino mutation has been placed onto the B6 background by an outcross followed by 14 generations of backcrosses, after which a brother–sister mating regime is begun and followed for eight more generations. The offspring produced at this stage would be considered to be members of the $N_{15}F_8$ generation. When generational information is incorporated into the name of a congenic strain, the numbers are no longer subscripted. So, in this example, the complete name for the congenic animals at the stage indicated would be B6.BALB-c (N15F8).

### 3.3.3.4 Consomic strains

Consomic strains are a variation on congenic strains in which a whole chromosome—rather than one local chromosomal region—is backcrossed from a donor strain onto a recipient background. In almost all cases, the donor chromosome is the Y. Like congenics, consomics are produced after a minimum of 10 backcross generations. Backcrossing to obtain consomics for the Y chromosome must be carried out in a single direction—males that contain the donor chromosome are always crossed to inbred females of the recipient strain. For example, to obtain a B6 strain consomic for the *M. m. castaneus* Y chromosome, one would start with an outcross between a B6 female and a *castaneus* male. $F_1$ males, and those from all subsequent generations, would also be mated with B6 females. After ten generations, the genetic background would be essentially B6, but the Y chromosome would be *castaneus*. This strain could be symbolized as B6-Y$^{CAS}$.

### 3.3.3.5 Conplastic strains

Conplastic strains are another variation on the congenic theme, except that in this case, the donor genetic material is the whole mitochondrial genome, which is placed into an alternative host. Since the mitochondrial genomes carried by all of the classical inbred strains are indistinguishable, conplasticity makes sense only in the context of interspecific or inter-subspecies crosses. Conplastic lines are generated by sequential backcrossing of females from the donor strain to recipient males; this protocol is reciprocal to the one used for the generation of Y chromosome consomics. For example, to obtain a B6 strain conplastic for the *M. m. castaneus* mitochondrial genome, one would start with an outcross between a B6 male and a *castaneus* female. $F_1$ females, and those from all subsequent generations, would also be mated with B6 males. After ten generations, the nuclear genome would be essentially B6 with the same statistics that hold for congenic production (Fig. 3.5), but all mitochondria would be derived from castaneus. This strain could be symbolized as B6-mt$^{CAS}$.

## 3.3.4 Recombinant inbred and related strains

### 3.3.4.1 Recombinant inbred strains

Recombinant inbred (RI) strains are formed from an initial cross between two different inbred strains followed by an $F_1$ intercross and 20 generations of strict brother–sister mating. This breeding protocol allows the production of a family of new inbred strains with special properties relative to each other that are discussed fully in Section 9.2. Different RI strains derived from the same pair of original inbred parents are considered members of a set. Each RI set is named by joining an abbreviation of each parental strain together with an "X." For example, RI strains derived from a C57BL/6J (B6) female and a DBA/2J male are members of the BXD set, and RI strains derived from AKR/J and C57L/J are members of the AKXL set. A complete listing of commonly used RI sets is given in Table 9.3. Each RI strain in a particular set is distinguished by appending a hyphen to the series name followed by a letter or number. Thus, BXD-15 is a particular RI strain that has been formed from an initial cross between a B6 female and a DBA male. At any point in time, it is always possible to add a new strain to a particular set through an outcross between the same two progenitor strains followed by 20 generations of inbreeding. The RI strains represent an important tool in the arsenal available for linkage studies of newly defined DNA loci.

### 3.3.4.2 Recombinant congenic strains

Recombinant congenic strains (abbreviated as RC strains) are a variation on the recombinant inbred concept (Demant and Hart, 1986). As with RI strains, the initial cross is between two distinct inbred strains. However, the next two generations are generated by backcrossing, without selection, to one of the parental strains. This sequence is followed by brother–sister mating for at least 14 generations. Whereas standard RI strains have genomes that are a mosaic of equal parts derived from both parents (as detailed in Section 9.2.2), RC strains will have mosaic genomes that are skewed in the direction of the parent to which the backcrossing occurred such that a random 7/8 fraction of the genome will be derived from this parent, and a random 1/8 fraction will be derived from the other parent. Sets of RC strains have some interesting properties in terms of limiting the amount of the genome that has to be searched for multiple genes involved in quantitative traits. However, with the new PCR-based methods for genotyping highly polymorphic loci discussed in Section 8.3, the advantages of the RC strains appear to have been superseded and they have not been used widely by the mouse genetics community.

## 3.4 STANDARDIZED NOMENCLATURE

### 3.4.1 Introduction

Mouse genetics is, by its very nature, a collaborative field of scientific investigation. This is because the interpretation of data collected by any one scientist is often highly dependent on data collected by others. High-resolution genetic maps are often formed through the integration of results obtained in many individual studies, and as each new result is published, it can be swept up into a system of databases (see Appendix B). Large-scale integration has been possible only because all mouse geneticists speak the same language. The definition of this language is provided by the International Committee on Standardized Nomenclature for Mice, which has been in existence since 1939. This committee is charged with the task of establishing and updating rules and guidelines for genetic nomenclature. The continued functioning of this committee is critical because, as the analysis of the genome becomes ever more sophisticated, new genetic entities become apparent, and these must be named in a standard fashion.

For a complete description of the "Rules and Guidelines for Gene Nomenclature," one should consult the Lyon and Searle book (Committee on Standardized Genetic Nomenclature for Mice, 1989), and updates published regularly in *Mouse Genome*. Here, I will briefly review the salient features of this nomenclature system with a focus on the naming of newly defined genes and loci. Once an investigator has chosen a new name and symbol for a locus, the chair of the Committee should be contacted for confirmation that the rules have been followed properly, and the names do not conflict with others already in use.

### 3.4.2 Strain symbols

There is no rhyme or reason to the names given to the original inbred strains derived at the beginning of the century. The name of the famous BALB/c line was derived by co-joining the name of the investigator (Bagg) with the color of the mouse (albino).

Bagg's ALBino became BALB. Other famous strains have names based on animal numbers; for example, female no. 57 (from Abbie Lathrop's farm) gave rise to both the C57BL/6 and C57BL/10, strains which are commonly abbreviated as B6 and B10, respectively. New inbred strains can be named freely by their originators as long as certain rules are followed: the name should be brief, and it should begin with a capital letter followed by other capital letters (preferably) or  numbers.[18] Strains with a common origin that have been separated prior to the $F_{20}$ generation must be given separate symbols, although these symbols can indicate their relationship to each other. All names should be registered with the appropriate contact person, who is indicated prominently in the current issue of *Mouse Genome.*

Substrains can arise whenever two or more colonies of an established inbred strain are maintained in isolation from each other for a sufficient period of time to allow detectable genetic differences to become fixed. There are three specific situations where substrain formation can be considered to have occurred: (1) when branches of an inbred strain are separated before the $F_{40}$ generation, when residual heterozygosity is still likely; (2) when a branch has been maintained separately from other branches for 100 or more generations; and (3) when genetic differences from other branches are uncovered. Such differences can be caused by any one or more of three factors: residual heterozygosity at the time of branching, mutation, or contamination.

Substrains are indicated by appending a slash (/) to the strain symbol followed by an appropriate substrain symbol, for example, DBA/1 and DBA/2. A laboratory registration code is often included within the substrain designation, for example, C57BL/6J and C57BL/10J are two substrains of C57BL that are both maintained at the Jackson Laboratory (indicated with a J). On the other hand, a different nomenclature has been formulated recently to distinguish the same strain maintained without any apparent genetic differences by two or more laboratories. In this case, the "@" character is appended to the strain symbol followed by the laboratory registration code. For example, the SJL strain maintained by the Jackson Laboratory would be symbolized as SJL@J.

### 3.4.3  Locus names and symbols

In the first set of rules for distinguishing gene symbols laid down by the Committee on Mouse Genetics Nomenclature in 1940, it was stated that "the initial letter of the mutant gene symbol shall be the same as the initial letter of the mutant gene, e.g., *d* for dilution. Additional letters shall be added to the initial letter *if necessary* to distinguish it from symbols already in use" (Snell, 1941, p.242).[19] With over 3,000 independent loci identified as of 1993, the necessity of using symbols that contain more than one letter is now obvious. In fact, the recent explosion in gene and locus identifications in the mouse has brought about a re-evaluation of the entire basis for the naming of chromosomal entities. At the time of this writing, a final consensus has not yet been reached. Thus, investigators are cautioned to contact members of the International Committee on Standardized Nomenclature for Mice before settling on a name for a new genetic entity.

Each mouse locus is given a unique name and a unique symbol. In devising new names, investigators should consider their suitability for inclusion into electronic databases. Thus, names should be limited in length to fewer than 40 characters (including spaces), and should not include Greek letters or Roman numerals. The symbol is a highly abbreviated version of the name. In published articles, locus symbols (but not names) are always set in *italic* font. Symbols always begin with a letter followed by any combination of letters or Arabic numbers without internal white space. In the past, symbols were typically 3–8 characters in length. Today, database considerations set a preferred maximum number of characters at 10, although this rule is frequently broken.

Loci that are members of a related series of some kind are given the same primary stem and symbol followed by a distinguishing number or letter. Thus, the third esterase gene to be defined is named "Esterase 3," with the symbol *Es3*, and the second homeo box gene cluster to be identified is named "homeo box B cluster" with the symbol *Hoxb*. In the past, a hyphen was often used to separate the numeral designation from the body of the gene symbol, e.g., *Es-3*. This practice has now been discontinued and hyphens have been deleted from all symbols except in the special cases discussed just below.

When one member of a series has been further duplicated into a closely linked cluster of related genes, a number can be appended to the cluster name; individual genes in the Hoxb cluster will be named homeo box B1, homeo box B2, etc., with symbols *Hoxb1, Hoxb2*, etc. For clusters that were initially named with an appended number—like *Lamb1*—the individual symbols can be named by appending the cluster name with a hyphen followed by a number to obtain the symbols *Lamb1–1, Lamb1–2*, etc.

All loci can be broadly separated into two classes. The first class includes loci known to be functional or homologous to functional loci. With few exceptions, these loci are genes or pseudogenes. The second class includes sequences identified solely on the basis of DNA variation. Members of this latter class are referred to as anonymous loci because their function or lack thereof is unknown. The rules for naming each of these classes of loci follow below.

*Gene names* should convey in a concise form, and as accurately as possible, the character by which the gene is recognized. Genes can be named according to an expressed phenotype (retinal degeneration or shiverer), an enzyme or protein name or function (glyoxalase-1, hemoglobin alpha chain, or octamer binding transcription factor 1), a pattern of expression (t-complex testes-expressed-1), a combination of these (myosin light chain alkali-fast skeletal muscle), or by homology to genes characterized in other organisms (homeo box A, B, etc.; retinoblastoma). Except in the case of genes that are first characterized through a recessive mutation, names and symbols should begin with an upper case letter. With all symbols, all letters that follow the initial character should be lower case.

Whenever a mouse gene is characterized based on homology to a gene already named in another species, the mouse homolog should be given essentially the same name and symbol. Of course, one should always check the mouse gene databases (see Appendix B) to be certain that the symbol has not already been assigned. In the translation from human to mouse symbols, characters beyond the first should be converted from upper to lower case.

All *pseudogenes* are defined by homology to known genes. Their symbol is a combination of the known gene name (as a stem), and the pseudogene designation (*ps*) followed by a serial number. Thus, the third alpha globin pseudogene has been given the name "hemoglobin alpha 3 pseudogene", with the symbol *Hba-ps3*.

When new loci are uncovered by hybridization with known genes, the functionality of the new locus is usually unknown. In these cases, where the locus could be either a functional gene or a pseudogene, it should be named with a "related sequence" symbol (*rs*). Thus, if a new locus is uncovered by cross-hybridization with a probe for the plasminogen gene (symbolized *Plg*), it would be named "plasminogen-related sequence-1" and would be symbolized as *Plg-rs1*. If a new locus is uncovered with a probe for one member of a series of loci, the *rs* symbol is appended without the hyphen. Thus, a locus related to *Ela1* would be symbolized as *Ela1rs1*.

*Anonymous DNA loci* are named in a straightforward manner. The symbol should begin with the character '*D*' (for DNA), followed by an integer representing the chromosomal assignment, followed by a 2–3 letter registration code representing the laboratory or scientist that described the locus, followed by a unique serial number given to the locus to distinguish it from others on the same chromosome described by the same investigator. For example, the 23rd anonymous locus mapped to chromosome 14 by the Pasteur Institute would be given the symbol *D14Pas23*. The name for this locus would include all of this information in longhand form—"DNA segment chromosome 14 Pasteur 23." This DNA locus nomenclature system should be used for all loci defined only as DNA segments including, but not limited to, microsatellites, minisatellites and restriction fragment length polymorphisms (see Chapter 8). To obtain a unique laboratory or investigator registration code, please contact the Institute for Laboratory Animal Resources, USA National Academy of Sciences, Washington, DC.

*Mouse homologs of anonymous DNA loci first mapped in humans* are named in a somewhat different format in order to allow the connection between the two species to be perfectly transparent.

In these cases, the symbol should still begin with the character "*D*" and the mouse chromosomal assignment, but this should now be followed by the character "*h*," the chromosomal assignment of the human homolog and its identification number. For example, a probe to the human locus D17S111—the 111th single copy (S) anonymous locus mapped to human chromosome 17—is used to identify a mouse homolog on Chr 1. This corresponding mouse homolog will now be named *D1h17S111*.

### 3.4.4 Alleles

In the case of a gene defined initially by a mutant phenotype, the symbol for the first defined mutant allele becomes both the gene symbol and the symbol for that allele. The corresponding wild-type allele is indicated by a + sign. For example, an animal heterozygous at the *tf* locus with a wild-type and the defining mutant allele would have a genotype symbolized as $+/tf$. In this case, the context is sufficient to indicate the association of the $+$ symbol with the *tf* locus. When the context is not sufficient to indicate association, the wild-type allele of a specific locus should have the locus symbol appended to it as a superscript. Thus, the wild-type allele at the *tf* locus can also be designated as $+^{tf}$.

In all other cases, alleles are designated by the locus symbol followed by an allele-defining symbol that is usually one or a two characters in length and set in superscript, with the entire expression set in italics. This rule also applies to mutant alleles beyond the first one that are uncovered at a phenotypically defined locus. For computer presentation with only ASCII (text format) code, the allele designation can be set off from the locus symbol by prefixing it with a * or with angular brackets; for example, $Hbb^d$ becomes Hbb*d or Hbb < d > .

The simplest means for assigning allele names is through a series of lower case letters, beginning with *a*. Thus, the *Hba-ps4* gene has alleles $Hba\text{-}ps4^a$, $Hba\text{-}ps4^b$, etc. In many cases, it can be useful to provide information within the allele symbol. For example, a *M. spretus*-specific allele may be given the designation *s* as in $DXPas4^s$.[20] This type of nomenclature can be extended to alleles associated with the common inbred strains such as B6 (signified by the *b* allele) and DBA (signified by the *d* allele) as well as the subspecies *musculus (m), castaneus (c), and domesticus (d)*.

New "random" mutations—at previously characterized genes are denoted by a superscript *m* followed by a serial number and the 1–3 letter code representing the laboratory or scientist that described the new allele. When specific mutations are generated by the gene targeting technologies (see Section 6.4), the same nomenclature applies except that the superscript *m* is preceded by a superscript *t*. Thus, the third knockout allele created at Princeton University by gene targeting at the *cftr* locus would be designated as $cftr^{tm3Pri}$. To obtain a unique laboratory or investigator registration code, please contact the Institute for Laboratory Animal Resources, USA National Academy of Sciences, Washington, DC.

### 3.4.5 Transgene loci

The experimental introduction of foreign DNA into the germ line of a mouse results in the creation of a new *transgene* locus at the site of integration. The official symbol for a transgene locus has five parts. First is the designation *Tg* for transgene. Second is a letter indicating the mode by which the transgene was inserted; *N* is used for non-homologous insertion, *R* for insertion with a retroviral vector, and *H* for homologous recombination. With the standard production of transgenic mice by embryo injection, *N* would be used; with homologous recombination in embryonal stem cells followed by chimera formation to rescue the transgene into the germ line, *H* would be used; for transgenic animals produced by retroviral infection of embryos, *R* would be used.

The third part of the symbol contains a mnemonic, of six characters or fewer, that describes the salient features of the transgene insert written within parentheses. If the insert includes a defined gene, the gene symbol should be incorporated into the mnemonic without hyphens. Other standard abbreviations for use within the mnemonic include: *An* for anonymous sequence; *Nc* for non-coding sequence; *Rp* for reporter sequence; *Et* for enhancer trap; *Pt* for promoter trap; and *Sn* for synthetic sequence. The fourth part of the symbol is an investigator-assigned 1–5 digit number. The fifth and last part is the laboratory code. An example of this nomenclature is as follows. Castle has injected mouse embryos with a construct containing the *Pgk2* coding sequence as a reporter. He names the transgene locus present in the fourth line that he recovers *TgN(RpPgk2)4Cas*.

When the insertion of a transgene at a particular site results in a new mutation through the disruption of a gene present normally in the genome, this mutation should be named independently of the transgene locus itself. The rationale for this rule is that the contents of the transgene are independent of the locus uncovered through insertional mutagenesis. However, the mutant allele associated with the transgene should incorporate the transgene symbol as the superscripted allele designation. For example, if Castle's construct became inserted into the *Hbb* locus in the fifth line that he derived, the new *Hbb* allele that was created would be called $Hbb^{TgNrpPgk25Cas}$. Notice that the parentheses have been removed from the allele symbol. If a new mutation has been induced at a previously unidentified locus, the mutant phenotype should be used to name the new locus.

### 3.4.6 Further details

In this section, I have only touched upon those issues of nomenclature that will be of most concern to the majority of molecular biologists involved in studies of the mouse genome. In fact, the nomenclature rules developed for the mouse are rather extensive and are discussed in much greater detail in the Lyon and Searle compendium (Committee on Standardized Genetic Nomenclature for Mice, 1989), with additions and changes published regularly in *Mouse Genome*. As a final note, one must keep in mind that mouse genetic nomenclature will continue to evolve with the field as a whole. As new types of genetic elements and inter-relationships are uncovered, it will be the charge of the Nomenclature Committee to keep the rules internally consistent and up to date.

## 3.5 STRATEGIES FOR RECORD-KEEPING

### 3.5.1 General requirements

A breeding mouse colony differs significantly from a static one in the type and complexity of information that is generated. In a non-breeding colony, there are only the animals and the results obtained from observations and experiments on each one. In a breeding colony, there are animals, matings, and litters, with specific genetic connections among various members of each of these data sets. Classical genetic analysis is based on the transmission of information between generations, and as a consequence, the network of associations among individual components of a colony is as important as the components are in and of themselves.

An ideal record-keeping system would allow one to keep track of: (1) individual animals, their ancestors, siblings and descendants; (2) matings between animals; (3) litters born from such matings, and the individuals within litters that are used in experiments or to set up the next generation of matings; and (4) experimental material (tissues and DNA samples) obtained from individual animals. Ideally, one would like to maintain records in a format that readily allows one to determine the relationship, if any, that exists between any two or more components of the colony, past or present.

Based on these general requirements, two different systems for record-keeping have been developed by mouse geneticists over the last 60 years. The "mating-unit" system focuses on the mating pair as the primary unit for record-keeping. The "animal/litter" system treats each animal and litter as a separate entity. As discussed below, there are advantages and disadvantages to each approach.

### 3.5.2 The mating unit system

With this system, each mating unit is assigned a unique number and is given an individual record. When record-keeping is carried out with a notebook and pencil, each mating pair is assigned a page in the book. The cage that holds the mating pair can be identified with a simple card on which the record number is indicated; this provides immediate access to the corresponding page in the record book.

When litters are born, they are recorded within the mating record. Each litter is normally given one line on which the following information is recorded in defined columns: (1) a number indicating whether it is the first, second, third, or a subsequent litter born to the particular mating pair; (2) the date of birth; (3) the number of pups; and (4) other characteristics of the litter that are of importance to the investigator. Individual mice within any litter can be identified uniquely with a code that includes the mating unit number, followed by a hyphen, the litter number, and a letter that distinguishes siblings from each other. For example, the fourth pup in the third litter born to mating unit 7371 would be numbered 7371–3d. This system provides for the individual numbering of animals in a manner that immediately allows one to identify siblings and parents.

At the outset, parental numbers are incorporated into each mating record, and since these are linked implicitly to the litters from which they come, it becomes possible to trace a complete pedigree back from any individual. It also becomes possible to trace pedigrees forward if, as a matter of course, one cross-references all new matings within the litter records from which the parents derive. For example, if one sets up a new mating unit that is assigned the number 8765 with female 5678–2e and male 5543–1c, the number 8765 could be inscribed on appropriate lines in records 5678 and 5543.[21]

There are several important advantages to a record-keeping system based on the mating unit: (1) only a single set of primary record numbers is required; (2) one can easily keep track of the reproductive history of each mating pair; and (3) information on siblings can be readily viewed within a single location.

Furthermore, it is easy to incorporate this system of record-keeping into a simple spreadsheet file that can be maintained on a desktop computer or fileserver. This can be accomplished most readily by having each row represent an individual litter (or even individuals within a litter) with columns for: (1) mating unit number; (2) father's number; (3) mother's number; (4) litter number; (5) birth date; (6) number of pups; and (7) further information. To record information on individuals within particular litters, new rows having the same format can be formed but with the litter number–sibling letter combination used in place of the litter number alone in column 4.

New litters can be recorded initially as they are born in empty rows at the bottom of the file. By re-sorting the database according to the first column, an investigator would be able to view all litters born to a particular mating pair in sequential rows. Upon sorting according to birth date, the list of litters could be displayed according to age. With search or find commands, it would be possible to identify ancestors, descendants, and siblings related to each animal.

The major disadvantage to this form of record-keeping is that records are focused on mating pairs and litters rather than individual animals. Thus, it is not well-suited for investigators who need to record and retrieve animal-specific information. It is also less than ideal for situations where the mating unit is not sacrosanct and animals are frequently moved from one mate to another. Under these circumstances, the animal/litter system described below is more appropriate for record-keeping.

### 3.5.3 The animal/litter system

In a second system developed originally by one of the earliest mouse geneticists, L. C. Dunn, there are two primary units for record-keeping—the individual animal and the individual litter. Each breeding animal is assigned a unique sequential number (at the time of weaning) that is associated with an individual record occupying one row across facing pages within an "animal record book" or a spreadsheet file. Each animal record contains the numbers of both parents and through these it is possible to trace back pedigrees. A separate "litter record book" or spreadsheet file is used to keep track of litters, which are also assigned unique sequential numbers attached to one row in the database. In this record-keeping system, animal numbers and litter numbers are assigned independently of each other.

A third independent set of numbers are those assigned to individual cages. Cage numbers can be assigned in a systematic manner so that related matings are in cages with related numbers. For example, different matings that derive from the same founder of a particular transgenic line may be placed in cages numbered from 2311 to 2319. A second set of matings that carry the same transgene from a different founder could be placed into cages numbered 2321 to 2329, and so on. Thus, the cages between 2300 and 2399 would all have animals that carried the same transgene, however, different sets of ten would be used for different founder lines. For matings of animals with a second transgene, you might choose to use the cages numbered 2400 to 2499. This type of numbering allows one to classify cages—which represent matings—in a hierarchical manner. Although at any point in time, every cage in the colony will have a different number, once a particular cage is dismantled, its number can be re-assigned to a new mating. Cage cards from dismantled matings can be saved for future reference.

When a litter is born, the litter record is initiated with entries into a series of columns for: (1) an identifying number; (2) the birth date; (3 & 4) the numbers of the parents; (5) the number of the cage in which the litter was born; (6) the number of pups born; and (7) any other information of importance to the investigator. In addition, the litter number can be inscribed on the cage card (which may or may not have additional information about the mating pair). When an animal is weaned from a litter for participation in the breeding program, an animal record is initiated. The most important information in the animal record is the number of the litter from which it came, the cage that it goes into, and the date of that move (all entered into predefined columns). The cage number is particularly important in allowing one to trace pedigrees forward from any individual at a future date. If an animal is moved from one cage to another at some later date, this can be added to the record in another column.

### 3.5.4 Comparison of record-keeping systems

With three unrelated sets of numbers and the need for extensive cross-referencing, the animal/litter system is complex, and implementation on paper is labor intensive. However, it does provide the investigator with additional power for analysis. For example, by choosing cage numbers wisely and saving cage cards in numerical order, it becomes possible to go back at any point in the future and look at all of the litters born to a particular category of matings over any period of time. With the mating unit system, this could only be accomplished by using different files for different categories of matings. However, it then becomes very difficult to keep track of matings formed with animals taken from different files.

Another difference between the mating unit system and the animal/litter system is the ease with which it is possible to keep track of animals that are moved from one mating unit to another. The mating unit system is most effective for colonies where "animals are mated for life." The animal/litter system is effective for colonies of this type as well, but is also amenable to those where animals are frequently switched from one mate to another.

### 3.5.5 A computer software package for mouse colony record-keeping

The animal/litter system of record-keeping has been incorporated into more extensive computer software packages that greatly facilitate data entry, with automatic cross-referencing and extensive error checking. These software packages, called the "Animal House Manager" or AMAN and MacAMAN, are specialized database programs that allow users to record and retrieve information on animals, litters, cages, tissue and DNA samples, and restriction digests generated from a breeding mouse colony (Silver, 1993b) . Data are entered through a series of queries and answers (for AMAN) or pointing and clicking (for MacAMAN). With automatic cross-referencing and specialized protocols, the same information never has to be entered more than once. Hard-copy printouts can be obtained for cage cards, individual records, or sets of records (in abbreviated form) that have been recovered through searches for positive or negative matches to particular words or parameters. Search protocols are highly versatile; for example, it is possible to print out a cage-ordered list of litters that are old enough for weaning on a particular date, or a list of live mice from a particular set of breeding cages that are ordered according to the cage in which they were born. AMAN and MacAMAN provide investigators with the ability to maintain control over a complex breeding program with instant access to each record, current and past. They can store up to 100,000 records in each of four files for animals, litters, DNA/tissue samples, and restriction digests.

For licensing and other information, please refer to appendix B.

# 4

# Reproduction and Breeding

## 4.1 REPRODUCTIVE PERFORMANCE: COMPARISON OF INBRED STRAINS

Reproductive performance can be measured according to several different parameters including age at first mating, number of litters sired, number of pups per litter, and the frequency with which a strain has productive matings. Table 4.1 shows the values obtained for these different parameters with the most commonly used inbred strains of mice. As the table shows, inbred strains vary widely in their reproductive fitness.

The first important measure of reproductive fitness is the frequency with which a mating pair will produce any offspring at all.[22] With some strains, such as C3H/HeOuJ, CBA/CaJ and FVB/N, over 90% of all matings that are set up

**Table 4.1**  Reproductive characteristics of some important inbred strains

| Strain | Productive matings $1$ | Weeks at first mating | Litter size $2$ | Number of litters $3$ | Relative fecundity $1 \times 2 \times 3$ | Response to superovulation |
|---|---|---|---|---|---|---|
| 129/SvJ | 75% | 7.9 | 5.9 | 4.1 | 18.1 | High |
| A/J | 65% | 7.6 | 6.3 | 2.9 | 11.9 | Low |
| AKR/J | 84% | 6.6 | 6.1 | 2.2 | 11.3 | — |
| BALB/cJ | 47% | 8.0 | 5.2 | 3.8 | 9.3 | Low |
| C3H/HeJ | 86% | 6.7 | 5.7 | 2.9 | 14.2 | Low |
| C3H/HeOuJ | 99% | 5.9 | 6.4 | 3.7 | 23.4 | — |
| C57BL/6J | 84% | 6.8 | 7.0 | 4.0 | 23.5 | High |
| C57BL/10SnJ | 67% | 7.7 | 6.3 | 2.8 | 11.8 | — |
| CBA/CaJ | 96% | 6.4 | 6.9 | 2.7 | 17.9 | High |
| DBA/2J | 75% | 7.4 | 5.4 | 3.9 | 15.8 | Low |
| FVB/N | > 90% | — | 9.5 | 4.8 | 41.0 | Moderate |
| SJL/J | 72% | 7.4 | 6.0 | 3.1 | 13.4 | High |

Matings are considered "productive" if any offspring at all are born. "Weeks at first mating" refers to the age at which the first "productive" mating occurred; it is estimated by subtracting 3 weeks from the age of the parents at the first birth. "Relative fecundity" is obtained as the following product: (productive mating frequency) × (litter size) × (number of litters); the value obtained is a measure of the overall fecundity of the strain. The "response to superovulation" is based on information extracted from *Manipulating the Mouse Embryo: A Laboratory Manual* (Hogan et al., 1994); the significance of this parameter is discussed in Section 6.2. Data for all strains other than FVB/N are extracted from the *Handbook of Genetically Standardized JAX mice* (Green and Witham, 1991). Data for the FVB/N strain are extracted from a publication by Taketo and collegues (1991).

will produce offspring. The C3H/HeOuJ strain is at the extreme end of this group with a 99% frequency of productive matings. At the opposite extreme, among the most well-characterized strains, is BALB/cJ with a frequency of non-productive matings that is over 50%. A second measure of fitness is the age at which females first become pregnant. This can vary from an early 5.9 weeks for C3H/HeOuJ to a late 8.0 weeks for BALB/cJ. The third measure of reproductive fitness is litter size. Once again, BALB/cJ performs the worst in this category with an average litter size of just 5.2. All but one of the remaining inbred strains have average litter sizes in the range of 5.4–7.0. The one strain that outperforms all others in this category is FVB/N with a much larger average litter size of 9.5. The final measure of reproductive fitness is the average number of litters that a single female can produce in a lifetime. This varies from a low of 2.2 litters with AKR/J females to a high of 4.8 litters with FVB/N females.

Three of the easily quantified measures of reproductive performance— frequency of productive matings, litter size, and number of litters—have been multiplied together to give a sense of the overall fecundity associated with any one inbred strain in comparison to the others (Table 4.1). Far and away the highest fecundity value (41.0) is associated with the relatively new FVB/N strain. It is for this reason, as well as others, that FVB/N has become the strain of choice for use in the production of transgenic animals (see Section 6.2). Among the traditional inbred strains, C57BL/6J (B6) and C3H/HeOuJ show fecundity values (23.5 and 23.4) that are significantly above all others. The lowest fecundity (9.3) is associated with the BALB/cJ strain.

The fecundity of female mice declines with both age and number of prior pregnancies. Few inbred females of any strain, with the exception of FVB/N, will produce more than five litters (Green and Witham, 1991). Irrespective of their past reproductive history, most inbred females exhibit greatly reduced fecundity by the age of 8–10 months. Male mice, like male humans, can remain fertile throughout their lives. However, older males that have become obese or sedentary are unlikely to breed.

Reproductive performance is among the characteristics most affected by inbreeding. Outbred animals and $F_1$ hybrids of all types will routinely surpass the inbred strains in all of the categories listed in Table 4.1 as a consequence of "hybrid vigor." With non-inbred animals, the frequency of productive matings is close to 100%, the age of first mating can be as early as 5 weeks, and litters can have as many as 16 pups. Finally, non-inbred females can sometimes remain fertile up to 18 months of age, and bring as many as 10 litters successfully to weaning.

## 4.2 GERM CELL DIFFERENTIATION AND SEXUAL MATURATION

### 4.2.1 Males

Male germ cell differentiation occurs continuously in the seminiferous tubules of the testes throughout the life of a normal animal. This process has been very well characterized in the mouse (Bellvé et al., 1977; Eddy et al., 1991), and only its salient features will be summarized here. Spermatogenic cells at different stages are classified into four broad categories—*spermatogonia, spermatocytes, spermatids*, and *spermatozoa*—with numerous substages defined within each category. All premeiotic cells are called spermatogonia; these include regenerating stem cells as well as those that have taken the path to terminal differentiation. With the commencement of meiosis, germ cells are called spermatocytes, and subsequent to meiosis, haploid cells are called spermatids. Finally, with the release of the morphologically mature product, the germ cells are called spermatozoa or, more simply, just sperm.

The timing of the stages of spermatogenesis in the mouse was described originally by Oakberg (1956a, 1956b). At birth, the testis contains only undifferentiated type A1 spermatogonia, which will serve as a self-renewing stem cell population throughout the life of a male mouse. By day 3, differentiation has begun through a series of mitotic divisions into more advanced spermatogonial stages ($A_2$, $A_3$, $A_4$, intermediate and type B spermatogonia). By 8–10 days, spermatocytes are observed for the first time in the leptotene phase of meiosis (Nebel et al., 1961). The meiotic phase is relatively long, extending over a 13 day period. When the male has reached 17–19 days of age, approximately 50% of the seminiferous tubules are found to contain cells in the late pachytene stage. The earliest postmeiotic cells—round spermatids—are not observed until after 20 days (Nebel et al., 1961). During the next 13 days, the round spermatids differentiate into elongating spermatids in which the sperm tail forms and the nucleus condenses. At the end of this process, morphologically mature sperm are released into the fluid-filled lumen.

The entire process of differentiation from stem cell to released spermatozoa is called *spermatogenesis*. The term *spermiogenesis* refers specifically to the final morphological differentiation of haploid cells into sperm. At the time of release into the lumen of the seminiferous tubules, sperm cells are still not physiologically mature. After leaving the testes, they pass through the *epididymis* where they undergo further biochemical changes. From the epididymis, they go to the *vas deferens*, where they are stored until ejaculation. The final stage of sperm maturation—known as *capacitation*—is required for fertilizing activity and does not occur until after contact has been made with the milieu of the female reproductive tract.

### 4.2.2 Females

Female germ cell differentiation operates under a two-phase time course dramatically different from that found in the male.[23] By the 12th or 13th day after conception, the primordial oocytes within the fetal ovary have undergone their last mitotic division and are referred to as oogonia. At this point, the young female, still not born, has produced all of the germ cells that she will ever have; the total number is somewhere between 30,000 and 75,000. All of these oogonia progress into meiosis, and by 5 days after birth, they reach the diplotene stage of prophase of the first meiotic division. At this point, also called the dictyate stage, the oogonia become arrested and remain quiescent until sexual maturation. As they move into the dictyate stage, all primordial oocytes acquire a coat of follicle cells; the complete coat surrounding each oocyte is called a follicle.

With the onset of puberty, the ovaries become activated by hormone stimulation, and every 4 days, a new group of oogonia are stimulated to proceed forward toward their ultimate differentiated state. This second phase of differentiation occurs over a period of 20 days. During this entire period—until a few hours prior to ovulation—the oocytes still remain fixed in the dictyate stage of meiosis, but they become highly active metabolically and increase greatly in size from 15 to 80 µm. The size of each follicle also increases through the addition of follicle cells up to a total of 50,000 per ovum. At 20 days after activation, oocytes have become competent for ovulation, which occurs in response to the correct hormonal cues during the estrus cycle described in the next section. During each natural cycle, only 6–16 oocytes are stimulated to undergo ovulation. The stimulated follicles swell with fluid and move to the periphery of the ovary where they burst out to begin their journey into the oviduct and further down the reproductive tract.

Stimulation to ovulate also releases the oocyte (now also called an egg) from its state of arrest and induces it to continue through meiosis. The first meiotic division is completed and the first polar body is formed prior to release from the ovary. The second meiotic division begins immediately but stops at metaphase, where the oocyte remains arrested until fertilization. Penetration by the sperm triggers completion of the final meiotic division and the formation of the second polar body.

Surprisingly, at least 50% of the oocytes present at birth degenerate before the mouse reaches 3 weeks of age. The vast majority of the remaining oocytes are never ovulated—many degenerate throughout the life of the animal, and all that remain are eliminated at the time of mouse menopause, which occurs at approximately 12–14 months of age.

## 4.3 MATING AND PREGNANCY

### 4.3.1 Puberty

The onset of puberty—when ovulation first occurs in a female, and when males have achieved full spermatogenic activity—is variable even among different animals within the same inbred strain. Although it is possible for some outbred females to reach puberty by the age of 4 weeks, the majority of females from most inbred strains first ovulate naturally between 6 and 8 weeks after birth (Table 4.1). Numerous environmental factors appear to have an effect on the timing of this event (Whittingham and Wood, 1983). Exposure to adult males or their urine can bring it on sooner, whereas adult females or their urine may retard its onset. Furthermore, 3–6-week-old females can be induced to ovulate with a specific regimen of hormone treatment as described in section 6.2.2.1. The onset of male puberty in most laboratory strains usually occurs between 34 and 38 days, however, it is sometimes possible for non-inbred males to reach sexual maturity by 30–32 days after birth. Thus, if one does not want littermates to mate with each other, they should be separated according to sex before the appropriate age is reached.

### 4.3.2 The estrus cycle

The normal estrus cycle of a laboratory mouse is 4–6 days in length. The cycle has been divided into four phases, which are distinguished by changes in physiology, morphology, and behavior. (1) The *proestrus* portion of the cycle begins when a new batch of eggs reach maturity within ovarian follicles that are ripe and large. External examination of the female will usually show a bloated vulva with an open vagina. (2) *Estrus* begins with the ovulation of fully mature oocytes. The vulva remains in an extended state with an open vagina, and females are maximally receptive to male advances. When mice are maintained on a standard light-dark cycle, the estrus phase will usually begin soon after midnight and last for 6–8 hours. (3) The *metestrus* phase follows, when mature eggs move through the oviducts and into the uterus. The vulva is no longer bloated, and the vagina is now closed.

At the end of metestrus, a physiological branch point occurs with the direction to be taken dependent on whether a successful copulation has occurred. The act of successful copulation induces hormonal changes that prepare the uterus for a pregnancy, which will ensue under normal circumstances. However, a sterile copulation—one that does not lead to fertilization—can induce a state of pseudopregnancy (see Section 6.2.3.2). A pseudopregnancy can extend the *metestrus* phase by as long as 10–13 days.

(4) If pregnancy does not occur, the metestrus phase is ultimately followed by the last phase of the estrus cycle, *diestrus*. Unfertilized eggs are eliminated, the vagina and vulva are at a minimum size, and new follicles begin to undergo a rapid growth for the next ovulation. (The proestrus and estrus phases together constitute the *follicular* phase; the metestrus and diestrus phases together constitute the *luteal* phase.)

### 4.3.3 Mating

Once animals have been together for more than a few days, mating will be restricted to the late proestrus/early estrus portion of the female cycle. It is only during this period that a female will be receptive and that a male will normally be interested. (However, in some instances, when a new couple is first brought together in a cage, the male will rape his partner, irrespective of her estrus phase.) Mating typically occurs over a period of 15–60 minutes with clear strain-specific differences: DBA males are quick (20 minutes) and BALB/c males are slow (one hour) according to Wimer and Fuller (1966). The male first examines the female genitalia, and then mounts her and withdraws from one to 100 times until ejaculation occurs during a final mounting. The male is quiet for a short period of time and then resumes normal activity. Although a full sperm count is not built up again for 2 days, it is possible for a male, especially an outbred one, to mate with up to three females in a single night, causing all to become pregnant. Different inbred strains have very different average times for recovery of libido, defined operationally as the time between attempted matings. DBA/2 mice can mate again within 1 hour, whereas B6 males usually wait for 4 days (Wimer and Fuller, 1966).

In some instances, one may want to maximize the rapid output of offspring from a single male. This situation could arise with rare genotypes, such as new mutants or first-generation transgenic founders. For this purpose, a single male can be rotated among sets of females (two or three per cage) in three or four cages. The factors that play a role in the length of each rotation have just been discussed: the length of the estrus cycle, the time it takes for a male to recover a full sperm count, and the libido recovery time. Together, these factors suggest an optimal rotation period of 4 days in each cage. For full optimization of offspring output, a male should receive two new, 8-week-old, virgin females in his cage, every 4 days.

### 4.3.4 Fertilization

Fertilization takes place in the upper reaches of the oviduct (a region referred to as the ampulla). The egg remains viable for 10–15 hours after ovulation, although a gradual aging process slowly reduces the probability that fertilization will occur. Fertilization causes an immediate activation of the egg and induces the completion of the second meiotic division, which leads to the formation of the second polar body within 2 hours.

The actual process of fertilization can be divided into a series of highly ordered steps that lead ultimately to the joining of a single sperm cell with an ovulated egg (Wassarman, 1993). The first step in this process occurs with the binding of multiple spermatozoa to the zona pellucida, a thick extracellular coat that surrounds the egg. The association between the zona and the sperm surface triggers the acrosome reaction that affects an elongated sperm-specific membrane-bound organelle just below the surface, which contains a specialized protease called acrosin. The acrosome reaction is a form of exocytosis that results in the complete loss of the plasma membrane overlying the acrosome in hybrid vesicles along with the outer acrosomal membrane. The acrosomal contents are released, and these allow the resulting "acrosome-reacted" sperm to use protease to digest its way through the zona pellucida to reach the perivitelline space between the zona and the egg plasma membrane. Finally, fusion occurs between the egg plasma membrane and the plasma membrane overlying the equatorial region of a single sperm cell. Fusion leads to the activation of the egg and the initiation of embryonic development.

The ultimate fusion reaction is not species-specific and can occur between heterologous gametes when the zona pellucida is first removed from the egg. Thus, in general, the main biochemical barrier to cross-species fertilization appears to lie within the initial interaction between the sperm plasma membrane and the egg zona pellucida. The specificity of this interaction implicates the existence of specific complementary molecules on egg and sperm, referred to respectively as the "sperm receptor" and the "egg binding protein" or EBP. The sperm receptor has been identified as a specific zona protein called ZP3

(Wassarman, 1990). The identity of the sperm surface EBP is still under investigation with multiple candidates described to date.

### 4.3.5 Determination of copulation and pregnancy

After a successful copulation has been completed, particular components of the male ejaculate will coagulate to form a hard *plug* that occludes the entrance to the vagina. The plug is a coagulum of fluids derived from both the vesicular and coagulating glands, and as such, it can be produced even by a vasectomized male. Usually the plug is visible through a simple visual examination of the vulva. In some instances, a probe will be required to detect a plug located further back in the vagina. The most common probe used for this purpose is a simple dental tool with a blunt end. "Plugging" should be performed as early as possible in the morning after a potential mating. By noon, some inbred strain plugs will begin to disappear, however, most will persist for 16–24 hours after copulation. Plugs formed by outbred mice can persist for several days.

Later in the pregnancy, from 10–12 days postconception and beyond, it becomes possible to feel the maturing fetuses within the uterus by simple palpation. Pregnancy palpation is most readily carried out on older, multiparous females who have looser skin and are more accustomed to being handled. Right-handed workers should hold the female in the left hand (left-handers should hold the mouse in the right hand), with the thumb and forefinger grasping the skin behind the neck, and the smallest finger holding back the tail. The other fingers of this hand should be brought in behind the mouse to arch her forward. When the female is securely held by one hand and relatively calm, one should use the other hand to close down firmly on the abdomen close to the spine on one side at a time with the forefinger and, thumb and then gently move the fingers out. Initially, a pregnant female will seem to have a string of beads on each side of her body. As development proceeds, these "beads" will mature into larger, more defined shapes. With experience, this method can be used to determine the gestational stage of a pregnancy to within a single day.

It is possible to identify a state of pregnancy in young females by a simple visual inspection that does not even require one to handle the animal. The gestational day at which this becomes possible is greatly dependent on a number of factors including the age of the female, the number of fetuses inside, and whether she has given birth previously. For first-time pregnant females carrying large litters, tell-tale bulges from the center of her body can be detected by day 15. At the opposite extreme, older multiparous females with small litters never "show" in this way. Fortunately, these older animals are easier to palpate when a prenatal determination is required.

### 4.3.6 The gestational period

The gestation period for the mouse ranges from 18 to 22 days. Different strains have different averages within this range but even within a single strain, and even for a single female, there can be significant differences from one pregnancy to the next. Many factors can have an effect on the length of pregnancy. For example, larger litters tend to be born earlier (Rugh, 1968) , as is the case with humans as well. Non-inbred females tend to have shorter pregnancies then inbred ones, but this may be simply because they tend to produce larger litters as well as larger pups. Birth occurs most frequently between the hours of midnight and 4.00 a.m. when animals are maintained under a standard light–dark cycle; however, it can occur anytime of the day or night.

The gestation period can be greatly extended when the pregnant mother continues to nurse a previous litter. Prolongation up to 7 days is not uncommon, and birth can sometimes be pushed back by as many as 16 days (Grüneberg, 1943; Bronson et al., 1966). This fact should be kept in mind when trying to count back from the day of birth to the day of conception in order to determine paternity for females in contact with sequential males.

### 4.3.7 Effects of a foreign male on pregnancy and pup survival

The fertilized embryo is a free-floating entity in the female reproductive tract for the first 4.5–5 days of development. It is during this pre-implantation period that external events can play a role in determining whether a successful implantation will occur. Obvious disturbances to the mental health of the pregnant female— such as erratic lighting, extremes in temperature or humidity, high noise levels, or insufficient food and water—can cause a failure to implant. In addition, there is one other less obvious disturbance that is highly significant in the eyes of the female—the introduction into her cage of a male other than the one with whom she had mated. If the foreign male is not genetically identical to her partner, he can cause a premature termination of the pregnancy through a mechanism that is almost certainly a hormonally induced block to implantation (Bruce, 1959, 1968). This pregnancy block is also known as the *Bruce effect* (after its discoverer) and it provides an obvious selective advantage by ensuring that females will use their resources only to raise offspring who carry the genes of the intruding male (who is presumably more fit since he has displaced the original mating male). With the previous pregnancy terminated, the female can quickly become pregnant again with her new partner.

It is interesting that females do *not* recognize males from the same inbred strain as foreign (Bruce, 1968). On the other hand, a pregnancy block is induced in nearly all other cases. These findings indicate that one or more genetic differences are responsible for the distinction between the original and the intruder male, but in addition, they clearly show that the genetic recognition system is highly polymorphic. Further studies with congenic and coisogenic strains have demonstrated conclusively that a major component of this recognition system is the highly polymorphic class I family of genes in the major histocompatibility complex (Yamazaki et al., 1986).

The Bruce effect has important implications for the management of a breeding mouse colony. Quite simply, if a mating event has occurred and one wants to recover live-born offspring from this mating, for the first 5 days that follow, the pregnant female should not be placed either into a cage with a foreign male or in contact with bedding that has been soiled by a foreign male. After this initial stage, there is no longer any problem. In fact, if one wishes to quickly set up a new mating pair, one should be sure to do it before the litter is born. If a foreign male is in the cage at the time of birth, he will normally accept the newborn pups. (Presumably, he "thinks" these pups are his own.) On the other hand, if a male is placed into a new cage that already has newborn pups, he is likely to kill them ("knowing" that he could not possibly have been the father).

## 4.4 THE POSTNATAL PERIOD

### 4.4.1 Postnatal development

A mouse is born naked with closed ears and eyes, and if a female, with a closed vagina. Hair begins to appear at 2– days, ears open at 3–5 days, and eyes open at about 14 days. Typically, the vagina opens at 24–28 days of age, but it can be delayed in some mice until they are 35–40 days old. As soon as the eyes are fully functional, at about 16 days, pups will begin to eat solid food. However, nursing can continue to at least the end of the third week, and sometimes a week or more longer. By the end of the third week of life, a young mouse resembles the adult in every aspect other than size and sexual differentiation.

### 4.4.2 Determination of sex

The sex of the newborn mouse can be determined from both the distance that separates the genital papilla and the anal opening, and from the general appearance of the urogenital–anal region. The genital–anal distance in newborn males is generally 50% greater than in newborn females. In addition, the male genitalia are often more prominent, and in the prescrotal region below, a dark pigmentation is often visible. If a litter is large enough, it is likely to have both males and females. The simplest way to become adept at distinguishing gender is through pairwise comparisons of the pups—in each hand, a newborn pup can be held gently, but firmly, between the index finger and the thumb in an upside-down position. As neonates age, gender determination becomes somewhat more difficult. It becomes easier again at 8–10 days with the appearance of nipples along the ventral side of the female, and at the age of weaning (18–28 days), when the penis has developed more fully in the male.

### 4.4.3 Lactation, culling, and supplementing litters

The most important factor in the growth of infant mice is the amount of milk available for suckling. Thriving newborns will begin to nurse immediately after birth, and within a matter of hours, it is possible to clearly see the milk in their stomachs through their translucent bodies. The amount of milk present is an excellent gauge of the likelihood with which a young pup will become a vibrant, healthy adult. When little or no milk is present by 6 hours after birth, it is almost certainly the case that something is wrong with either the pup or the mother.

For litters of four or more pups, the amount of milk produced by a lactating mother will increase with the number of young, but the increase will not be proportional (Grüneberg, 1943). However, if there are two or more mature females in a cage, all can be induced to lactate in response to a single litter. This phenomenon makes sense from an evolutionary point of view since females living together in a deme are likely to be related—often as sisters—and thus child-care sharing serves to enhance the survival of the common gene pool. For the mouse geneticist, however, it means that when there are two or more females in a breeding cage, one can not use a state of lactation as a means to distinguish the birth mother.

When the number of pups is eight or more, a single inbred mother will not be able to provide the nourishment required for the optimal growth and development of all. Thus, when it is not detrimental to the experimental protocol, it makes sense to cull litters soon after birth. In some cases, one will wish to select pups according to sex (as described in the previous section), or other visible phenotypes such as eye or coat color (at day 2–4), or gross morphological characters. To ensure optimal growth of selected animals, litters should be culled to 5–6 pups during the first days after birth. A further reduction to 3–4 youngsters can be carried out between days 10 and 14.

With the common strains of mice, it is not often the case that only one or two pups will be born in a litter. However, this situation can arise more frequently with the breeding of animals that carry embryonic lethal mutations, and it is also problematic. Especially for first-time mothers, 1–2 pups may not provide the level of suckling stimulation required to stimulate milk production effectively. When this problem arises (as indicated by an insufficient level of milk in the stomach), the simplest solution is to supplement the litter with 2–3 age-matched pups that are clearly distinguishable from those born to the mother.

### 4.4.4 Foster mothers

When newborn animals are not receiving sufficient amounts of milk and it is likely that the mother is the problem, one can consider fostering as a last resort. The foster mother should be an experienced female with her own newborn litter. This entire litter should be removed from the foster mother's cage and placed onto a clean surface. The pups to be fostered can then be added to this group and an equivalent (or greater) number of the foster mother's pups can be removed; when there is a choice, the largest and best-fed of these pups should be eliminated. The new mixed litter can now be placed back into the foster mother's cage. Obviously, it is critical to be able to easily distinguish the pups of the foster mother from those that have been added. This is most readily accomplished by known coat color differences between the two litters that have been mixed together.

### 4.4.5 Age of weaning

Mice can be weaned from their mothers when they are as young as 18 days old. However, especially for fragile inbred strains or animals that carry deleterious mutations, it is best to wait until they are 4 weeks of age. When young are kept with their mothers for a longer period, they are more likely to thrive as adults.

### 4.4.6  Postpartum estrus

Amazingly, within 28 hours of giving birth, a nursing mother will normally go into a postpartum estrus that can allow her to become pregnant again immediately. There is a tendency to ovulate 12–18 hours after the time of birth, but this can be countered by the tendency to ovulate nocturnally (Bronson et al., 1966). The level of postpartum estrus fertility is reduced somewhat relative to that achieved during a normal estrus cycle. A postpartum pregnancy can have negative consequences for the litter already born as well as the one on the way. Since the mother is forced to split her resources between two sets of "progeny," her milk production will fall off more quickly than would otherwise be the case. In addition, the duration of the postpartum pregnancy can be extended for up to 2 weeks. Finally, when the second litter is born, there will be competition between the new pups and the older ones (if they are not yet weaned), and the new ones can suffer malnourishment or death.

### 4.4.7  Genetically controlled variation in the adult mouse

Although it is possible to make general statements about the gross characteristics of all laboratory mice—they reach adult weights of 22–40 g, they have life spans of 1–3 years, they have a gestation period of 18–22 days, and an average litter size of 5–10 pups—much more discrete numbers can be obtained for individual inbred strains. It is not surprising that the members of an inbred strain show much less variance in these numbers since they are, for all practical purposes, genetically identical. Statistical evaluations of the growth and reproductive characteristics of the older, established inbred lines have been compiled in two different handbooks. One is published by the Federation of American Societies for Experimental Biology, abbreviated FASEB (Altman and Katz, 1979). The second is published by the Jackson Laboratory in regularly updated editions (Green and Witham, 1991). The Jackson Laboratory handbook is available without charge and has detailed information on each of the inbred strains that are sold to investigators.

A quick survey of the information provided in these books provides ample evidence of the wealth of genetic variation that exists among the classical inbred strains in terms of gross morphological and physiological characters. Although most of this variation shows a high degree of heritability, it is polygenic and, as a consequence, it was not readily accessible to the types of genetic analyses carried out in the past. However, with the new genetic markers described in Section 8.3, polygenic traits are no longer beyond reach, and it is only matter of time before many of the common forms of variation in mice—which often have human counterparts—will be linked to individual loci and, ultimately, to cloned genes (see Section 9.5).

One note of caution is the possibility that genetic drift within an inbred line can lead to a drift in gross phenotype. This is a concern because mutations occur constantly and inbred lines are continuously rederived through two-member population bottlenecks, which can lead to the rapid fixation of new genetic variants. In general, gross phenotypic features are controlled by multiple genes,

# 5

# The Mouse Genome

## 5.1 QUANTIFYING THE GENOME

Even before the discovery of the structure of DNA, it was clear that the fertilized mammalian egg could contain only a finite amount of genetic information, and that this information was all that was needed to define something as complicated as a whole mouse or human being. However, with the demonstration of the double helix and the unraveling of the relationships that exist between basepairs, codons, genes, and polypeptides, it became possible to determine just how finite the total sum of genetic information actually is. But the problem that still looms large is an understanding of the *essential* genetic information needed to make a mammal. Is it the total amount of DNA in a haploid set of chromosomes, just that portion of DNA that does not include repeated sequence copies, transcription units, coding and regulatory regions, or only those genes required for viability? In some cases, it seems possible to distinguish among what is essential, what is nice to have but not essential, and that which serves no useful function at all. However, in many cases, the distinctions are still not clear. This section addresses the quantification of the genome at various levels of analysis.

### 5.1.1 How large is the genome?

Quantitative DNA-specific staining can be achieved with the use of the Feulgen reagent. Through microphotometric measurements of the staining intensity in individual sperm nuclei, it is possible to determine the total amount of DNA present in the haploid mouse genome (Laird, 1971). These measurements indicated a total haploid genome content of 3 pg, which translates into a molecular weight of $1.8 \times 10^{12}$ daltons (Da).

The smallest unit of genetic information is the basepair (bp), which has a molecular weight of $\sim 600$ Da. By dividing this number into the total haploid DNA mass, one arrives at an approximate value for the total information content in the haploid genome: three billion bp, which can also be written as three million kilobasepairs (kb) or 3,000 megabasepairs (mb). All eutherian mammals have genomes of essentially the same size.

It is instructive to consider the size of the mammalian genome in terms of the amount of computer-based memory that it would occupy. Each basepair can have one of only four values (G, C, A, or T) and is thus equivalent to two bits of binary

with changes in each having small additive effects. The extent of genetic drift should be roughly linear with time; thus, the longer the period that has elapsed since a study was performed, the more likely it is that a gross characteristic can change in a statistically significant manner. Genetic drift is much more likely to occur in outbred laboratory stocks, which are heterogeneous to begin with, but also pass through narrow population bottlenecks (Papaioannou and Festing, 1980).

The anatomy of the laboratory mouse is described with numerous illustrations by Cook (1983). The average body weight of a full-grown adult member of the standard inbred strain C57BL/6J is 30 g for males and 25 g for females (Altman and Katz, 1979). The largest of the commonly available inbred strains is AKR/J with males that reach 40 g in weight. The smallest is 129/J with full-grown males and females that weigh 27 g and 22 g respectively. Hybrids between inbred strains are usually larger than either parent.

The life span of a mouse is highly strain dependent. At one extreme, the AKR/J mouse has a mean life span of only 10 months due to its propensity to develop lymphatic leukemia; at the opposite extreme, the life span of the B6 mouse is among the longest of the common inbred lines. B6 animals have a median life span of 27–28 months—somewhat over 2 years (Zurcher et al., 1982; Green and Witham, 1991). The longevity of inter-strain hybrids tends to be greater than either inbred parent. The hybrid formed by a cross between B6 and DBA/2J (called B6D2F1) has a median life span of over 2½ years and some animals survive as long as 3½ years (Green and Witham, 1991).

The genetic factors responsible for longevity have been studied by a number of different investigators. In one study, the second generation F2 offspring derived from an outcross-intercross between the inbred strains B6 and DBA/2J were analyzed for correlations between longevity and genotype at three different autosomal loci—H-2, b (brown), and d (dilute)—as well as the two sex chromosomes. Statistically significant correlations were observed between longevity and particular genotypes for each of the loci analyzed. This does not mean that the tested loci, in and of themselves, have any bearing on longevity, but rather that genes in their vicinity do. The fact that an effect was observed with all of the loci tested points to a strong likelihood that the number of genes involved in this polygenic trait is many. This result should not be unexpected, since one can imagine that many different phenotypic characteristics will have an indirect effect on life span.

## 4.5 ASSISTED REPRODUCTION FOR THE INFERTILE CROSS

### 4.5.1 Artificial insemination

Although artificial insemination is a critical tool for reproductive biologists working with other species (including humans), it is not often used by mouse geneticists. Its major use in other species is to initiate a successful pregnancy when, for any of a number of reasons, the male cannot or should not, be directly involved in the process of mating. Artificial insemination has been a boon to the cattle industry because the semen from one good bull can be shipped around the world to impregnate unlimited numbers of females. Male mice are somewhat smaller than bulls and, as a consequence, the whole animal can be shipped for a cost that is likely to be the same (or less) than one would pay for frozen semen alone. Furthermore, obtaining semen from a mouse is a "one-shot" deal. Since assisted masturbation of the male mouse is not practical, sperm must be recovered from the epididymis after the animal has been sacrificed.

There are some special cases where artificial insemination can be used as an experimental tool for the study of the mouse. One example is in those cases where, for behavioral reasons, males of a particular strain refuse to mate with selected females of another strain. This scenario is most likely to occur when the males and females are members of different *Mus* species. West and colleagues (1977) used artificial insemination to overcome this problem in order to determine the viability of various hybrid embryos formed between distantly related members of the *Mus* genus.

Another use of artificial insemination is in those cases where one wants to alter the composition of the sperm pool. For example, Olds-Clarke and Peitz (1985) were able to analyze the relative fertilizing potentials associated with sperm obtained from two different males by mixing equal numbers together before insemination. Finally, there will always be the case where a one-of-a-kind male—such as a first generation transgenic or another new mutant—refuses to participate in the mating process. As a last resort, one can recover epididymal sperm from such an animal for a single chance at achieving a pregnancy. Detailed protocols for sperm recovery and artificial insemination have been described elsewhere (Rugh, 1968; West et al., 1977; Olds-Clarke and Peitz, 1985).

When a choice is possible, females to be inseminated should not be inbred; $F_1$ hybrids and random-bred animals will always have higher levels of fertility. A successful fertilization can only occur when the inseminated female is in the late proestrus/early estrus stage of the estrus cycle. Appropriately staged females can be obtained either by visual inspection of naturally cycling animals (as described earlier in this chapter) or through superovulation (see Section 6.2). The implantation of fertilized embryos will occur only in females that have been stimulated into a state of pseudopregnancy (Section 6.2.3). If the investigator intends to use a sterile stud male for this purpose, the mating should be performed after the insemination (within 0.5–2 hours) so that the vaginal plug does not interfere with the protocol (Olds-Clarke and Peitz, 1985). If pseudopregnancy is to be induced manually, it should be accomplished in the fully alert female prior to the insemination protocol (see Section 6.2.3 for details).

### 4.5.2 Transplantation of ovaries

In a small number of instances, females that express certain mutations may be fertile in the sense that they are able to produce functional oocytes but infertile in the sense that they are physically unable to bring offspring to term. Such females may not be able to mate, they may not be fit enough to allow gestation to proceed properly or they may be unable to give birth to live offspring.

In such cases, it is possible to transplant the ovaries from these incapac[itated] females into healthy females of another strain as means for obtaining germ[line] transmission (Russell and Hurst, 1945; Stevens, 1957). This protoc[ol is] commonly used at the Jackson Laboratory to maintain several mutant strai[ns of] mice, including those that carry the obese mutation, the dwarf mutation, [and] dystrophia muscularis (muscular dystrophy) mutation (Green and Wi[tham,] 1991).

### 4.5.3 In vitro fertilization

A third method of assisted reproduction entails fertilization outside the f[emale] reproductive tract. There are two general types of circumstances when in[vitro] fertilization becomes useful: when the *female* partner of a cross is unable to [carry] litters to term for one reason or another; and when an investigator wan[ts to] establish the timing of fertilization to a more precise degree and/or wan[ts to] synchronize the development of a batch of embryos for later recovery [and] analysis. A detailed discussion of this procedure and all other aspects of em[bryo] manipulation are provided in the manual by Hogan and her colleagues (1[986]).

code information (with potential values of 00, 01, 10, 11). Computer information is usually measured in terms of bytes that typically contain 8 bits. Thus, each byte can record the information present in 4 bp. A simple calculation indicates that a complete haploid genome could be encoded within 750 megabytes of computer storage space. Incredibly, small lightweight storage devices with such a capacity are now available for desktop computers. Of course, the computer capacity required to actually interpret this information will be many orders of magnitude larger.

## 5.1.2  How complex is the genome?

Another method for determining genome size relies upon the kinetics of DNA renaturation as an indication of the total content of different DNA sequences in a sample. When a solution of double-stranded DNA is denatured into single strands, which are then allowed to renature, the time required for renaturation is directly proportional to the *complexity* of the DNA in the solution, if all other parameters are held constant. Single-stranded and double-stranded molecules are easily distinguished by various physical, chemical, and enzymatic procedures.

Complexity is a measure of the information contained within the DNA. The maximal information possible in a solution of genomic DNA purified from one animal or tissue culture line is equivalent to the total number of basepairs present in the haploid genome.[24] The information content of a DNA solution is independent of the actual amount or concentration of DNA present. DNA obtained from one million cells of a single animal or cell line contains no more information than the DNA present in one cell. Furthermore, if sequences within the haploid genome are duplicates of one another—repeated sequences—the complexity will drop accordingly.

The effect of complexity on the kinetics of renaturation can be understood by viewing the system through the eyes of a single strand of DNA, randomly diffusing through a solution, looking for its complementary partner. For example, imagine two DNA solutions, both $2 \mu g/ml$ in concentration, but one from a genome having a complexity of $3 \times 10^9$ bp, and the second from a genome having a complexity of only $3 \times 10^8$ bp. In the second solution, with the same quantity of DNA but ten-fold less complexity, each segment of DNA sequence will be represented ten times as often as any particular segment of DNA sequence in the first solution. Thus, a single strand will be able to find its partner ten times more quickly in the second solution as compared to the first solution. The speed with which a DNA sample renatures can be expressed in the form of a *Cot curve*, which is a graphic representation of the fraction of a sample that has renatured (along the $Y$ axis) as a function of the single stranded DNA concentration at time zero ($C_0$) multiplied by the time allowed for renaturation ($t$) shown on the $X$ axis. The $C_0 t$ value attained at the midpoint of renaturation—when half of the molecules have become double-stranded—is called $C_0 t_{1/2}$ and is used as a indicator of the complexity of the sample being measured. Different $C_0 t_{1/2}$ values can be compared directly to allow a determination of complexity in a new sample relative to a calibrated control.

Renaturation analysis of mouse DNA reveals an overall complexity of approximately $1.3-1.8 \times 10^9$ bp. This value is only 40–60% of the size of the complete haploid genome and it implies the existence of a large fraction of repeated sequences. In fact, a careful analysis of the renaturation curve indicates that 5% of the genome renatures almost one million times faster than the bulk of the DNA. This "low complexity class" of sequences represents the satellite DNA, which is discussed in detail in Section 5.3.3. After renaturation of the satellite DNA class comes a very broad class of repeated sequences (whose copy number varies from several hundred thousand to less than ten), which merges into the final bulk class of "unique" sequences. With the advent of DNA cloning and sequencing, the "repeated sequence" class of mouse DNA has been divided into a number of functionally and structurally distinct subclasses, which are also discussed more fully in Sections 5.3 and 5.4. It was originally assumed that nearly all of the protein-coding genes would be present in the final renaturation class of unique copy sequences. However, we now know that the situation is not that simple and that many genes are members of gene families that can have anywhere from two to 50 similar, but non-identical, cross-hybridizing members.

### 5.1.3 What is the size of the mouse linkage map?

The genome size of any sexually reproducing diploid organism can actually be measured according to two semi-independent parameters. There is the physical size measured in numbers of basepairs, as just discussed, and there is recombinational size measured in terms of the cumulative linkage distances that span each chromosome (discussed fully in Section 7.1). The size of the whole mouse linkage map can be arrived at by a number of different approaches. First, one can perform a statistical test on the frequency with which new loci are found to be linked to previously identified loci. In 1954, Carter used this test on then-available data for 43 loci to estimate the size of the complete mouse linkage map at $1,620 \pm 352 \, cM$[25] (Carter, 1954).

A second estimate is based on counting the number of chiasmata that appear in spreads of chromosomes prepared from germ cells undergoing meiosis and viewed under the microscope. A chiasma (the singular of chiasmata) represents the cytological manifestation of crossing over; it is seen as a visible connection between non-sister chromatids at each site where a crossover event has occurred between the maternally and paternally derived chromosomes of the animal that provided the sample.[26] Chiasma formation occurs after the final round of DNA replication when each of the two homologs contains two identical sister chromatids—the genomic content of cells at this stage is represented by the notation "4N." Each crossover event involves only two of the four chromatids present. Thus, there is only a 50% chance that any one crossover event will be segregated to any one haploid (1N) gamete and so the total number of crossovers segregated into any one gamete genome will be approximately half the number of chiasmata present within 4N meiotic cells. Thus, one can derive an estimate of total linkage distance by multiplying the average number of chiasmata observed per meiotic cell by the expected interchiasmatic distance (100 cM) and dividing by two. This analysis provided the basis for a whole mouse genome linkage size of 1,954 cM (Slizynski, 1954).

With the generation of high-density whole genome linkage maps based on the segregation of hundreds of loci, it is now possible to determine map size directly from the distance spanned by the set of mapped loci. In two cases, this calculation was performed for data generated within the context of single crosses: the resulting map sizes were 1,424 cM for a B6 × M. *spretus* intercross–backcross (Copeland and Jenkins, 1991) and 1,447 cM for an $F_2$ intercross between B6 and M. m. *castaneus* (Dietrich et al., 1992). Two other direct estimates of 1,468 cM and 1,476 cM are based on whole genome consensus maps formed by the incorporation of data from large numbers of different crosses that used overlapping sets of markers for mapping (Hillyard et al., 1992; Lyon and Kirby, 1992).

All of these estimates are remarkably consistent with each other and yield a simple average value of 1,453 cM.[27] This consistency is remarkable because, in isolated regions of the genome, linkage distances are highly strain-dependent with differences that vary by as much as a factor of two (see Section 7.2.3). Nevertheless, the accumulated data suggest that the *overall* level of recombination is predetermined in the *Mus* genus and maintained from one cross to another through compensatory changes so that suppression of recombination in one region will be offset by an increase in recombination in another region of the genome.

One can derive an average equivalence value between the two metrics of genome measurement described in this section—kilobases and centimorgans—of approximately 2,000 kb per centimorgan. As mentioned above and discussed in Section 7.2.3.3, the actual relationship between linkage distance and physical distance can vary greatly in different parts of the genome as well as in crosses between different strains of mice.

### 5.1.4 What proportion of the genome is functional?

Bacterial species are remarkably efficient at packing the most genetic information into the smallest possible space. In one analysis of a completely sequenced 100 kb region of the E. *coli* chromosome, it was found that 84% of the total DNA content was actually used to encode polypeptides (Daniels et al., 1992). Most of the remaining DNA is used for regulatory purposes, and only 2% was found to have no recognizable function.

In higher eukaryotes of all types, the situation has long been known to be quite different. The early finding that some primitive organisms had haploid genome sizes which were many times larger than that of mammals[28] led to the realization that large portions of higher eukaryotic genomes might be "non-functional." However, to answer the question posed in the title to this section, one must first define what is meant by functional. Are entire transcription units considered functional even though, in most cases, 80% or more of the transcript will be spliced away before translation begins? Are both copies of a perfectly duplicated gene considered functional even though the organism could function just as well without one. What about the twilight class of pseudogenes which, in some cases, may be functional in some individuals but not others, and may serve as a reservoir for the emergence of new genetic elements in a future generation? Finally, comparative sequence analysis over long regions of the mouse and human genomes shows evolutionary conservation over stretches of sequence that do not have coding potential or any *obvious* function (Koop and Hood, 1994). However, sequences can only be conserved when selective forces act to maintain their integrity for the benefit of the organism. Thus, conservation implies functionality, even though we may be too ignorant at the present time to understand exactly what that functionality might be in this case.

Taking all of these caveats into consideration, and defining functional sequences as those with coding potential or with potential roles in gene regulation or chromatin structure, one can come up with a broad answer to the question posed in this section based on a synthesis of the data described in the next section. The fraction of the mouse genome that is functional is likely to lie somewhere between 5% and 10% of the total DNA present.

### 5.1.5 How many genes are there?

#### 5.1.5.1 Gene density estimates

How many genes are in the genome? A truly accurate answer to this question will be a long time in coming. The complete sequence of the genome will almost certainly provide a means for uncovering most genes, however, an unknown percentage will probably still remain hidden from view. But, in the absence of a complete sequence, one is forced to make multiple assumptions in order to come up with just a broad estimate of the final number.

One approach to estimating gene number is to derive an average gene size and then determine how many average genes can fit into a 3,000 megabase pair space. Unfortunately, the sizes of the genes characterized to date do not form a nice discrete bell curve around some mean value. The first mammalian gene to be characterized—*Hbb*—encodes the beta globin polypeptide; the *Hbb* gene has a length of the order of 2 kb. The alpha globin gene is even smaller with a length of less than 1 kb. At the opposite extreme is the mouse homolog of the human Duchene's muscular dystrophy gene (called *mdx* in the mouse); at 2,000 kb, the size of *mdx* is three orders of magnitude larger than *Hbb*. Nevertheless, a survey spanning all of the hundreds of mammalian genes characterized to date would seem to suggest that *mdx* is an extreme example, with most genes falling into the range of 10–80 kb, with an median size in the range of 20 kb. This estimate must be considered highly qualified, since size could play a role in determining which genes have been cloned and characterized.

Interestingly, a similar estimate of median gene size is obtained by viewing complete cellular polypeptide patterns on two-dimensional gels where the highest density of proteins appears in the 50–70,000 $M_r$ window for all cell types. A "typical" polypeptide in this range will have an amino acid length of ~600, encoded within 1,800 nucleotides, that will "typically" be flanked by another 200 nucleotides of untranslated regions on the 3′ and 5′ ends of a 2 kb mRNA that has been "typically" spliced down from an original 20 kb transcript that included 18 kb of intronic sequences.[29] If one assumes an average inter-gene distance of 10 kb—including gene regulatory regions and various non-essential repetitive elements to be discussed later—one obtains an average density of one gene per 30 kb. When this number is divided into the whole 3,000 mb genome, one derives a total gene number of 100,000.

Actual validation of a gene density in the range just estimated has been obtained for the major histocompatibility (MHC) region of the human and mouse genomes. With intensive searches for all transcribed sequences present within portions of the 4 mb MHC region, a gene density of one per 20 kb has been found (Milner and Campbell, 1992). Direct extrapolation of this gene density to the whole genome would yield a total of 130,000 genes. However, such an extrapolation is probably not valid since the average gene size in the MHC appears to be significantly smaller than the average overall. Another problem is that a significant proportion of the "genes" in the MHC (and elsewhere as well) are non-functioning pseudogenes. If one assumes a pseudogene rate of 20%, the value of 130,000 is reduced down to near 100,000.

A serious problem with all estimates made from the extrapolation of "average" genes is that genomic regions containing smaller, more densely packed genes will contribute disproportionately to the total number. As an example only, if 1% of the genome was occupied by (mostly uncharacterized) short genes that were only 500 bp in length and were packed at a density of one per kilobase, this class alone would account for 30,000 genes to be added onto the previous 100,000 estimate.

### 5.1.5.2 Number of transcript estimates

A very different approach to placing boundaries on the total gene number is to estimate the number of different transcripts produced in various cell types. Estimates of this type can be made, in a manner similar to that described for Cot studies, by analyzing the kinetics of mRNA–cDNA renaturation for a determination of the "complexity" of transcript populations in single cell types or tissues. This approach allowed Hasties and Bishop (1976) to estimate the presence of 12,000 different transcripts (representing the products of 12,000 different genes) in each of the three tissues analyzed—liver, kidney, and brain. However, as these investigators indicate, the brain in particular is a very complex tissue with millions of cells that are likely to have different patterns of gene expression. Genes expressed at low levels in a small percentage of cells will go undetected in a broad-tissue analysis of the type described. Thus, the actual level of gene expression in the brain, and other complex "tissues" like the developing embryo, could be much greater than the number derived experimentally. A revised complexity estimate of 20–30,000 has been suggested for the brain.

The only way in which estimates of transcript number could provide an estimate of total gene number is if every tissue in the body was analyzed at every developmental stage, and the number of cell- or stage-specific transcripts was determined apart from the number that were expressed elsewhere. A comprehensive analysis of this type is impossible even today, but some simple estimates can be made. For example, by analysis of cross-hybridization between sequences from different tissues, an overlap in expression of 75–85% has been estimated (Hasties and Bishop, 1976). This would suggest that perhaps 3,000 genes may be uniquely expressed in any one tissue relative to any other. However, when the data from many tissues are brought together, the actual number of tissue-specific transcripts is likely to be further reduced. On the other hand, it is also the case that some genes are likely to function in some tissue types only during brief periods of development.

Interpretation of the accumulated data provides a means only for estimating the minimum number of transcribed genes that could be present in the genome. By adding the brain estimate of ~25,000 to ~1,000 unique genes for each of 25 different tissue types, one arrives at a minimum estimate of 50,000.

### 5.1.5.3 Vital function estimates

Another independent, and very old, method for estimating gene number is to first saturate a region of known length with mutations that cause homozygous lethality, then count the number of lethal complementation groups and extrapolate from this number to the whole genome. The assumption one makes with an approach of this type is that the vast majority of genes in the genome will be essential to the viability of the organism. If one eliminates the expression of a "vital" gene through mutagenesis, the outcome will be a clear homozygous phenotype of prenatal or postnatal lethality.

It has long been clear that the early assumption of vitality associated with most genes is incorrect. Even genes thought to play critical roles in cell cycling and growth, such as p53, can be "knocked out" and still allow the birth of normal-looking viable animals (Donehower et al., 1992). Whole genomic regions of 550 kb in length can be eliminated with a resulting phenotype not more severe than short ears and subtle changes in skeletal structures (Kingsley et al., 1992). Observations of this type can be interpreted in two ways. First, there is likely to be some redundancy in many genetic pathways so that, if the absence of one gene prevents one pathway from being followed, another series of unrelated genes may provide compensation (sufficient for viability) through the use of an alternative pathway. Second, genes need not be vital to be maintained in the genome. If a gene provides even the slightest selective advantage to an animal, it will be maintained throughout evolution.

Although vital genes will represent only a subset of the total in the genome, it is still of interest to determine the size of this particular subset. In two saturation mutagenesis experiments in regions from different chromosomes, estimates of 5,000–10,000 vital genes were derived (Shedlovsky et al., 1988; Rinchik et al., 1990b). This range of values is likely to represent only 5–10% of the total functional units in the mouse genome. However, it is interesting that the number of vital genes in mice is not very different from the number of vital genes in *Drosophila melanogaster*, where the genome size is an order of magnitude smaller. This suggests that genes added on to the genome later in evolutionary time are less likely to be vital to the organism and more likely to help the organism in more subtle ways.

### 5.1.5.4 Overview

From the discussion presented in this section, it seems fair to say, with a high level of confidence, that the actual number of genes in the mammalian genome will be somewhere between 50,000 and 150,000. As of 1993, fewer than 10% of these genes have been characterized at any level from DNA to phenotype, and many fewer still are fully understood in terms of their effect on the organism and their interaction with other genes. Although the efforts to clone and sequence the entire human and mouse genomes will provide an entry point into many more genes, an understanding of the relationship between genotype and phenotype, in nearly every case, will still require much more work with the organism itself. Thus, the need to breed mice is likely to remain strong for many years to come.

## 5.2 KARYOTYPES, CHROMOSOMES, AND TRANSLOCATIONS

### 5.2.1 The standard karyotype

#### 5.2.1.1 Chromosome number and banding patterns

All of the *Mus musculus* subspecies (*domesticus, musculus, castaneus,* and *bactrianus*) as well as the closely related species *M. spretus, M. spicilegus* and *M. macedonicus* have the same "standard karyotype" with 20 pairs of chromosomes, including 19 autosomal pairs and the X and Y sex chromosomes as shown in Fig. 5.1. The correct chromosome number was first established by Painter in 1928. Surprisingly, all of the 19 autosomes as well as the X chromosome appear to be telocentric, with a centromere at one end and a telomere at the other.[30] The biological explanation for this uniformity in chromosome morphology is entirely unknown; however, it makes the task of individual chromosome identification much more difficult than it is with human karyotypes. Nevertheless, trained individuals can distinguish chromosomes on the basis of reproducible banding patterns that are accentuated with the use of various staining protocols. The most common of these includes a mild trypsin treatment followed by staining with the dye Giemsa to produce dark Giemsa-stained bands—called G bands—that alternate with Giemsa-negative bands—called R bands for *reverse* G-bands. A variety of other staining protocols have been developed—called R, Q, and T banding—that are all based on the same principal of chromatin denaturation and/ or mild enzymatic digestion followed by staining with a DNA-binding dye (Craig and Bickmore, 1993). In general, all of these different protocols produce the same pattern of bands and interbands observed with Giemsa staining, although in some cases, the dark and light regions are reversed.

The reproducibility of the alternative pattern of G and R bands observed with many different staining protocols implies an underlying difference in the structure of chromatin which, in turn, suggests an underlying heterogeneity in the long-range structure of the genome. In fact, numerous differences have been found in the DNA associated with the two types of bands. G-band DNA condenses early, replicates late and is relatively A:T rich; in contrast, R-band DNA condenses late, replicates early and is relatively G:C rich (Bickmore and Sumner, 1989). All housekeeping genes are located in R bands, while tissue-specific genes can be located in both G and R bands. Each band type is also associated with a different class of dispersed repetitive DNA elements: G bands

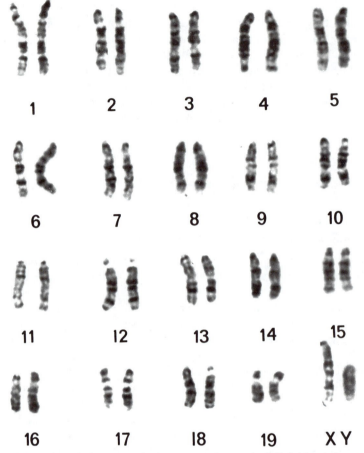

**Figure 5.1**   Normal *Mus musculus* karyotype. A complete diploid set of metaphase chromosomes from the laboratory mouse is shown. This karyotype was kindly provided by Dr Muriel Davisson (Jackson Laboratory, Bar Harbor).

contain LINE-1 elements whereas R bands contain SINE elements (see Section 5.4 for a detailed discussion of these elements).

With all of these contrasting properties, it becomes an interesting problem to distinguish between cause and effect in the generation of the two major types of chromosomal domains. In other words, is there a particular DNA element that defines the G or R bands, and somehow contributes to the preferential association, or disassociation, of all other DNA elements that contribute to the characteristics of the band type? Further research will be necessary to unravel this problem.

### 5.2.1.2 Idiograms and band names

As a mechanism for facilitating data presentation and for comparing results obtained by different investigators, the light and dark bands observed in a raw karyotype are usually converted into idiograms, which are black and white drawings of idealized chromosomes as shown in Fig. 5.2. Autosomes are numbered from 1 to 19, in descending order of length. Major bands (alternating dark and light regions) within each autosome are designated with a capital letter starting from A at the centromere and ascending in alphabetical order. With an increase in resolution, most major bands can be resolved into a series of smaller bands, which are numbered sequentially from 1 starting at the *proximal*—or centromeric—side of the major band and ending at the *distal*—or telomeric— side. Finally, when increased resolution allows the visualization of multiple minor bands within a single previously defined sub-band, these are designated with a number (in sequence from 1) demarcated with a decimal point. As an example of the use of this nomenclature, the designation 17E1.3 represents (in reverse order), the third minor band within the first sub-band within the fifth major band (all in order from the centromere) on the mouse chromosome ranked 17th in size (illustrated in Fig. 7.1).

### 5.2.1.3 Chromosome length and DNA content

The amount of DNA present in each chromosome can be estimated by measuring its length—cytologically—relative to the sum of the lengths of all 20 chromosomes and multiplying this fraction by the total genome length of 3,000 kb (Evans, 1989). From these measurements, one finds that the largest chromosome (1) has a DNA length of approximately 216 mb and the smallest chromosome (19) has a DNA length of 81 mb, with all others following in a near continuum between these two values (see Table 9.4 for estimates of the centimorgan lengths of individual chromosomes).

## 5.2.2 Robertsonian translocations

### 5.2.2.1 Presence in natural populations

Since 1967, there have been numerous reports of wild-caught house mice with karyotypes containing fewer than 20 sets of chromosomes. The first report described a karyotype with 13 sets of chromosomes (seven metacentrics and six telocentrics) in mice captured from the "Valle di Poschiavo" in southeastern Switzerland (Gropp et al., 1972). The assumption was made that animals with such a grossly different karyotype could not possibly be members of the *M. musculus* species, and as a consequence, these Swiss mice were classified as belonging to a separate species named *Mus poschiavinus* and informally referred to as the "tobacco mouse." In subsequent years, additional populations of animals from the alpine regions of both Switzerland and Italy were found with a variety of non-standard karyotypes having anywhere from one to nine metacentrics. Further studies of wild house mice by other investigators have led to the discovery of additional non-standard karyotypes in house mice from other regions of Europe as well as South America and Northern Africa (Adolph and Klein, 1981; Wallace, 1981; Searle, 1982).

When the "*M. poschiavinus*" animals and others with non-standard karyotypes were subjected to a variety of tests—both morphological and genetic—to determine their relatedness to *M. m. domesticus*, investigators were surprised to find that no characteristics, other than karyotype, distinguished these populations

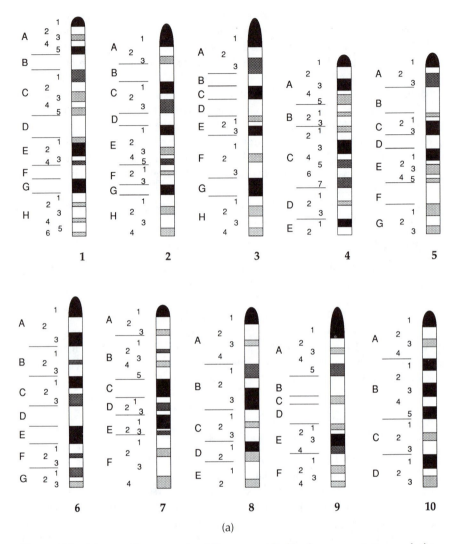

**Figure 5.2** Mouse chromosomes idiograms. Idealized representations of the Giemsa banding patterns associated with each chromosome in a normal karyotype. (a) Chromosomes 1–10; (b) chromosomes 11–19, and the X and Y. This figure was formed with electronic versions of individual chromosome drawings developed by Dr David Adler (University of Washington, Seattle) and used here with his permission. Interested investigators can download these electronic files directly from the Internet Server maintained by Dr Adler. The Server address is: Larry.Pathology-Washington.Edu 70. See Appendix B for further information on data access over the Internet.

from each other. In particular, phylogenetic studies place the *"M. poschiavinus"* animals securely within the *M. m. domesticus* fold; thus the *M. poschiavinus* species name is inappropriate and should not be used.

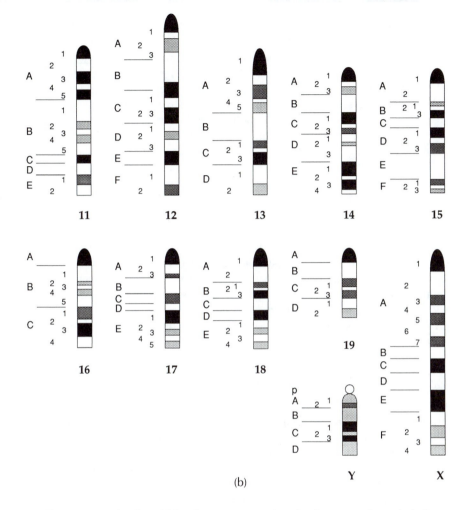

(b)

How can animals within the same species (and even subspecies) have karyotypes that have diverged apart so radically in what has to be a very short period of evolutionary time? The first point to consider is that the karyotypes are actually not as different from each other as they might appear to be at first glance. When subjected to staining and banding analysis, each arm of every metacentric chromosome uncovered to date has been found to be identical to one of the chromosomes present in the standard *M. musculus* karyotype. Thus, it would appear that all of the non-standard karyotypes have arisen by simple fusion events, each of which resulted in the attachment of two standard mouse chromosomes at their centromeres. These centromeric fusions, also referred to as whole arm translocations, have been given the formal name of Robertsonian translocations, because W. R. B. Robertson was the first to identify such chromosomes in the grasshopper.

A three-part nomenclature is used to describe each individually isolated Robertsonian chromosome. First, the "Rb" symbol indicates a Robertsonian;

second, the two chromosomes that have fused together are separated by a dot and listed within parenthesis (with the lower numbered chromosome first); and third, the laboratory number and symbol are indicated. Thus, the 23rd Robertsonian uncovered at the Institute for Pathology in Lubeck, Germany, that resulted from a fusion between chromosomes 10 and 15, would be designated Rb(10.15)23Lub.

Why does the standard mouse karyotype contain no metacentric chromosomes, and at the same time, why do multiple metacentric chromosomes become fixed so rapidly in unrelated populations from isolated geographical regions? In the latter case, genetic drift alone does not appear to provide a satisfactory answer since metacentric fixation requires an intermediate stage during which animals must be karyotypically heterozygous; heterozygosity for one or more metacentric chromosomes can result in decreased fertility as a consequence of non-disjunction.[31] Thus, spontaneously arising Robertsonians cannot be expected to survive in a population (let alone reach fixation) unless they engender a selective advantage to the animal within which they reside. Based on the limited, but scattered, occurrence of populations that contain Robertsonians, it would appear that they can only provide a selective advantage under certain environmental conditions, whereas under other conditions, mice are better served with only telocentric chromosomes. The mechanisms by which such selective pressures would operate on chromosome structure remain totally obscure at the present time.

### 5.2.2.2 Experimental applications

Robertsonian translocations are useful as genetic tools in two types of experimental applications. Like other translocations, they can catalyze non-disjunction events in heterozygous meiotic cells, which—in this case—can lead to the genetic transmission of both homologs of the affected chromosome through individual gametes. This phenomenon will be discussed in the following section. In addition, Robertsonians are especially valuable as visible genetic markers in somatic cells. This usefulness is peculiar to the mouse where all chromosomes other than the Robertsonians will be telocentric. Under microscopic examination, Robertsonians can be easily distinguished as the only chromosomes with two arms.

The advantages of using Robertsonians as genetic markers can be best exploited within animals that contain a single pair of such homologs on a standard karyotypic background. Numerous strains carrying single Robertsonian pairs have been generated through selective breeding between wild and laboratory animals. Each of the standard mouse autosomes is available within the context of a Robertsonian in one or more strains of this type, which can be purchased from the Jackson Laboratory.

The most useful Robertsonians are those with two chromosome arms that differ significantly in length. For example, if one is interested in the analysis of chromosome 2, it would make sense to work with a strain that carries the Rb(2.18)6Rma fusion in which the longer arm of the metacentric (Chr 2) will be easily distinguished under the microscope from the shorter arm (Chr 18).

Robertsonians can be used as somatic markers for both analytical and preparative purposes. Analysis of a particular chromosome by *in situ* hybridization or other staining protocols is greatly aided by the ability to easily identify the relevant chromosome in all metaphase plates. Preparative microdissection for the purpose of generating subchromosome-specific DNA libraries (Section 8.4) is less tedious and more rapidly accomplished by this easy identification (Röhme et al., 1984). Finally, Robertsonians can be more easily distinguished from all other mouse chromosomes by fluorescence-activated cell sorting (FACS) methods for chromosome identification and purification (Bahary et al., 1992).

## 5.2.3  Reciprocal translocations
### 5.2.3.1  Derivation and genetics

Although crossovers normally occur between homologous sequences present on sister chromatids, in rare instances, aberrant crossovers will occur between sequences that are non-allelic. When the two non-allelic sequences that partake in a crossover event of this type come from different chromatids in the same pair of homologs (as illustrated in Fig. 5.5), the result is a pair of reciprocal recombinant products that are "unequal" with one having a duplication and the other having a deletion of the material located between the two breakpoints. Intra-chromosomal unequal crossover events are discussed at length in Section 5.3.2.

When crossing over occurs between sequences located on entirely different chromosomes, the result is even more dramatic. As shown in Fig. 5.3, inter-chromosomal crossing over results in the production of two reciprocal translocation chromosomes. Although inter-chromosomal crossover events occur even less often than intrachromosomal events, the former are more much readily detected (in most cases) for two reasons. First, reciprocal translocations cause the swapping of entire distal portions of two different chromosomes. Since the portions being swapped are usually not equal in size and are always associated with different banding patterns, each resultant translocation chromosome will usually look quite different from any normal chromosome. So long as the breakpoints are not exceedingly close to the centromeres or telomeres, these aberrant chromosomes will be easily recognized through karyotypic analysis. Second, reciprocal translocations usually cause a significant reduction in fertility as a consequence of the unusual pairing that must occur during synapsis and the production of a high frequency of unbalanced gametes through adjacent-1 segregation discussed in more detail below and illustrated in Fig. 5.3. Unbalanced gametes derived from reciprocal translocation heterozygotes give rise to embryos that are partially trisomic or monosomic, and in some cases, these do not survive to birth.

Unlike Robertsonian fusions, reciprocal translocations are not found in wild populations of mice. They can arise spontaneously in laboratory animals and they are recovered at a higher frequency in offspring of males that have been subjected to chemical mutagenesis or irradiation treatment (discussed in Section 6.1). A large number of translocations have been recovered to date (Searle, 1989) and strains homozygous for many can be purchased from the Jackson Laboratory.

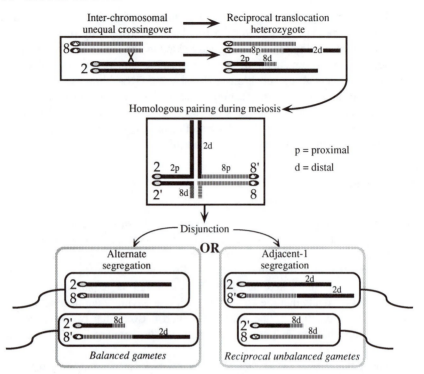

**Figure 5.3**  Chromosome segregation from a reciprocal translocation heterozygote. The box at the top illustrates the process by which unequal crossing over between sequences on different chromosomes can lead to the production of reciprocal translocation heterozygotes. The unequal crossover event will have actually occurred in a meiotic cell that gave rise to a balanced translocation gamete (as shown in the lower part of the Figure) that participated in the formation of the zygote from which the animal derived. The box in the center illustrates the process of chromosome pairing and synapsis that will occur for just the two chromosome pairs involved in the translocation. Through the process of normal disjunction, each gamete will receive just one homolog of each chromosome pair. At the bottom right is an illustration of the two gametes (sperm in this case) produced by "adjacent-1 segregation" in which the adjacent homologs of chromosomes 2 and 8 assort together. At the bottom left is an illustration of the two gametes produced by "alternate segregation" in which non-adjacent homologs of the two chromosomes assort together. A third much less frequent possibility, not illustrated here, is "adjacent-2 segregation;" in this case of *non-disjunction*, both chromosome 2 centromeres would assort to one gamete, and both chromosome 8 centromeres would assort to the brother gamete.

As shown in Fig. 5.3, translocations will cause genetic linkage between chromosomal regions that assort independently in animals with normal karyotypes. Eva Eicher (1971) was the first to use this correlation between genetic linkage and karyotypic linkage to make a specific chromosomal assignment for a particular linkage group, and by the end of the 1970s, all 19 autosomal linkage groups and chromosomes had been paired together (Miller and

Miller, 1975). Higher resolution studies that compared genetic and cytological breakpoint positions provided a means for further mapping of genes to particular chromosome bands; these data also provided a means for determining the centromeric and telomeric ends of each linkage map (Searle, 1989).

### 5.2.3.2 Chromosome segregation

The main contemporary use of reciprocal translocations is as a tool to generate animals that receive both homologs of a chromosomal region from a single parent. To understand the genetic basis for this outcome, you can follow the process of chromosome segregation during meiosis for the fictitious reciprocal translocation heterozygote shown in Fig. 5.2. In this example, mouse chromosomes 2 and 8 have exchanged material.

During the anaphase I stage of the first meiotic division, the two homologs of every chromosome "disjoin" from each other and are pulled to opposite poles by spindles that attach to the centromeric regions. This *disjunction* of chromosomes is the physical basis for the genetically observed segregation of alleles according to Mendel's first law. In mice with a normal karyotype, the segregation of any one pair of homologs will not affect the segregation of any other pair of homologs. Thus, individual homologs of different chromosomes that came into the animal together from one parent will go out into the offspring in an independent manner. This is the physical basis for Mendel's second law of independent assortment.

In animals with a normal karyotype, chromosome disjunction will always lead to the production of gametes that are "balanced" with a complete haploid genome—no more, no less. However, the same is not true with animals heterozygous for a reciprocal translocation. As shown in Fig. 5.3, there are two equally likely outcomes called "alternate segregation" and "adjacent-1 segregation." With the alternate segregation pathway, one gamete class will receive one chromosome (Chr 2) homolog and one Chr 8 homolog {2, 8}, just like all gametes produced by mice with a normal karyotype. The other gamete class will receive both translocated chromosomes called 2' and 8' in this example {2', 8'}; although the genetic material is rearranged, one complete haploid genome is present, and thus these gametes are considered to be "balanced." If a balanced {2', 8'} gamete joins together with a normal gamete during fertilization, the resulting animal will be a balanced, reciprocal translocation heterozygote just like the original parent.

With the adjacent-1 segregation pathway, the two gamete classes are unbalanced in a reciprocal fashion. One will have a normal Chr 2 and a translocated Chr 8' {2, 8'}; this gamete is deleted for sequences at the distal end of the normal Chr 8 (8d) and duplicated with both homolog copies of sequences from the distal end of Chr 2 (2d). The other {2', 8} will be deleted for distal Chr 2 sequences (2d) and duplicated for distal Chr 8 sequences (8d).

### 5.2.3.3 Partial trisomies and uniparental disomies

The special consequences of chromosome segregation from reciprocal translocation heterozygotes have been exploited with two types of breeding protocols. In the first, translocation heterozygotes are bred to animals with a normal karyotype. Adjacent-1 segregation will give rise to animals that are partially trisomic (for the distal end of one translocated chromosome) and partially monosomic (for the distal end of the other). Thus, by choosing appropriate translocations, it becomes possible to construct animals that are deleted or duplicated for particular genes of interest. By breeding-in mutations at these loci, it becomes possible to construct genotypes of the { +/+/m} and { +/m/m} variety (where + and m are wild-type and mutant alleles, respectively) as a means toward a better understanding of gene dosage effects, and dominance and recessive relationships (Agulnik et al., 1991; Ruvinsky et al., 1991).

In the second type of breeding protocol, animals heterozygous for the same pair of reciprocal translocations are mated to each other. The most interesting offspring to emerge from such unions are those formed through the fusion of complementary unbalanced gametes that represent the two different products of adjacent-1 segregation (Fig. 5.3). Although the resulting zygotes have fully balanced genomes—they are not deleted nor duplicated for any sequences— they carry two subchromosomal regions in which both homologs came from only one or the other parent respectively. In other words, for one chromosomal region, these animals are maternally disomic and paternally nullisomic; for the other chromosomal region, the opposite holds true.

Uniparental disomy can also be obtained, albeit with lower frequency, in the offspring of matings between animals heterozygous for the same Robertsonian translocation. Pairing between the Robertsonian and the homologous acrocentric chromosomes can lead to *non-disjunction* with gametes that contain either two copies or no copies of one homolog represented within the Robertsonian. Once again, the fusion of two complementary nondisjunction gametes will lead to zygotes with fully balanced genomes but with whole chromosome uniparental disomy. Both whole and partial chromosome disomy provide powerful genetic tools for the analysis of genomic imprinting which is discussed later in this chapter (Cattanach and Kirk, 1985).

## 5.3 GENOMIC ELEMENTS, GENOME EVOLUTION, AND GENE FAMILIES

### 5.3.1 Classification of genomic elements

#### 5.3.1.1 Functional and non-functional sequences

Sequences within the genome can be classified according to a number of criteria. The most important of these is functionality and the largest class of functional DNA elements consists of coding sequences within transcription units. Transcription units usually contain exons and introns, and are usually associated with flanking regulatory regions that are necessary for proper expression. For the most part, transcription units correspond one-to-one with Mendelian genes and they usually function on behalf of the organism within which they lie. However, mammalian genomes also contain transcribable elements that do not benefit the organism and whose sole function appears to be self-propagation. Such sequences are referred to as *selfish* DNA or *selfish genes* and will be described at length in Section 5.4. Although these sequences may undergo transcription, they cannot be detected, in and of themselves, in terms of traditional Mendelian phenotypes. The functional class of DNA elements also includes a number of specialized sequences that play roles in chromosome structure and transmission. The best characterized structural elements are associated with the centromeres and telomeres.

Most of the genome *appears* to consist of DNA sequences that are entirely non-functional. This non-functional class includes pseudogenes that derive from, and still share homology with, specific genes but are not themselves functional with a lack of transcription or translation. However, for the most part, non-functional DNA is present in the context of long lengths of apparently random sequence—located between genes and within their introns—with origins that have long since become indecipherable as a consequence of constant "genetic drift."

### 5.3.1.2 Single-copy and repeated sequences

Both functional and non-functional sequences can be distinguished by a second criterion—copy number. Sequences in a genome that do *not* share homology with any other sequences in the same genome are considered *unique* or *single copy*. This single-copy class contains both functional and non-functional elements. Sequences that *do* share homology with one or more other genomic regions are considered to be *repeated* or *multicopy*.

At one homology extreme, two sequences can show 100% identity to each other at the nucleotide level. At the other extreme, homology may be recognized only through the use of computer algorithms that show a level of identity between two sequences that is unlikely to have occurred by chance. In the case of many gene families, individual members are not identical—in fact, they are likely to have evolved different functions—yet a probe from one will cross-hybridize with sequences from the others. Cross-hybridization provides a powerful tool for the identification of multicopy DNA elements by simple Southern blot analysis, and for their characterization by library screening and cloning.

Homologies among more distantly related functional sequences that do not show cross-hybridization can sometimes be uncovered through the use of the polymerase chain reaction (PCR). The rationale behind this approach—which has been used successfully with a number of different gene families—is that specific short regions of related gene sequences may be under more intense selective pressure to remain relatively unchanged due to functional constraints on the encoded peptide regions. These highly conserved regions may not be long enough to allow detectable cross-hybridization under blotting conditions, but the constrained peptide sequences that they encode can be used to devise two degenerate oligonucleotides for use as primers to identify additional members of the gene family through PCR amplification from either genomic DNA or tissue-specific cDNA.

All sequences that are partially identical to each other—as recognized by hybridization, PCR, or sequence comparisons—are considered to be members of

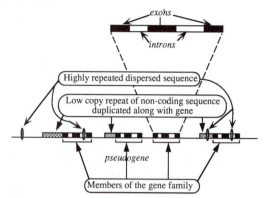

**Figure 5.4** Hypothetical region from the mouse genome. Illustration of a genomic region containing a four-member gene family. Each of these genes has the same underlying exon–intron structure as expected from duplication by unequal crossing over. One member of this family is no longer transcribed and is, thus, considered a pseudogene. Adjacent to three of the four members of this gene family is a non-functional sequence that must have come along with two of the duplication events. Given a sufficient period of evolution, these non-functional sequences will diverge into random unique sequence. Selfish elements have inserted into several sites including two that flank genes and two within introns.

the same DNA element family. Families of functional elements are called gene families. Families of non-functional elements have been referred to simply as "repeat families" or "DNA element families." Multicopy DNA families—both functional and non-functional—can be further classified according to copy number, element size, and distribution within the genome. Related sequences can be found closely linked to each other in a cluster, they can be unlinked to each other and dispersed to different chromosomes, or they can have a combination of these two arrangements with multiple clusters dispersed to different sites.

From a distance, the genome appears to be a chaotic mixture of sequences from all of these classes thrown together without any structure or order, like craters, one overlapping the next, on the surface of the moon. However, on closer examination, it becomes possible to make sense of the genome, the relationship of different genomic elements to each other, and the mechanisms by which they have evolved as indicated for the hypothetical genomic region shown in Fig. 5.4.

### 5.3.2 Forces that shape the genome

#### 5.3.2.1 Genomic complexity increases by gene duplication and selection for new function

Mice, humans, the lowly intestinal bacterium *E. coli*, and all other forms of life evolved from the same common ancestor that was alive on this planet a few billion years ago. We know this is the case from the universal use of the same molecule—DNA—for the storage of genetic information, and from the nearly universal genetic code. But *E. coli* has a genome size of 4.2 mb, while the mammalian genome is nearly a thousand-fold larger at ~ 3,000 mb. If one assumes that our common ancestor had a genome size that was no larger than that of the modern-day *E. coli*, the obvious question one can ask is where did all of our extra DNA come from?

The answer is that our genome grew in size and evolved through a repeated process of duplication and divergence. Duplication events can occur essentially at random throughout the genome and the size of the duplication unit can vary from as little as a few nucleotides to large subchromosomal sections that are tens, or even hundreds, of megabases in length. When the duplicated segment contains one or more genes, either the original or duplicated copy of each is set free to accumulate mutations without harm to the organism, since the other good copy with an original function will still be present.

Duplicated regions, like all other genetic novelties, must originate in the genome of a single individual and their initial survival in at least some animals in each subsequent generation of a population is, most often, a simple matter of chance. This is because the addition of one extra copy of most genes—to the two already present in a diploid genome—is usually tolerated without significant harm to the individual animal. In the terminology of population genetics, most duplicated units are essentially neutral (in terms of genetic selection) and thus, they are subject to genetic drift, inherited by some offspring but not others derived from parents that carry the duplication unit. By chance, most neutral genetic elements will succumb to extinction within a matter of generations. However, even when a duplicated region survives for a significant period of time, random mutations in what were once-functional genes will almost always lead to non-functionality. At this point, the gene becomes a *pseudogene*. Pseudogenes will be subject to continuous genetic drift with the accumulation of new mutations at a pace that is so predictable (~ 0.5% divergence per million years) as to be likened to a *molecular clock*. Eventually, nearly all pseudogene sequences will tend to drift past a boundary where it is no longer possible to identify the functional genes from which they derived. Continued drift will act to turn a once functional sequence into a sequence of essentially random DNA.

Miraculously, every so often, the accumulation of a set of random mutations in a spare copy of a gene can lead to the emergence of a new functional unit—or gene—that provides benefit and, as a consequence, selective advantage to the organism in which it resides. Usually, the new gene has a function that is related to the original gene function. However, it is often the case that the new gene will have a novel expression pattern—spatially, temporally, or both—which must result from alterations in cis-regulatory sequences that occur along with codon changes. A new function can emerge directly from a previously functional gene or even from a pseudogene. In the latter case, a gene can go through a period of non-functionality during which there may be multiple alterations before the gene comes back to life. Molecular events of this class can play a role in "punctuated evolution" where, according to the fossil or phylogenetic record, an organism or evolutionary line appears to have taken a "quantum leap" forward to a new phenotypic state.

### 5.3.2.2 Duplication by transposition

With duplication acting as such an important force in evolution, it is critical to understand the mechanisms by which it occurs. These fall into two broad categories: (1) *transposition* is responsible for the dispersion of related sequences; (2) *unequal crossing over* is responsible for the generation of gene clusters. Transposition refers to a process in which one region of the genome relocates to a new chromosomal location. Transposition can occur either through the direct movement of original sequences from one site to another or through an RNA intermediate that leaves the original site intact. When the genomic region itself (rather than its proxy) has moved, the "duplication" of genetic material actually occurs in a subsequent generation after the transposed region has segregated into the same genome as the originally positioned region from a non-deleted homolog. In theory, there is no upper limit to the size of a genomic region that can be duplicated in this way.

A much more common mode of transposition occurs by means of an intermediate RNA transcript that is reverse-transcribed into DNA and then inserted randomly into the genome. This process is referred to as retrotransposition. The size of the retrotransposition unit—called a retroposon—cannot be larger than the size of the intermediate RNA transcript. Retrotransposition has been exploited by various families of selfish genetic elements (described in Section 5.4), some of which have been copied into 100,000 or more locations dispersed throughout the genome with a self-encoded reverse transcriptase. However, examples of functional, intronless retroposons—such as *Pgk2* and *Pdha2*—have also been identified (Boer et al., 1987; Fitzgerald et al., 1993). In such cases, functionality is absolutely dependent upon novel regulatory elements either present at the site of insertion or created by subsequent mutations in these sequences.

### 5.3.2.3 Duplication by unequal crossing over

The second broad class of duplication events result from unequal crossing over. Normal crossing over, or recombination, can occur between equivalent sequences on homologous chromatids present in a synaptonemal complex that forms during the pachytene stage of meiosis in both male and female mammals. Unequal crossing over—also referred to as illegitimate recombination—refers to crossover events that occur between non-equivalent sequences. Unequal crossing over can be initiated by the presence of related sequences—such as highly repeated retroposon-dispersed selfish elements—located nearby in the genome (Fig. 5.5). Although the event is unequal, in this case, it is still mediated by the homology that exists at the two non-equivalent sites.

So-called non-homologous unequal crossovers can also occur, although they are much rarer than homologous events. I say so-called because even these events may be dependent on at least a short stretch of sequence homology at the two sites at which the event is initiated. The initial duplication event that produces a two-gene cluster may be either homologous or non-homologous, but once two units of related sequence are present in tandem, further rounds of homologous unequal crossing over can be easily initiated between non-

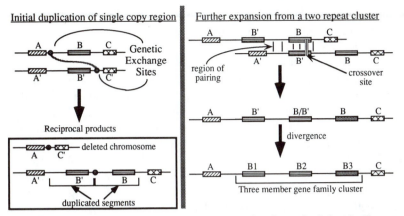

**Figure 5.5**  Unequal crossing over generates gene families. The left side illustrates an unequal crossing over event and the two products that are generated. One product is deleted and the other is duplicated for the same region. In this example, the duplicated region contains a second complete copy of a single gene (*B*). The right side illustrates a second round of unequal crossing over that can occur in a genome that is homozygous for the original duplicated chromosome. In this case, the crossover event has occurred between the two copies of the original gene. Only the duplicated product generated by this event is shown. Over time, the three copies of the *B* gene can diverge into three distinct functional units (B1, B2, and B3) of a gene family cluster.

equivalent members of the pair as illustrated in Fig. 5.5. Thus, it is easy to see how clusters can expand to contain three, four, and many more copies of an original DNA sequence.

In all cases, unequal crossing over between homologs results in two reciprocal chromosomal products: one will have a duplication of the region located between the two sites and the other will have a deletion that covers the same exact region (Fig. 5.5). It is important to remember that, unlike retrotransposition, unequal crossing over operates on genomic regions without regard to functional boundaries. The size of the duplicated region can vary from a few basepairs to tens or even hundreds of kilobases, and it can contain no genes, a portion of a gene, a few genes, or many.

### 5.3.2.4 Genetic exchange between related DNA elements

There are many examples in the genome where genetic information appears to flow from one DNA element to other related—but non-allelic—elements located nearby or even on different chromosomes. In some special cases, the flow of information is so extreme as to allow all members of a gene family to co-evolve with near-identity as discussed in Section 5.3.3.3. In at least one case—that of the class I genes of the major histocompatibility complex (*MHC* or *H2*)—information flow is unidirectionally selected, going from a series of 25–38 non-functional pseudogenes into two or three functional genes (Geliebter and Nathenson, 1987). In this case, intergenic information transfer serves to increase dramatically the level of polymorphism that is present at the small number of functional gene members of this family.

Information flow between related DNA sequences occurs as a result of an alternative outcome from the same exact process that is responsible for unequal crossing over. This alternative outcome is known as intergenic gene conversion. Gene conversion was originally defined in yeast through the observation of altered ratios of segregation from individual loci that were followed in tetrad analyses. These observations were fully explained within the context of the Holliday model[32] of DNA recombination, which states that homologous DNA duplexes first exchange single strands that hybridize to their complements and migrate for hundreds or thousands of bases. Resolution of this "Holliday intermediate" can lead with equal frequency to crossing over between flanking markers or back to the *status quo* without crossing over. In the latter case, a short single-strand stretch from the invading molecule will be left behind within the DNA that was invaded. If an invading strand carries nucleotides that differ at any site from the strand that was replaced, these will lead to the production of heteroduplexes with basepair mismatches. Mismatches can be repaired (in either direction) by specialized "repair enzymes" or they can remain as they are to produce non-identical daughter DNAs through the next round of replication.

By extrapolation, it is easy to see how the Holliday model can be applied to the case of an unequal crossover intermediate. With one resolution, unequal crossing over will result; with the alternative resolution, gene conversion can be initiated between non-allelic sequences. Remarkably, information transfer— presumably by means of gene conversion—can also occur across related DNA sequences that are even distributed to different chromosomes.

### 5.3.3 Gene families and superfamilies

#### 5.3.3.1 Origins and examples

Much of the functional DNA in the genome is organized within gene families and hierarchies of gene superfamilies. The superfamily term was coined to describe relationships of common ancestry that exist between and among two or more gene families, each of which contains more closely related members. As increasingly more genes are cloned, sequenced, and analyzed by computer, deeper and older relationships among superfamilies have unfolded. Complex relationships can be visualized within the context of branches upon branches in evolutionary trees. All of these superfamilies have evolved out of combinations of unequal crossover events that expanded the size of gene clusters and transposition events that acted to seed distant genomic regions with new genes or clusters.

A prototypical small-size gene superfamily is represented by the very well-studied globin genes illustrated in Fig. 5.6. All functional members of this superfamily play a role in oxygen transport. The superfamily has three main families (or branches) represented by the beta-like genes, the alpha-like genes and the single myoglobin gene. The duplication and divergence of these three main branches occurred early during the evolution of vertebrates and, as such, all

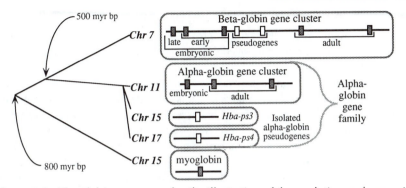

**Figure 5.6** The globin gene superfamily. Illustration of the evolution and genomic organization of all mouse globin-like genes. The members of each of the two individual clusters are more closely related to each other than any are to other members of the superfamily. This suggests that each cluster was seeded by a single transposed gene that duplicated later by unequal crossing over into all other members of the local family. Dark boxes represent functional genes and open boxes represent pseudogenes.

three are a common feature of all mammals. The products encoded by genes within two of these branches—alpha-globin and beta globin—come together (with heme cofactors) to form a tetramer which is the functional hemoglobin protein that acts to transport oxygen through the blood stream. The product encoded by the third branch of this superfamily—myoglobin—acts to transport oxygen in muscle tissue.

The beta-like branch of this gene superfamily has duplicated by multiple unequal crossing over events, and diverged into five functional genes and two beta-like pseudogenes that are all present in a single cluster on mouse chromosome 7 as shown in Fig. 5.6 (Jahn et al., 1980). Each of the beta-like genes codes for a similar polypeptide, which has been selected for optimal functionality at a specific stage of mouse development: one functions during early embryogenesis, one during a later stage of embryogenesis, and two in the adult. The alpha-like branch has also expanded by unequal crossing over into a cluster of three genes—one functional during embryogenesis and two functional in the adult—on mouse chromosome 11 (Leder et al., 1981). The two adult alpha genes are virtually identical at the DNA sequence level, which is indicative of either a very recent duplication event or concerted evolution (see Section 5.3.3.3).

In addition to the primary alpha-like cluster are two isolated alpha-like genes (now non-functional) that have transposed to dispersed locations on chromosomes 15 and 17 (Leder et al., 1981). When pseudogenes are found as single copies in isolation from their parental families, they are called "orphons." Interestingly, one of the alpha globin orphons (*Hba-ps3* on Chr 15) is intronless and would appear to have been derived through a retrotransposition event, whereas the other orphon (*Hba-ps4* on Chr 17) contains introns and may have been derived by a direct DNA-mediated transposition. Finally, the single myoglobin gene on chromosome 15 does not have any close relatives either nearby or far away (Blanchetot et al., 1986; Drouet and Simon-Chazottes, 1993). Thus, the globin gene superfamily provides a view of the many different mechanisms that can be employed by the genome to evolve structural and functional complexity.

The *Hox* gene superfamily provides an alternative prototype for the expansion of gene number as illustrated in Fig. 5.7. In this case, the earliest duplication events (which predate the divergence of vertebrates and insects) led to a cluster of related genes that encoded DNA-binding proteins used to encode spatial information in the developing embryo. The original gene cluster has been duplicated *en masse* and dispersed to a total of four chromosomal locations (on Chrs 2, 6, 11, and 15) each of which contains 9–11 genes (McGinnis and Krumlauf, 1992). Interestingly, because of the order in which the duplication events occurred—unequal crossing over to expand the cluster size first, transposition *en masse* second—an evolutionary tree would show that a single "gene family" within this superfamily is actually splayed out physically across all of the different gene clusters as shown in Fig. 5.7. Some gene additions and subtractions within individual clusters have occurred by unequal crossing over since the *en masse* duplication so that differences in gene number and type can be seen within a basic framework of homology among the different whole clusters.

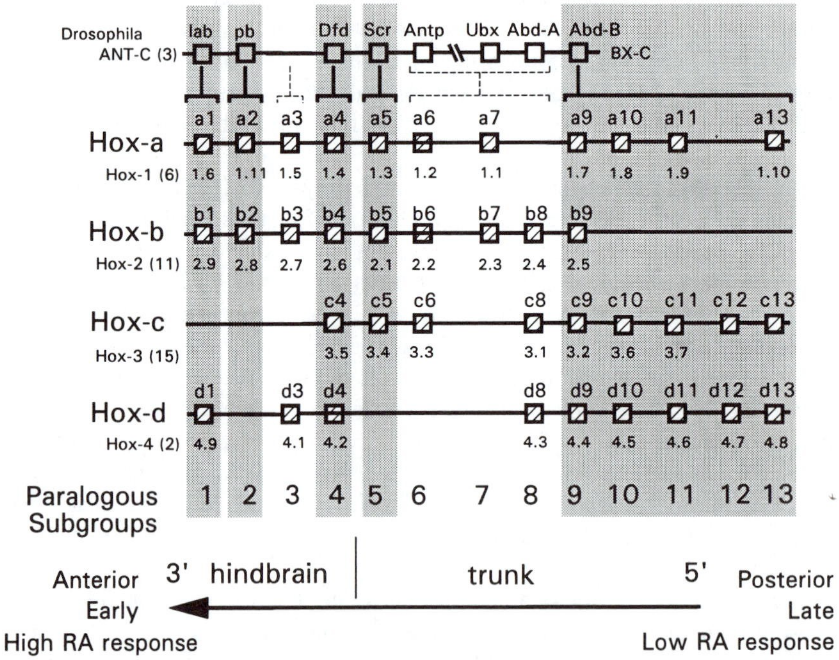

**Figure 5.7**   The *Hox* gene superfamily. Illustration of the four mouse homeobox gene clusters and the relationship of genes within each cluster to those in other clusters and *Drosophila* homologs in the HOM-C cluster (McGinnis and Krumlauf, 1992). Related genes from different clusters, so-called paralogous subgroups, are shown together in the same shaded region. The general spatial pattern of expression of each gene along the anterior–posterior axis within the developing embryo is indicated below. This illustration was kindly provided by Robert Krumlauf (Mill Hill, England).

A final example of a gene super-superfamily is the very large set of genes, which contain immunoglobulin (Ig)-like domains and function as cell-surface or soluble receptors involved in immune function or other aspects of cell–cell interaction. This set includes the Ig gene families themselves, the major histocompatibility genes (called *H2* in mice), the T-cell-receptor genes and many more (Hood et al., 1985). There are dispersed genes and gene families, small clusters, large clusters, and clusters within clusters, tandem and interspersed. Dispersion has occurred with the transposition of single genes that later formed clusters and with the dispersion of whole clusters *en masse*. Furthermore, the original Ig domain can occur as a single unit in some genes, but it has also been duplicated intragenically to produce gene products that contain two, three, or four domains linked together in a single polypeptide. The Ig superfamily, which contains hundreds (perhaps thousands) of genes, illustrates the manner in which the initial emergence of a versatile genetic element can be exploited by the forces of genomic evolution with a consequential enormous growth in genomic and organismal complexity.

### 5.3.3.2 Does gene order or localization matter?

Does the chromosome on which a gene lies matter to its function? Is gene clustering significant to function or is it simply a remnant of the fact that duplicated genes are most often generated by unequal crossover events? One can gain insight toward the answers to these questions by comparing the positions of homologous genetic information in different species: specifically mice and humans. Whole genome comparisons quickly demonstrate that the question of conservation to a particular chromosome only makes sense in the context of the X and Y. This is because every autosome from one species contains significant stretches of homology with two or more autosomes in the other species. Thus, the question of autosome conservation is meaningless. The X and Y chromosomes are a different story for three interrelated reasons. First, as a pair, they play a special role in sex determination. Second, they are the only chromosomes that can appear in a hemizygous state in normal genomes. Third, the X chromosome alone is subject to stable inactivation in all normal female mammals. With few exceptions, X-linked and Y-linked genes have remained in the same linkage groups throughout mammalian evolution as originally proposed by Ohno (1967), although various intrachromosomal rearrangements have occurred (Bishop, 1992; Brown et al., 1992; Foote et al., 1992).

The second question asked at the head of this section can be restated as follows: do fine-structure genetic maps have functional significance? The answer is that in at least some cases, the integrity of genes within a clustered family is clearly important to function. This was first illustrated in the case of the beta-globin gene family with its five members arranged in a 70 kb array (Fig. 5.6). Although beta-globin was used in the first transgenic experiments conducted in 1980 and many subsequent experiments, it was never possible for researchers to achieve full expression of the transgene at the same level as the endogenous gene. The problem was that all of the members of the endogenous gene family are dependent for expression on a *locus control region* (or LCR) that maps outside of the gene cluster and appears to play a role in "opening-up" the chromatin structure of the entire cluster in hematopoietic cells so that individual family members can then be regulated in different temporal modes (Talbot et al., 1989; Townes and Behringer, 1990). When transgene constructs are produced with the beta-globin LCR linked to the beta-globin structural gene, full endogenous levels of expression can be obtained (Grosveld et al., 1987). In recent years, evidence has accumulated for the role of LCRs in the global control of other gene clusters as well.

There is not only a requirement for some genes to remain in their ancestral cluster, but in some cases, the precise order of genes is conserved as well. Actual gene order has been observed to play roles in two different patterns of expression. Transgenic experiments indicate that for the beta-globin cluster, the temporal sequence of expression appears to be directly encoded (to a certain extent) in the order in which the genes occur (Hanscombe et al., 1991). In the Hox gene clusters, the order of genes correlates with the pattern of spatial expression along the anterior–posterior axis of the developing embryo (McGinnis and Krumlauf, 1992) Fig. 5.7).

There are also a few examples of genes and clusters that are unrelated by sequence, but which map together in a small chromosomal region and have a common arena of function. The best example of this phenomenon is the major histocompatibility complex, which contains various gene families that diverged from a common immunoglobulin-like domain ancestor, but also unrelated genes that play a role in antigen presentation and other aspects of immune function. This conjunction of immune genes has been conserved in all mammalian species that have been examined. Is this significant? Farr and Goodfellow (1992) quote Sydney Brenner in likening gene mappers to astronomers boldly mapping the heavens and conclude that "Seeking meaning in gene order may be the equivalent of astrophysics—or it might be astrology." I think it is safe to bet that sometimes it will be one and sometimes it will be the other. The problem will be to distinguish between the two.

### 5.3.3.3 Tandem families of identical genetic elements

A limited number of multicopy gene families have evolved under a very special form of selective pressure that requires all members of the gene family to maintain essentially the same sequence. In these cases, the purpose of a high copy number is not to effect different variations on a common theme but rather to supply the cell with a sufficient amount of an identical product within a short period of time. The set of gene families with identical elements includes those that produce RNA components of the cell's machinery within ribosomes and as transfer RNA. It also includes the histone genes, which must rapidly produce sufficient levels of protein to coat the new copy of the whole genome that is replicated during the S phase of every cell cycle.

Each of these gene families is contained within one or more clusters of tandem repeats of identical elements. In each case, there is strong selective pressure to maintain the same sequence across all members of the gene family because all are used to produce the same product. In other words, optimal functioning of the cell requires that the products from any one individual gene are *directly* interchangeable in structure and function with the products from all other individual members of the same family. How is this accomplished? The problem is that once sequences are duplicated, their natural tendency is to drift apart over time. How does the genome counteract this natural tendency?

When ribosomal RNA genes and other gene families in this class were first compared both between and within species, a remarkable picture emerged: between species, there was clear evidence of genetic drift with rates of change that appeared to follow the molecular clock hypothesis described earlier. However, within a species, all sequences were essentially equivalent. Thus, it is not simply the case that mutational changes in these gene families are suppressed. Rather, there appears to be an ongoing process of "concerted evolution" that allows changes in single genetic elements to spread across a complete set of genes in a particular family. So the question posed previously can now be narrowed down further: how does concerted evolution occur?

Concerted evolution appears to occur through two different processes (Dover, 1982; Arnheim, 1983). The first is based on the expansion and contraction of gene family size through sequential rounds of unequal crossing over between homologous sequences. Selection acts to maintain the absolute size of the gene family within a small range around an optimal mean. As the gene family becomes too large, the shorter of the unequal crossover products will be selected; as the family becomes too small, the longer products will be selected. This cyclic process will cause a continuous oscillation around a mean in cluster size. However, each contraction will result in the loss of divergent genes, whereas each expansion will result in the indirect "replacement" of these lost genes with identical copies of other genes in the family. With unequal crossovers occurring at random positions throughout the cluster and with selection acting in favor of the least divergence among family members, this process can act to slow down dramatically the continuous process of genetic drift between family members.

The second process responsible for concerted evolution is intergenic gene conversion between "non-allelic" family members. It is easy to see that *different* tandem elements of nearly identical sequence can take part in the formation of Holliday intermediates, which can resolve into either unequal crossing over products or gene conversion between non-allelic sequences. Although the direction of information transfer from one gene copy to the next will be random in each case, selection will act upon this molecular process to ensure an increase in homogeneity among different gene family members. As discussed above, information transfer—presumably by means of gene conversion—can also occur across gene clusters that belong to the same family but are distributed to different chromosomes.

Thus, with unequal crossing over and interallelic gene conversion (which are actually two alternative outcomes of the same initial process) along with selection for homogeneity, all of the members of a gene family can be maintained with nearly the same DNA sequence. Nevertheless, concerted evolution will still lead to increasing divergence between whole gene families present in different species.

### 5.3.4 Centromeres and satellite DNA

In the early days of molecular biology, equilibrium sedimentation through $CsCl_2$ gradients was used as a method to fractionate DNA according to buoyant density.

Genomic DNA prepared from animal tissues according to standard protocols is naturally degraded by shear forces into fragments that are, on average, smaller than 100 kb. When a solution of genomic DNA fragments is subjected to high-speed centrifugation in $CsCl_2$, each fragment will move to a position of equivalent density in the $CsCl_2$ concentration gradient that forms. DNA buoyant density is related to the molar ratio of G:C basepairs to A:T basepairs by a simple linear function.[33] The greater the G:C content, the higher the density. When mouse DNA is subjected to $CsCl_2$ fractionation, the bulk of the DNA (90%) is distributed within a narrow bell-shaped curve having an average density of 1.701 $g/cm^3$, equivalent to a G:C content of 42%.

In addition to this "main band" of DNA, a second "satellite band" was observed with an average density of 1.690 $g/cm^3$ equivalent to a G:C content of 31% (Kit, 1961). Approximately 5.5% of the total mouse genome is found within this band and the DNA within this fraction was given the name "satellite DNA" (Davisson and Roderick, 1989). It was not until 1970 that Pardue and Gall used their newly invented technique of *in situ* hybridization to demonstrate the localization of satellite DNA sequences to the centromeres of all mouse chromosomes except the Y (Pardue and Gall, 1970). Centromeres are highly specialized structural elements that function to segregate eukaryotic chromosomes during mitosis and meiosis (Rattner, 1991).

When DNA recovered from the satellite band was subjected to renaturation analysis, as described earlier in this chapter, the $C_0t_{1/2}$ value obtained indicated a complexity of only ~ 200 nucleotides. This result showed that the satellite DNA fraction was composed of a simple sequence that was repeated over and over again, many times. Modern cloning and sequence analysis has demonstrated a basic repeating unit with a size of 234 bp (Hörz and Altenburger, 1981).[34] One can calculate the copy number of this basic repeat unit by dividing the proportion of the genome devoted to satellite sequences (5.5% $\times$ 3 $\times$ $10^9$ bp = 1.65 $\times$ $10^8$ bp) by the repeat size (234 bp) to obtain 700,000 copies. If these copies were distributed equally among all chromosomes, each centromere would contain 35,000 copies having a total length of 8 mb.

Although the original definition of "satellite" DNA was based on a density difference observed in $CsCl_2$ gradients, the meaning of the term has expanded to describe all highly repeated simple sequences found in the centromeres of chromosomes from higher eukaryotes. In many species, satellite sequences do not have G:C contents that differ significantly from that of the bulk DNA.

The *M. musculus* genome has a second family of satellite sequences present in only 50–100,000 copies (Davisson and Roderick, 1989). This "minor satellite" is also localized to the centromeres and appears to share a common ancestry with the major satellite. It is of interest that the relative proportion of the two satellites in *M. spretus* is the reverse of that found in *M. musculus*. The *M. spretus* genome has only 25,000 copies of the "major satellite" and 400,000 copies of the "minor satellite." This difference can be exploited to allow the determination of centimorgan distances between centromeres and linked loci in interspecific crosses as discussed in Section 9.1.2 (Matsuda and Chapman, 1991).

The satellite sequences in the distant *Mus* species *M. caroli*, *M. cervicolor*, and *M. cookii* (Fig. 2.2) have diverged so far from the *musculus* sequences that cross-hybridization between the two is minimal. This qualitative difference can be exploited, once again, by *in situ* hybridization, to mark differentially cells from each species in interspecific chimeras (Rossant et al., 1983). A satellite DNA marker is useful for cell lineage studies because it is easy to detect by hybridization of tissue sections, and it is present in all cells irrespective of gene activity or developmental state.

The term satellite has been incorporated as a suffix into a number of other terms (microsatellite, minisatellite, midisatellite, etc.) that are used to describe DNA sequences formed from basic units that have become amplified by multiple rounds of tandem duplication. Some of these sequence classes are described in Section 5.4.5 and Chapter 8.

## 5.4 REPETITIVE "NON-FUNCTIONAL" DNA FAMILIES

In the preceding section, we examined several different classes of DNA families with members that carry out a variety of tasks necessary for the survival of the organism. This section surveys a final major class of DNA families whose members in and of themselves do not function for the benefit of the animal in which they lie. This class can be subdivided further into individual families that are actively involved in their own dispersion—the so-called *selfish genes*—and those that consist of very simple sequences that appear to arise *de novo* at each genomic location. The selfish gene group can be further divided—somewhat arbitrarily—into subclasses based on copy number in the genome. Each of the resulting subclasses of repetitive DNA families will be discussed in the subsections that follow.

### 5.4.1 Endogenous retroviral elements

Retroviruses are RNA-containing viruses that can convert their RNA genome into circular DNA molecules through a viral-associated reverse transcriptase, which becomes activated upon cell infection. The resultant DNA "provirus" can integrate itself into a relatively random site in the host genome. The genetic information present in the retroviral genome is retained within the integrated provirus and, under certain conditions, the provirus can be activated to produce new RNA genomes along with the associated proteins—including reverse transcriptase—that can come together to form new virus particles that are ultimately released from the cell surface by exocytosis. However, in many cases, stably integrated retroviral elements appear not to be active.

Once it has become integrated into a chromosome, the provirus will become replicated with every round of host replication irrespective of whether the provirus itself is active or silent. Furthermore, proviruses that integrate into the germ line—through the sperm or egg genome—will segregate along with their host chromosome into the progeny of the host animal and into subsequent generations of animals as well. In certain hybrid mouse strains, new proviral integrations into the germ line can be observed to occur at abnormally high frequencies (Jenkins and Copeland, 1985).

All strains of mice as well as all other mammals have endogenous proviral elements. These elements can be classified and subclassified according to the type of retrovirus from which they derived (ecotropic, MMTV, xenotropic, and others). Ecotropic elements are generally present at 0–10 copies (Jenkins et al., 1982), MMTVs are present at 4–12 copies (Kozak et al., 1987), and non-ecotropic elements are present at 40–60 copies (Frankel et al., 1990). Loss and acquisition of new proviral sequences is an ongoing process and, as a consequence, the genomic distribution of these elements is highly polymorphic. Thus, these elements can be very useful as genetic markers as discussed in Section 8.2.4.

In addition to the DNA families clearly related to known retroviruses, there are a number of additional families that are retroviral-like in structure but are not clearly related to any known virus strain in existence today. The intracisternal A particle (IAP) DNA family is defined by homology to RNA sequences that are actually present within non-functional retroviral-like particles found in the cytoplasm of some types of mouse cells. The IAP family is present in ~ 1,000 copies (Lueders and Kuff, 1977) but very few of these copies can actually produce transcripts. Another retroviral-like DNA family is called VL30, which stands for viral-like 30S particles (Carter et al., 1986); there are approximately 200 copies of this element in the mouse genome (Courtney et al., 1982; Keshet and Itin, 1982). There is no reason to expect that additional retroviral-like families will not be uncovered through further genomic studies.

It is of evolutionary interest to ask the question: from where do retroviruses come? Retroviruses cannot propagate in the absence of cells but cells can propagate in the absence of retroviruses. Thus, it seems extremely likely that retroviruses are derived from sequences that were originally present in the cell genome. The first retrovirus must have been able to free itself from the confines of the cell nucleus through an association with a small number of proteins that allowed it to coat, and thus protect, itself from the harsh extracellular environment. Of course, the protein most critical to the propagation of the retrovirus is the enzyme that allows it to reproduce—RNA-dependent DNA polymerase, commonly referred to as reverse transcriptase. But where did this enzyme come from? Reverse transcriptase catalyzes the production of single-stranded complementary DNA molecules from an RNA template. This enzymatic activity does not appear to be required for any normal cellular process known in mammals! How could such an activity—without any apparent benefit to the host organism—arise *de novo* in a normal cell? One possible answer is that reverse transcriptase did not evolve for the benefit of the organism itself but, rather, for the benefit of selfish DNA elements *within* the *genome* that utilize the enzyme to propagate themselves *within the confines of the genome* as described in the next section.

### 5.4.2 The LINE-1 family

The mouse genome contains three independent families of dispersed repetitive DNA elements—called B1, B2, and LINE-1[35] (or L1)—that are each present at more than 80,000 chromosomal sites (Hasties, 1989). The general name coined for genomic elements of this type that disperse themselves through the genome by means of an RNA intermediate is *retroposon*. Of the three major retroposon families, it is only L1 that appears to be derived from a full-fledged selfish DNA sequence with a self-encoded reverse transcriptase. The mouse L1 DNA family is very old and homologous repetitive families have been found in a wide variety of organisms including protists and plants (Martin, 1991). Thus, LINE-related elements, or others of a similar nature, are likely to have been the source material that gave rise to retroviruses.

Full-length L1 elements have a length of 7 kb; however, the vast majority ( > 90%) of the ~ 100,000 L1 elements have truncated sequences that vary in length down to 500 bp (Martin, 1991). However, of the ~ 10,000 full-length L1 elements, only a few retain a completely functional reverse transcriptase gene, which has not been inactivated by mutation. Thus, only a very small fraction of the L1 family members retain "transposition competence," and it is these that are responsible for dispersing new elements into the genome.

Dispersion to new positions in the germ-line genome presumably begins with the transcription of competent L1 elements in spermatogenic or oogenic cells. The reverse transcriptase coding region on the L1 transcript is translated into enzyme that preferentially associates with and utilizes the transcript that it came from as a template to produce L1 cDNA sequences (Martin, 1991). For reasons that are unclear, it seems that the reverse transcriptase usually stops before a full-length copy is finished. These incomplete cDNA molecules are, nevertheless, capable of forming a second strand and integrating into the genome as truncated L1 elements that are forever dormant.

The L1 family appears to evolve by repeated episodic amplifications from one or a few progenitor elements, followed by the slow degradation of most new integrants—by genetic drift —into random sequence. Thus, at any point in time, a large fraction of the cross-hybridizing L1 elements in any one genome will be more similar to each other than to L1 elements in other species. In a sense, episodic amplification followed by general degradation is another mechanism of concerted evolution.

A large percentage of the mouse L1 elements share two *Eco*RI restriction sites located at a 1.3 kb distance from each other near the 3' end of the full-length sequence. With its very high copy number, this 1.3 kb fragment is readily observed in—and, in fact, diagnostic of—*Eco*RI digests of total mouse genomic DNA that has been separated by agarose gel electrophoresis and subjected to staining by ethidium bromide. This high copy number *Eco*R1 fragment was originally given other names, including MIF-1 and 1.3RI, before it was realized to be simply a portion of L1.

### 5.4.3 The major SINE families: B1 and B2

The two other major families of highly repetitive elements in the mouse—B1 and B2—are both of the SINE type with relatively short repeat units of ~ 140 bp and ~ 190 bp in length, respectively. The significance of this short repeat length is that it does not provide sufficient capacity for these elements actually to encode their own reverse transcriptase. Nevertheless, SINE elements are able to disperse themselves through the genome, just like LINE elements, by means of an RNA intermediate that undergoes reverse transcription. Clearly, SINEs are dependent on the availability of reverse transcriptase produced elsewhere, perhaps from L1 transcripts or endogenous retroviruses.

All SINE elements, in the mouse genome and elsewhere, appear to have evolved out of small cellular RNA species—most often tRNAs but also (in the case of mice and humans) the 7SL cytoplasmic RNA, which is one of the components of the signal recognition particle (SRP) essential for protein translocation across the endoplasmic reticulum (Okada, 1991). Unlike the LINE families, however, SINE families present in the genomes of different organisms appear, for the most part, to have independent origins. The defining event in the evolution of a functional cellular RNA into an altered-function self-replicating SINE element is the accumulation of nucleotide changes in the 3' region that lead to self-complementarity with the propensity to form hairpin loops. The open end of the hairpin loop can be recognized by reverse transcriptase as a primer for strand elongation. Since hairpin loop formation of this type is likely to be very rare among normal cellular RNAs, the SINE transcripts in a cell will be utilized preferentially as templates for the production of cDNA molecules that are able to integrate into the genome at random sites. Like the L1 family, the B1 and B2 families appear to be evolving by episodic amplification followed by sequence degradation.

The B1 element is repeated $\sim 150{,}000$ times, and the B2 element is repeated $\sim 90{,}000$ times (Hasties, 1989); together these elements alone account for $\sim 1.3\%$ of the material in the *Mus musculus* genome. The B1 element is derived from a portion of the 7SL RNA gene, whereas the B2 element appears to be derived (in a complicated fashion) from a tRNA$^{lys}$ gene. Human beings have just one family of SINE elements, referred to as the "Alu family," which is present in 500,000 copies and is also derived from 7SL RNA, although in an independent fashion from the mouse B1 element. Interestingly, the mouse genome does contain about 10,000 copies of a retroposon family that is closely related to the human Alu family (Hasties, 1989).

### 5.4.4 General comments on SINEs and LINEs

A number of other independent SINE families have been identified in the mouse genome but none are present in more than 10,000 copies. One such family of 80 bp tRNA-derived elements called ID was originally found in the rat genome at a copy number of 200,000; in the mouse genome, there are only 10,000 ID copies. In addition, there are probably other minor SINE families in the mouse genome that have yet to be well characterized.

The total mass of SINE and LINE elements probably accounts for less than 15% of the mouse genome. With the efficient means of self-dispersion that these elements employ, one can ask why they have not amplified themselves to even higher levels? The answer is almost certainly that if the amount of selfish DNA in a genome goes above a certain critical level, it will cause the host organism to be less fit and, thus, less likely to pass its selfish DNA load on to future generations. The existence of a critical ceiling means that the various SINE and LINE families are in direct competition for a limited amount of genomic real estate.

If one assumes that each of the major highly repetitive families—B1, B2 and L1—is dispersed at random, it is a simple matter to calculate that, on average, a member of each family will be present in every 20–30 kb of DNA. In fact, if one screens a complete genomic library with 15 kb inserts in bacteriophage lambda with a probe for each family, 80% of the clones are found to contain B1 elements, 50% are found to contain B2 elements, and 20% are found to contain the central portion of the full-length L1 element (Hasties, 1989). However, if one analyzes individual clones for the presence of both SINE and LINE elements, there is a significant negative correlation. In other words, SINE and LINE elements appear to prefer different genomic domains.

To better understand the non-random distribution of the three major mouse repetitive families and to investigate possible correlations with chromosome structure, the karyotypic distribution of each family was investigated by fluorescence *in situ* hybridization or FISH (Boyle et al., 1990; Section 10.2). Incredibly, the distribution of the LINE-1 elements corresponds almost precisely with the distribution of the Giemsa-stained dark G bands. In contrast, both SINE element families co-localize to the lightly stained chromosomal regions located between G bands (R bands). When the same type of experiment was performed on human karyotypes, the same result was obtained—the human SINE element Alu was found in R bands, whereas human LINE sequences co-localized with G bands (Korenberg and Rykowski, 1988). Since, essentially all of the SINE and LINE elements integrated into the mouse and human genomes subsequent to their divergence from a common mammalian ancestor, the implication is that preferential integration into different chromosomal domains is a property of each element class.

The correlation between G and R bands and LINE and SINE distribution respectively is not perfect. Some chromosomal regions are observed to have an overabundance or underabundance of the associated sequences, and a small fraction of elements are located outside the "correct" regions. One consistent exception to the general correlation is that centromeric heterochromatic regions, which normally stain brightly with Giemsa, do not have any detectable LINE elements in mice or humans. However, this exception can be easily understood in terms of the special structural role played by centromeric satellite DNA in chromosome segregation—any integration of a LINE element would disrupt this special DNA and its function, and this would be selected against evolutionarily.

As a final note, it should be mentioned that, although the SINE and LINE elements have amplified themselves for selfish purposes, they have, in turn, had a profound impact on whole genome evolution. In particular, homologous elements located at nearby locations can, and will, act to catalyze unequal but "homologous" crossovers that result in the duplication of single-copy genes located in between and the initiation of gene cluster formation as illustrated in Fig. 5.5.

### 5.4.5 Genomic stutters: microsatellites, minisatellites, and macrosatellites

With large-scale sequencing and hybridization analyses of mammalian genomes came the frequent observation of tandem repeats of DNA sequences, without any apparent function, scattered throughout the genome. The repeating unit can be as short as two nucleotides (CACACACA, etc.), or as long as 20 kb. The number of tandem repeats can also vary from as few as two to as many as several hundred. The mechanism by which tandem repeat loci *originate* may be different for loci having very short repeat units as compared to those with longer repeat units. Tandem repeats of short di- or trinucleotides can originate through random changes in non-functional sequences. In contrast, the initial duplication of larger repeat units is likely to be a consequence of unequal crossing over. Once two or more copies of a repeat unit (whether long or short) exist in tandem, unequal pairing followed by crossing over can lead to an increase in the number of repeat units in subsequent generations (see Fig. 8.4). Whether stochastic mechanisms alone can account for the rich variety of tandem repeat loci that exist in the genome or whether other selective forces are at play is not clear at the present time. In any case, tandem repeat loci continue to be highly susceptible to unequal crossovers and, as a result, they tend to be highly polymorphic in terms of overall locus size.

Tandem repeat loci are classified according to both the size of the individual repeat unit and the length of the whole repeat cluster. The smallest and simplest—with repeat units of 1–4 bases and locus sizes of less than 100 bp—are called microsatellites. The use of microsatellites as genetic markers has revolutionized the entire field of mammalian genetics (see Section 8.3.6). Next come the minisatellites with repeat units of 10 to 40 bp and locus sizes that vary from several hundred base pairs to several kilobases (see Section 8.2.3). Tandem repeat loci of other sizes do not appear to be as common but a great variety are scattered throughout the genome. The term midisatellite has been proposed for loci containing 40 bp repeat units that extend over distances of 250 to 500 kb, and macrosatellites has been proposed as the term to described loci with large repeat units of 3–20 kb present in clusters that extend over 800 kb (Giacalone et al., 1992). However, the use of arbitrary size boundaries to "define" these other types of loci is probably not meaningful, since it appears that, in reality, no such boundaries exist in the potential for tandem repeat loci to form in the mouse and other mammalian genomes.

## 5.5 GENOMIC IMPRINTING

### 5.5.1 Overview

From the birth of the field of genetics until a decade ago, it was generally assumed that the parental origin of a gene could have no effect on its function. In the vast majority of studies carried out during the last 90 years, this paradigm has appeared to hold true. However, with increasingly sophisticated genetic and embryological investigations in the mouse, important exceptions to this rule have been uncovered over the last decade. First, the results of nuclear transplantation experiments carried out with single-cell fertilized embryos have demonstrated an absolute requirement for both a maternally derived and a paternally derived pronculeus to allow full-term development (McGrath and Solter, 1983). Second, in animals that receive both homologs of certain chromosomes or sub-chromosomal regions from one parent and not the other (through the mating of translocation heterozygotes as described in Section 5.2.3), dramatic effects on development can be observed including enhanced or retarded growth, and outright lethality (Cattanach and Kirk, 1985). Third, either of two deletions that cover a small region of mouse chromosome 17 can be transmitted normally from a father to his offspring but these same deletions cause prenatal lethality when they are maternally transmitted (Johnson, 1974; Winking and Silver, 1984). Fourth, similar parent-of-origin effects have been observed on the phenotypes expressed by animals that carry a targeted knock-out allele at the *Igf2* locus (DeChiara et al., 1991). Finally, molecular techniques have been used to demonstrate directly the expression of transcripts from one parental allele and not the other at the *Igf2r* locus (Barlow et al., 1991) and the *H19* locus (Bartolomei et al., 1991).

The accumulated data indicate that a subset of mouse genes (on the order of 0.2%) will function differently in normal embryos depending on whether they have been inherited through the male or the female gamete, such that one allele will be expressed and the other will be silent. *Genomic imprinting* is the term that has been coined to describe this situation in which the phenotype expressed by a gene varies depending on its parental origin (Sapienza, 1989). Further experiments have demonstrated that, in general, the "imprint" is erased and regenerated during gametogenesis so that the function of an imprintable gene is fully determined by the sex of its progenitor alone, and not by earlier ancestors.

With the demonstration of genomic imprinting in the mouse, patterns of disease inheritance in humans have been investigated for the possibility of phenotypes determined by parent of origin in this mammal as well. To date, clear-cut parental effects have been uncovered in the transmission of the juvenile form of Huntington disease (Ridley et al., 1991), Prader-Willi and Angelman deletion syndromes (Nichols et al., 1989) and certain forms of juvenile familial carcinomas, such as multifocal retinoblastoma, Wilm's tumor, embryonal rhabdomyosarcoma and Beckwith–Wiedemann syndrome (Ferguson-Smith et al., 1990; Henry et al., 1991; Sapienza, 1991).

### 5.5.2 Why is there imprinting?

The first explanation for the existence of imprinting was as a mechanism to prevent the full-term development of parthenogenetic embryos. This explanation was never satisfactory because it did not account for the intricate control of imprinting at multiple well-bounded loci. An alternative hypothesis put forward by Haig and his colleagues is based on a tug of war between the sexes (Haig and Graham, 1991; Moore and Haig, 1991). According to this hypothesis, it is in the interest of a male to attempt to recover more maternal resources for his developing offspring in relation to offspring in the same mother that were sired by other males. This can be accomplished with a paternal imprint that down-regulates the expression of genes that normally act to slow down the growth of embryos. As a consequence, embryos that are sired by these males will grow more rapidly than half-siblings sired by other males. Although overgrowth may be beneficial to these offspring, it extracts a heavy reproductive cost from the mother. Consequently, it is in the interest of the mother to counteract this increased level of growth. She can do this with an imprint that down-regulates the relevant growth factor genes themselves. The evolutionary endpoint of this tug of war is the current day situation where genes that act to increase embryonic growth (such as *Igf2*) have inactivated maternal alleles and genes that act to limit growth (such as *Igf2r*) have inactivated paternal alleles.

The only other currently viable hypothesis to explain imprinting is that it results from the accidental, ectopic use of machinery that has evolved for the really important imprinting associated with X chromosome inactivation. According to this hypothesis, autosomal imprinting is a red herring whose study is unlikely to provide information of significance to an understanding of developmental genetics. The major strike against this hypothesis is dealt by selectionists who would contend that genetic accidents of this magnitude just do not happen and there must be something peculiar about mammals that has promoted the evolution of imprinting. In support of the selectionist view is the recent demonstration of monoallelic expression of the *H19* gene in humans (Zhang and Tycko, 1992). Conservation of imprinting during the evolution of both humans and mice from a common ancestor strongly suggests the existence of selective forces. Nevertheless, it is still possible that the Haig hypothesis is not entirely correct and that other reasons for imprinting lie hidden beneath the surface waiting to be uncovered.

### 5.5.3 The molecular basis for imprinting

The question "how does it happen?" can be easily separated from the question "why does it happen?". However, here again, our understanding is still quite rudimentary as of 1993. Figure 5.8 illustrates in a very general way the essential requirements of a paternal imprinting system. Both parents have one imprinted allele (derived from their fathers) and one active allele (derived from their mothers). During oogenesis, the imprint must be erased so that all eggs will contain equivalent alleles that can become activated in all offspring. In contrast, at the completion of spermatogenesis in the father, all sperm will contain alleles that are "marked" for imprinting. It is possible that the mark present on one of the father's alleles is erased and both copies are marked *de novo* in all spermatogenic cells, or the one imprinted copy may retain its mark, with *de novo* marking applied only at the second copy. In either case, the new embryo will receive one "marked" gene from the father and one non-marked gene from the mother.

The "mark" may itself be replicated faithfully along with its homolog, and the "mark" may itself be responsible for the actual repression of gene activity. On the

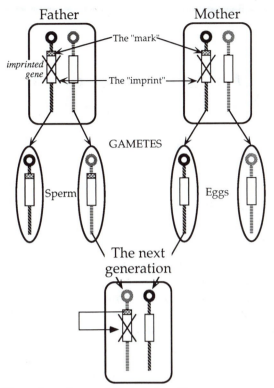

**Figure 5.8**   Illustration of genomic imprinting. The genotypes around an imprinted gene are shown for two parents, their gametes and their offspring. The gene is represented by an open box. The imprinting "mark" is represented by a stippled box adjacent to the gene, and the imprinted state itself is indicated with an X across the gene.

other hand, the "mark" may simply identify the paternal allele so that a separate imprinting machinery that acts to prevent gene repression can be laid down within the developing embryo. If there is a separate imprinting machinery, either it or the "mark" could be replicated along with the paternal homolog to maintain the imprint through each cell division.

It is still the case in 1993 that the nature of both the mark and the imprinting machinery (if it exists as a separate entity) are entirely unknown. Both could presumably entail direct chemical modifications of DNA and/or specific protein components (that might lead to changes in the local chromatin configuration). In addition, the specific DNA sequences that must be recognized by the gametogenic marker are also unknown at this time.

# 6

# Mutagenesis and Transgenesis

## 6.1 CLASSICAL MUTAGENESIS

### 6.1.1 The specific locus test

*Genetic variation*—the existence of at least two forms—is the essential ingredient present in all breeding experiments. Phenotypic variation, in particular, can be used as a means for uncovering the normal function of a wild-type allele. As discussed in the first chapter of this book, it was the availability of many variant phenotypes within the fancy mouse trade that made the house mouse such an ideal organism for studies by early geneticists. In a sense though, the house mouse won by default because, in the absence of domestication and artificial selection, variation in traits visible to the eye is extremely rare and thus, other small mammals were genetically intractable. Although the fancy mouse variants provided material for a host of early genetic studies, the number of different variants was still limited, and the rate at which new ones arose spontaneously in experimental colonies was exceedingly low: it is now known that, on average, only one gamete in 100,000 is likely to carry a detectable mutation at any single locus.

During the 1920s, several investigators began investigating the effects of X-rays on reproduction and development. In two laboratories, at least, new mutant alleles were recovered in the offspring of irradiated parents, but the investigators failed to make any connection between irradiation and the induction of these mutations (Little and Bagg, 1924; Dobrovolskaia-Zavadskaia, 1927). The connection was finally made by Muller who, in 1927, published his classic paper explaining the induction of heritable mutations by X-rays (Muller, 1927). Since that time, geneticists who study all of the major experimental organisms— from bacteria to mice—have used both ionizing irradiation and various chemicals as agents of mutagenesis to uncover novel alleles as tools for understanding gene function.

Large-scale mouse mutagenesis experiments were first begun at two government-based "atomic energy" laboratories: the Oak Ridge National Laboratory in Oak Ridge, Tennessee, in the US and the Medical Research Council Radiobiological Research Unit first at Edinburgh, Scotland, and then at Harwell, England, in the UK. Both of these experimental programs were begun initially after World War II as a means for quantifying the effects of various forms of radiation on mice and, by extrapolation, humans, to understand the consequences of detonating nuclear weapons better. The US effort was directed by W. L. Russell and the British effort was directed by T. C. Carter (Green and Roderick, 1966). Scientists at both laboratories quickly realized the potential of their newly created resource of mutant animals and both laboratories have since gone on to generate mutations by chemical agents as well. The very large-scale studies conducted at Oak Ridge and Harwell—where 10,000–60,000 first generation animals were typically analyzed in an experimental protocol—have provided most of the empirical data currently available on the mechanisms and rates at which mutations are caused by all well-characterized mutagenic agents in the mouse.

The experiments performed by Russell and Carter, and other colleagues who followed in their footsteps, were designed to obtain discrete values for the mutagenic potential of different radiation protocols. Rather than attempt to examine all animals for all effects of a particular irradiation protocol (as was common in earlier experiments), these mouse geneticists chose instead to look only at the small fraction of animals that were mutated at a small set of well-defined "specific" loci. The rationale for the "specific locus test" was that effects on individual loci could be more easily quantified and that the limited results obtained could still be extrapolated for an estimate of whole genome effects. Russell decided that mutation rates should be followed simultaneously at a sufficient number of loci to distinguish and avoid problems that might be caused by locus to locus variations in sensitivity to particular mutagens. He decided further that the same set of loci should be examined in each experiment performed. The seven loci chosen to be followed in the specific locus test were defined by recessive mutations with visible homozygous phenotypes that were easily distinguished in isolation from each other, and had no effect on viability or fertility. The seven loci are agouti (*a* is the recessive non-agouti allele), brown (*b*), albino (*c*), dilute (*d*), short-ear (*se*), pink-eyed dilution (*p*), and piebald (*s*). A special "marker strain" was constructed that was homozygous for all seven loci.

In its simplest form, the specific locus test is carried out by mating females from the special marker strain to completely wild-type males that have been previously exposed to a potential mutagen. In the absence of any mutations, offspring from this cross will not express any of the seven recessive phenotypes visible in the marker strain mother. However, if the mutagen has induced a mutation at one of the specific loci, the associated mutant phenotype will be uncovered. This test is very efficient because it only requires a single generation of breeding and visual examination is all that is required to score each animal.

Although recessive mutations at all loci other than the specific seven will go undetected in the first generation offspring from this cross, it is possible to detect a dominant mutation at any locus so long as it is viable and produces a gross alteration in heterozygous phenotype, such as a skeletal or coat-color change. One should realize that the most common effect of any undirected mutagen will be to "knock out" a gene and, in the vast majority of cases, the resulting null allele will be recessive to the wild type. There is, however, a very small class of loci at which null alleles will act in a dominant or semidominant fashion to wild type. These "haplo-insufficient" phenotypes are presumably caused by a developmental sensitivity to gene product dosage. Among the best characterized of the dominant null mutations are the numerous ones uncovered at the $T$ locus—which result in a short tail—and the $W$ locus—which result in white spotting on the coat.

### 6.1.2 Mutagenic agents

Mutations can be induced by both physical and chemical means. The physical means is through the exposure of the whole animal to ionizing radiation of one of three classes—X-rays, gamma rays, or neutrons (Green and Roderick, 1966). The chemical means is to inject a mutagenic reagent into the animal such that it passes directly into the gonads and into differentiating germ cells. The specific locus test has provided an estimate of the relative efficiency with which each reagent induces mutations. Under different protocols of exposure, X-irradiation was found to induce mutations at a rate of $13–50 \times 10^{-5}$ per locus, which is a 20–100-fold increase over the spontaneous frequency, but still not high enough to be used by any but the largest facilities as a routine means for creating mutations (Rinchik, 1991). The mutations created by irradiation are often large deletions or other gross lesions, such as translocations or complex rearrangements.

The class of known chemical agents that can induce mutations (known as mutagens) is very large and expanding all the time. However, two chemicals in particular—ethylnitrosourea (ENU) and chlorambucil (CHL)—have been found to be extremely mutagenic in mouse spermatogenic cells (W. L. Russell et al., 1979; L. B. Russell et al., 1989). Both of these chemicals produce much higher yields of mutations than any form of radiation treatment tested to date (Russell et al., 1989). Optimal doses of either ENU or CHL can induce mutations at an average per locus frequency that is greater than one in a thousand—$150 \times 10^{-5}$ with ENU and $127 \times 10^{-5}$ per locus with CHL (W. L. Russell et al., 1982; L. B.Russell et al., 1989). Although the rates at which ENU and CHL induce mutations are very similar, the types of mutations that are induced are quite different. In general, ENU causes discrete lesions that are often point mutations (Popp et al., 1983), whereas CHL causes large lesions that are often multilocus deletions (Rinchik et al., 1990a). Originally, it was thought that the basis for mutational differences of this type was the chemical nature of the mutagen itself (Green and Roderick, 1966), but this no longer appears to be the case. Rather, it now appears that the germ-cell stage in which the mutation arises is the major determinant of the lesion type (Russell, 1990). The correlation observed between chemical and lesion type is a result of the fact that different mutagens are active at different stages of spermatogenesis. Thus, ENU acts upon premeiotic spermatogonia, where mutations are likely to be of the discrete type, and CHL acts upon postmeiotic round spermatids where mutations are likely to be of the large lesion type (Russell et al., 1990).

ENU was the first chemical to be identified that was sufficiently mutagenic to be used by smaller laboratories in screens for mutations at particular loci or chromosomal regions of interest (Bode, 1984; Shedlovsky et al., 1988). ENU has also been used in screens for non-locus-specific phenotypic variants that could serve as models for various human diseases (McDonald et al., 1990). Several laboratories are beginning to use ENU for saturation mutagenesis of small chromosomal regions defined by deletions, as one tool (among several complementary ones) for obtaining a complete physical and genetic description of such a region (Shedlovsky et al., 1988; Rinchik et al., 1990b; Rinchik, 1991). The major limitation to the global use of this approach is the very small number of genomic regions in the mouse at which large deletions have been characterized.

The availability to *Drosophila* geneticists of deletions (or *deficiencies* as the fly people call them) that span nearly every segment of the fly genome has played a critical role in the identification and characterization of large numbers of genes and the production of both gross functional maps and fine-structure point mutation maps by the very approach just described above. Clearly, a method to accumulate a similar library of deletions for the mouse would be well received. The mutations induced by X-rays are often large-scale genomic alterations including translocations, inversions and deletions. Indeed, most mouse deletion mutations maintained in contemporary stocks were derived in this manner. However, the overall yield of X-ray-induced deletions is quite low, and because of other problems inherent in this approach, it is not ideal for global use.

In 1989, CHL was reported to be an attractive alternative to X-rays as an agent for the high-yield induction of deletion mutations in the mouse (Russell et al., 1989). The per locus mutation rate was found to be on the order of one in 700 in germ cells of the early spermatid class, and of the eight mutations induced at this stage that were analyzed, all were deleted for DNA sequences around the specific locus marker (Rinchik et al., 1990a). This study also showed that CHL-induced mutations were often associated with reciprocal translocations. This last finding is unfortunate because translocations can reduce fertility with consequent negative effects on strain propagation.

There is hope that CHL can be used as a means for generating sets of overlapping deletions that span entire chromosomes (Rinchik and Russell, 1990). Projects of this type will require very large animal facilities and support resources, and will consequently be confined to only a handful of laboratories. However, once mouse strains with deletions have been created and characterized, they can serve as a resource for the entire community.

### 6.1.3 Mouse mutant resources

An advantage to using the mouse as a genetic system is the strong sense of community that envelops most of the workers in the field, and it is in the context of this community that strains carrying many different mutations—both spontaneous and mutagen-induced—have been catalogued and preserved, and are made available to all investigators. A catalog containing detailed descriptions of all mouse mutations characterized as of 1989 has been compiled by Margaret Green and is included as the centerpiece chapter in the *Genetic Variants and Strains of the Laboratory Mouse* edited by Mary Lyon and Tony Searle (Green, 1989). This catalog is now available in an electronic form that is updated regularly (see Appendix B). Of course, many more mutant animals are found and characterized with the passing of each year, and an updated list is published annually in the journal *Mouse Genome*. This list contains information on the individual investigators that one should contact to actually obtain the mutant mice.

The largest collection of mutant mouse strains is maintained at the Jackson Laboratory under the auspices of the "mouse mutant resource" (MMR), which is currently maintained under the direction of Dr Muriel Davisson (Davisson, 1990). In 1990, over 250 mutant genes were maintained in this resource, accounting for two-thirds of all known mouse mutants alive at the time (Davisson, 1990). Each year, animal caretakers identify an additional 75–80 "deviant" animals among the two million mice that are produced by the Jackson Laboratory's Animal Resources colonies (Davisson, 1993). Approximately 75% of the deviant phenotypes are found to have a genetic basis and breeding studies are conducted on these to determine whether or not they represent mutations at previously characterized loci. If they do, DNA samples are recovered and the lines are discarded or placed into the frozen embryo repository. If a mutation is novel, its mode of transmission (autosomal/X-linked, dominant/recessive) is determined, the phenotypic effect of the mutation is characterized, and it is mapped to a specific chromosomal location with the use of breeding protocols to be described in Section 9.4 (Davisson, 1990, 1993). Descriptions of all newly characterized mutations are publicized, and mutant mouse strains are made available for purchase through the standard Jackson Laboratory catalog. In 1992, over 35,000 mice from the MMR were distributed to investigators throughout the world (Davisson, 1993).

Space limitations make it impossible for the MMR to maintain breeding stocks of mice that contain every known mutant gene, with the total number expanding each year. Fortunately, mutant stocks that are not currently in demand by investigators can be maintained (at minimal cost) in the form of frozen embryos. The importance of embryo freezing as a storage protocol cannot be over-emphasized. Time and time again, modern-day molecular researchers have reached back to use mutations described long-ago as critical tools in the analysis of newly cloned human and mouse loci.

## 6.2 EMBRYO MANIPULATION: GENETIC CONSIDERATIONS

### 6.2.1 Experimental possibilities

The basic technology required to obtain preimplantation embryos from the female reproductive tract, to culture them for short periods of time in Petri dishes, and then to place them back into foster mothers where they can grow and develop into viable mice, has been available since the 1950s (Hogan et al., 1994). Over the ensuing years, this basic technology has been used in a host of different types of experiments aimed at manipulating the process of development or the embryonic genome itself. Embryos can be dissolved into individual cells that can be recombined in new combinations to initiate the development of chimeric mice. Pronuclei and nuclei can be switched from one early embryo to another to examine the relative contributions of the cytoplasm and the genome to particular phenotypes, as well as to investigate aspects of genomic imprinting and parthenogenesis. Foreign DNA can be injected directly into pronuclei for stable integration into chromosomes, which can lead to the formation of transgenic animals. Finally, embryonic cells can be converted into tissue culture cells (called embryonic stem [ES] cells) where targeted gene replacement can be accomplished. Selected ES cells can be combined with normal embryos to form chimeric animals that can pass the targeted locus through their germ line. These experimental possibilities are discussed more fully later in this chapter. This section is concerned simply with genetic considerations involved in the choice of mice to be used for the generation and gestation of embryos for various experimental purposes.

## 6.2.2  Choice of strains for egg production

### 6.2.2.1  General considerations

A number of factors will play a role in the selection of an appropriate strain of females who will contribute the eggs to be used as experimental material. First, in all cases, it is important that the eggs are hardy enough to resist damage from the manipulations that they will undergo. Second, the particular experimental protocol may impose a need for eggs that have special genetically determined qualities. Third, in those cases where very large numbers of eggs are required, it will be important that the strain is one that responds well to superovulation, as discussed in the next section. Finally, there is a question of genetic restrictions on the offspring that will emerge from the manipulation.

As concerns this last criterion, for some experiments it will be important to maintain strict control over the genetic background of embryos to be used for genomic manipulation. In these cases, inbred embryos should be derived from matings between two members of the same inbred strain. If these embryos are used for germ-line introduction of foreign genetic material, the resulting transgenic animals will be truly coisogenic to the original inbred strain.

For other experiments, strict genetic homogeneity will not be required. In these cases, it is possible to use $F_2$ embryos from superovulated $F_1$ females who have been mated to $F_1$ males of the same autosomal genotype. This breeding protocol is often preferable to the use of either a strictly inbred approach or a random-bred approach. First, in contrast to the random-bred approach, one still maintains a certain degree of control over the genetic input, since only alleles derived from one or both of the inbred strains used to generate the $F_1$ parents will be present at any locus in each embryo. Second, in contrast to the inbred approach, the use of both females and embryos with heterozygous genotypes allows the expression of hybrid vigor at all levels of the reproductive process. In particular, heterozygous embryos are less likely to be injured by in vitro manipulations.

### 6.2.2.2  The FVB/N strain is ideal for the production of transgenic mice

One inbred strain that has been developed relatively recently from a non-inbred colony of mice with a long history of laboratory breeding at NIH has special characteristics of particular interest to investigators interested in producing transgenic mice: this strain is called FVB/N. The FVB/N strain is unique in several important ways (Taketo et al., 1991). First, its average litter size of 9.5 (with a range up to 13) is significantly higher than that found with any other well-known inbred strain (see Table 4.1). Second, fertilized eggs derived from FVB/N mothers have very large and visually prominent pronuclei; this characteristic is unique among the known inbred strains and greatly facilitates the injection of DNA. Finally, the fraction of injected embryos that survive into live born animals is also much greater than that observed with all other inbred strains. For these reasons, FVB/N has quickly become the strain of choice for use in the production of transgenic animals.

### 6.2.3 Optimizing embryo production by superovulation

Although one can recover of the order of 6–10 eggs directly from individual naturally mated inbred or $F_1$ females, it is possible to obtain much larger numbers—up to 60 eggs per animal—by inducing a state of *superovulation*. For many experiments, it is important to begin with a large number of embryos; with superovulation, one can drastically reduce the number of females required to produce this large number. Superovulation is induced by administering two precisely timed intraperitoneal injections of commercially available gonado-tropin reagents that mimic natural mouse hormones and initiate the maturation of an aberrantly large number of egg follicles. Superovulation, like normal ovulation, causes both a stimulation of male interest in mating as well as female receptivity to interested males. The protocol is described in detail in the mouse embryology manual by Hogan and colleagues (1994).

Not unexpectedly, the average number of eggs induced by superovulation is highly strain dependent. Appropriately aged females[36] of the strains B6, BALB/cByJ, 129/SvJ, CBA/CaJ, SJL/J and C58/J can be induced to ovulate 40–60 eggs (Hogan et al., 1994). At the other extreme, females of the strains A/J, C3H/HeJ, BALB/cJ, 129/J, 129/ReJ, DBA/2J, and C57L/J do not respond well to the superovulation protocol, producing only 15 or fewer eggs per mouse. The response of the FVB/N strain to superovulation falls between the two extremes with the production of 25 embryos or fewer per female (Taketo et al., 1991). For generating transgenic mice, however, this single negative feature of FVB/N is outweighed by the advantageous characteristics of this strain discussed above.

An interesting aspect of the high versus low response to superovulation is that in two cases, substrains derived from the same original inbred strain (BALB/cByJ versus BALB/cJ and 129/SvJ versus 129/J) express such clearly distinct phenotypes. This finding suggests that subtle changes in genotype can have dramatic consequences on the expression of this particular reproductive trait.

One critical finding of both practical and theoretical importance is that $F_1$ hybrid females do not always express a better response to superovulation than both of their inbred parents. For example, the commonly used $F_1$ hybrid B6D2F$_1$, which is formed by a cross between a high ovulator (B6) and a low ovulator (DBA/2J), expresses the low ovulator phenotype (Hogan et al., 1994). This observation goes against the grain of hybrid vigor, and it suggests that the genetic basis for this phenotype may be much more specific and limited than it is for other general viability and fertility phenotypes. In addition, this observation suggests that for the major genes involved, the "high ovulatory" alleles are recessive.

Two $F_1$ hybrids have been determined empirically to express a high level superovulation—[BALB/cByJ $\times$ B6] and [B6 $\times$ CBA/CaJ] (Hogan et al., 1994). It is also very likely that $F_1$ hybrids derived from matings between any of the high responders listed above will themselves be high responders as well. Many of these $F_1$ hybrids can be purchased directly from animal suppliers; however, in most cases, suppliers cannot provide an exact day of birth, which is necessary to determine the optimal time of use.

### 6.2.4 The fertile stud male

Females that have undergone ovulation—either naturally or induced—must be mated with a "fertile stud male" to produce zygotes that can be used for nuclear injection or other purposes. As discussed earlier, it is always preferable to use a fertile stud male with the same genotype as the female, whether it is inbred or an $F_1$ hybrid. Obviously, it is important to use visibly healthy animals in the prime of their life, between 2 and 8 months of age. In addition, past experience is often a good indicator of future performance. Males that have mated successfully on demand in the past (as indicated by a vaginal plug) are likely to do the same in the future; for this reason, records should be maintained on the performance of each male used for this purpose. For optimal results, one should place only one male and one female in each cage, and after a successful mating, the male should be given a rest of 2–3 days.

### 6.2.5 Embryo transfer into foster mothers
#### 6.2.5.1 Choice of strain

Once embryos have been manipulated in culture, they can be placed back into the reproductive tract of a *foster mother*, to allow their development into fully formed live-born animals. Since the foster mother contributes only a womb, and not genomic material, to the engineered offspring, her genetic constituency should not be chosen according to the same criteria used for animals in most other experimental protocols. Only two considerations are important in the choice of a foster mother. First, and most important, she should have optimal reproductive fitness and "mothering" characteristics. This can be accomplished with either an $F_1$ hybrid between two standard inbred strains [B6 $\times$ CBA is recommended (Hogan et al., 1994) but others will do as well] or with outbred strains available from various commercial breeders.

A second consideration is whether the investigator will be able to distinguish natural-born pups from those that have been fostered. This is only a factor when the foster mother has been mated to a sterile male in order to induce the required state of pseudopregnancy, and there is some question as to whether the male has been properly sterilized. The simplest method for distinguishing the two types of potential offspring is by a coat-color difference; for example, albino versus pigmented. If the experimental embryos are derived from parents without an albino allele, then both the foster mother as well as her sterile stud partner (discussed below) can be chosen from commercially available outbred albino strains such as CD-1 (Charles River Breeding Laboratories) or Swiss Webster mice (from Taconic Farms). When one is certain that the sterile stud male is really sterile, coat color differences are less critical, so long as well-defined DNA differences exist if the unexpected need does arise to distinguish the genotypes of potential natural-born offspring from experimentally transferred offspring.

### 6.2.5.2 Induction of pseudopregnancy and the sterile stud male

In human females, the uterine environment becomes receptive to the implantation of fertilized eggs as a direct consequence of the hormonal induction of ovulation. In mice and most other non-primate mammals, the uterine environment becomes receptive to implantation *only* in response to a sufficient degree of sexual stimulation.[37] In addition, this stimulation also causes hormonal changes, which alter the normal estrus cycle under the assumption that a pregnancy will ensue.[38] When a successful stimulatory response has occurred in the absence of implantation, the female is said to be in a state of "pseudopregnancy." Only pseudopregnant females will allow the successful implantation and development of fostered embryos. Pseudopregnancy can be achieved in one of two ways: (1)

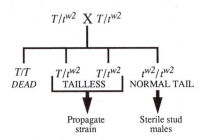

**Figure 6.1** A genetic strategy for generating sterile stud males. The breeding scheme shown is based on the interaction of two pseudo-allelic variants of the mouse *t* complex on Chr 17 (Silver, 1985). The *T* mutation is a recessive lethal allele. The $t^{w2}$ haplotype is a recessive sterile allele. Together, these two variants interact to produce doubly heterozygous animals that are fully viable and fertile, but have no tail. Matings between two $T/t^{w2}$ animals will generate offspring of two phenotypic classes. Tailless animals will have the same genotype as their parents and can be used to propagate the strain. Animals with normal tails must be homozygous for the $t^{w2}$ haplotype, and all such males will be unconditionally sterile. The percentage of sterile $t^{w2}/t^{w2}$ males produced from this intercross is actually greater than the 12.5% expected from simple Mendelian calculations because the ratio of transmission of the $t^{w2}$ haplotype is favorably distorted from heterozygous $t^{w2}$-bearing males (Silver, 1988, 1990). Although the ratio of transmission of the $t^{w2}$-bearing chromosome relative to the *T*-bearing homolog can be as high as 95:5, there is a counteracting, partially penetrant lethal phenotype associated with the $t^{w2}/t^{w2}$ genotype that reduces the percentage of live-born animals. Only 80% of these embryos are viable and, together with the distorted transmission ratio, this will lead to a final fraction of $t^{w2}/t^{w2}$ offspring of approximately 43% with half of these being males.

by mating to a sterile male; or (2) through the use of female masturbation tools such as vibrating rods inserted into the vagina (West et al., 1977). Most investigators have found that natural matings produce a higher percentage of pseudopregnancies than human surrogates.

Sterile males can be derived genetically or surgically. Genetic derivation can be accomplished with a breeding colony of mice that are doubly heterozygous for pseudo-allelic mutations on chromosome 17 in a region known as the $t$ complex (Silver, 1985). Animals with the genotype $T/t^{w2}$ (available from the Jackson Laboratory) are intercrossed for the purposes of both maintaining the strain as well as for the production of sterile males as shown in Fig. 6.1.

For those without the resources or personnel required to breed genetically sterile males, the only other choice is surgical vasectomy, which involves the severing of the vas deferens on both sides of the body (Hogan et al., 1994). The choice of mouse strain to use is based on criteria analogous to those set out for the choice of a foster mother, except that mating ability should be considered in place of mothering ability. The standard $F_1$ hybrids as well as "random-bred" animals can all be used with success. When it comes to choosing between individual males within a particular strain, one should use the same criteria described in Section 6.2.2.2 for the choice of fertile stud males. In addition, premating "sterile" males with fertile females serves to confirm the success of the vasectomy.

## 6.3 TRANSGENIC MICE FORMED BY NUCLEAR INJECTION

There are two problems inherent in all methods of classical mutagenesis. The first problem is that the process is entirely random. Thus, one must start out by designing a screening assay to allow the detection of mutations at the locus of interest, and then one must hope for the appearance of mutant animals at a frequency, which is at or above the usual per-locus rate. If a mutant allele fails to produce a phenotype that can be picked up by the screen, it will go undetected. Finally, even when mutant alleles are detected, the underlying lesion can usually not be ascertained without cloning and further molecular characterization.

The second problem with classical mutagenesis is that induced mutations are not tagged in any way to provide a molecular entry into a locus that has not yet been cloned. Thus, if a novel locus is uncovered by an induced mutation that causes an interesting phenotype, it can only be approached through candidate gene and positional cloning approaches in the same way as any other phenotypically defined locus. Furthermore, in the case of ENU-induced mutations, the mutant and wild-type alleles are likely to be molecularly indistinguishable with the exception of a single nucleotide that may or may not affect a restriction site.

One can imagine two types of mutagenic approaches that would be most ideal for the two different types of situations in which mutations can provide tools for molecular analysis of development and other aspects of mammalian biology. On the one hand, a random mutagenesis approach is fine for the elucidation of novel loci so long as the mutant allele is tagged to allow direct molecular access. On the other hand, to further analyze a locus that is already cloned and characterized, one would like to generate animals that misexpress the locus in some defined manner. The technologies of transgene insertion and gene targeting have provided geneticists with the tools needed to accomplish both of these goals.

### 6.3.1 Overview

In 1981, five independent laboratories reported the insertion of foreign DNA into the mouse germ line through the microinjection of one-cell eggs (Costantini and Lacy, 1981; Gordon and Ruddle, 1981; Harbers et al., 1981; E. F. Wagner et al., 1981; T. E. Wagner et al., 1981). Although the incorporation of exogenous DNA into the germ line through viral infection of embryos had been reported earlier (Jaenisch, 1976) , the 1981 reports implied for the first time that DNA from any source could be used to transform the mouse genome. The complexion of mouse genetics was changed forever with the development of this powerful tool. A strictly observational science was suddenly thrust into the realm of genetic engineering with all of its vast implications. The insertion of genetic material into the mouse germ line has now become sufficiently routine that the methodology is detailed in various "cookbooks" (Wassarman and DePamphilis, 1993; Hogan et al., 1994) and designer animals are even provided as a commercial service by a number of companies.

The term _transgenic_ has been coined to describe animals that have foreign sequences inserted stably into their genome through human intermediaries. Transgenic animals can be created by microinjection or viral infection of embryos, or through the manipulation in culture of embryonic-like "ES cells" that are subsequently incorporated back into the embryo proper for shepherding into the germ line. The latter technology will be discussed in a following section. Here I will focus on transgenic animals created by direct injection of DNA into embryos.

The initial animal that develops from each micromanipulated egg is called a _founder_. Even when multiple embryos have all been injected or infected with the same foreign DNA, the integration site—or _transgene locus_—in each founder will be different. However, all transgenic animals that descend from a single founder will share the same transgene locus. Protocols for the creation of transgenic mice, and extensive reviews of the technology and its uses have been described elsewhere (Palmiter and Brinster, 1986; Wassarman and DePamphilis, 1993; Hogan et al., 1994). Rules for naming transgene loci and transgenic animals are presented in Section 3.3.5.

With current protocols for the creation of transgenic mice by embryo microinjection, the site of integration is not predetermined and, for all practical purposes, should be considered random. Microinjection allows one to _add_, but not subtract genetic material in a directed manner; if a particular experiment leads to the insertion of an novel version of a mouse gene into the genome, this novel allele will be present in addition to the normal diploid pair. Consequently, only dominant, or codominant, forms of phenotypic expression will be detectable from the transgene.

1)
The embryo microinjection technology can be used to explore many different aspects of mouse biology and gene regulation. One class of experiments encompasses those aimed at determining the effects of expressing a natural gene product in an unnatural manner. By combining the gene of interest with regulatory regions chosen from other genes, one can cause transgenic mice to express the product at a higher than normal level, or in alternative tissues or developmental stages. The mutant phenotypes that result from such aberrant forms of expression can be used to elucidate the normal function of the wild-type gene. Experiments of this type can be used, for example, to demonstrate the capacity of some genes to induce specific developmental changes and the oncogenic nature of others when they are expressed aberrantly. Many other types of questions can be answered with this approach.

2)
In another class of experiments, one can dissect out the function of a regulatory region by forming constructs between it and a reporter gene whose expression can be easily assayed in the appropriate tissue(s). With a series of transgenic lines that have partially deleted or mutated forms of a regulatory region, one can pinpoint which DNA sequences are involved in the turning on and turning off of genes in different tissues or developmental stages.

3) A third class of experiments is aimed at correcting a genetic defect in a mutant mouse through the genomic insertion of a wild-type transgene. This use of the transgenic technology provides the most powerful means available to prove that a cloned candidate gene is indeed identical to the locus responsible for a particular mutant phenotype. Furthermore, the correction of genetic defects in model mammals is a necessary prelude to any attempt to perform similar studies in humans.

An important consideration in all transgenic experiments follows from the observation that the actual chromosomal location at which a transgene inserts can play a determining role in its expression. This will be readily apparent in cases where different founder lines with the same transgene show different patterns of transgene expression. The reason for such strain-specific differences is that some chromosomal regions are normally maintained in chromatin configurations that can act to suppress gene activity. Different transgene constructs will show different levels of sensitivity to suppression of activity when they land in such regions.

Another potential problem can result from the insertion of the transgene into a normally functioning endogenous locus with unanticipated consequences. In approximately 5–10% of all cases studied to date, homozygosity for a particular transgene locus has been found to cause lethality or some other phenotypic anomaly (Palmiter and Brinster, 1986). These recessive phenotypes are most likely due to the disruption of some normal vital gene. In less frequent cases, a transgene may land at a site that is flanked by an endogenous enhancer which can stimulate gene activity at inappropriate stages or tissues. This can lead to the expression of dominant phenotypes that are not strictly a result of the transgene itself.[39] For all of these reasons, it is critical to analyze data from three or more founder lines with the same transgene construct before reaching conclusions concerning the effect, or lack thereof, on the mouse phenotype.

In the vast majority of cases analyzed to date, the disruption of endogenous sequences caused by transgene integration has had no apparent effect on phenotype. However, the absence of a detectable phenotype does not necessarily mean that the transgene has integrated into a non-functional region of the genome. As discussed in Chapter 5, only a small subset of all mammalian genes are actually *vital*, and subtle effects on phenotype are likely to go unnoticed if one performs only a cursory examination of transgenic animals. Thus, the actual frequency of insertional mutagenesis resulting from embryo microinjection is likely to be significantly higher than the numbers imply.

## 6.3.2 Tracking the transgene and detecting homozygotes

Unless a particular transgenic insertion causes an easily detectable, dominant phenotype, the presence of the transgene in an animal is most readily determined through DNA analysis. For testing large numbers of mice, the best source of DNA is from tail clippings or ear punch-outs (Gendron-Maguire and Gridley, 1993). The presence or absence of the transgene can be demonstrated most efficiently with a method of PCR analysis that is based on transgene-specific target sequences.

The founder animal for a transgenic line will be heterozygous for the transgene insertion locus. The second homolog will be associated with a non-disrupted "wild-type" ( + ) allele at this locus, whereas the disrupted chromosome will carry a transgene (*Tg*) allele. As long as a heterozygous (*Tg/* + ) parent is used to transmit the transgene to offspring, it will be necessary to test each individual animal of each new generation for its presence. For this reason alone, it would be useful to generate animals homozygous for the transgene allele, since all offspring from matings between homozygous *Tg/Tg* animals would also be homozygous and there would be no need for DNA testing.

In rare cases, homozygous *Tg/Tg* animals will be phenotypically distinct from their *Tg/* + cohorts. This observation is usually a good indication that the transgene has disrupted the function of an endogenous locus through the process of integration. If the homozygous recessive phenotype is lethal, it will obviously be impossible to generate a homozygous line of animals. Otherwise, the phenotype may eliminate the need for DNA analysis. In the vast majority of cases, however, homozygous *Tg/Tg* animals will be indistinguishable in phenotype from heterozygous *Tg/* + animals, and without a recessive pheno-type, the identification of homozygous animals will not be straightforward.

One approach to confirming the genotype of a presumptive *Tg/Tg* animal is based on statistical genetics. In this case, confirmation is accomplished by setting up a mating between the presumptive homozygote and a non-transgenic + /+ partner. If the animal in question is only a *Tg/* + heterozygote, one would expect equal numbers of *Tg/* + and + /+ offspring. Through the method of chi-square analysis described in Section 9.1.3, one can calculate that if at least 13 offspring are born and all carry the transgene, the probability of a heterozygous genotype is less than one in a thousand.[40] If even a single animal is obtained without the transgene, the parent's genotype will almost certainly be *Tg/* + . Statistical testing of this kind must be performed independently for each presumptive *Tg/Tg* animal. Once homozygous *Tg/Tg* males and females have been confirmed, they can be remated to each other as the founders for a homozygous transgenic strain.

A second approach to demonstrating transgene homozygosity requires the cloning of a endogenous sequence from the mouse genome that flanks the transgene insertion site. This task is often not straightforward because transgenic material can be present in multiple copies that are intermingled with locally rearranged endogenous sequences. Nevertheless, with the cloning of any nearby endogenous sequence, one obtains a mapping tool that, in theory, can be used to distinguish both the disrputed and nondisrupted alleles at the transgene locus through the use of one of the various techniques described in Chapter 8 for detecting "codominant" DNA polymorphisms. With such a tool, and an associated assay, homozygosity for the transgene allele would be demonstrated by the absence of the non-disrupted wild-type allele. Unfortunately, this approach would require the generation of a separate endogenous clone for each and every transgenic line to be studied. Protocols for locating the transgene insertion site within the mouse linkage map are discussed in Section 7.3.2.

## 6.4 TARGETED MUTAGENESIS AND GENE REPLACEMENT

### 6.4.1 Overview

Although the transgene insertion technology described in the previous section provides a powerful tool for the analysis of gene action in the whole organism, it has one serious limitation in that it does not provide a mechanism for the directed generation of recessive alleles. This limitation can be overcome with a technology known variously as gene targeting, targeted mutagenesis, or gene replacement—the subject of this section. This powerful technology allows investigators to generate directed mutations at any cloned locus. These new mutant alleles can be passed through the germ line to produce an unlimited number of mutant offspring, and different mutations can be combined with variants at other loci to study gene interactions.

This ultimate tool of genetic engineering was born through the combination of several technologies that had developed independently over the preceding 10–20 years including embryonic stem cell culture and homologous recombination, with mouse embryo manipulation and chimera formation [Sedivy and Joyner (1992) provide an excellent review of all aspects of this field]. Two independent laboratories, headed by Oliver Smithies and Mario Cappechi, finally succeeded in bringing all of these various technologies together during the mid-1980s (Capecchi, 1989; Smithies, 1993).

One, although not the only, appeal of the gene targeting technology is the ability to create mouse models for particular human diseases (Smithies, 1993). But, in essence, gene targeting can provide investigators with powerful tools to study any cloned gene. While patterns of RNA and protein expression provide clues to the stages and tissues in which genes are active, it is only with mutations that a true understanding of function can be obtained (Chisaka and Capecchi, 1991).

After heaping such praises on gene targeting, it is important to forewarn potential users of this technology that its application is not problem-free. First, an investigator must achieve a high level of competence and experience with several distinct, technically demanding protocols; this requires a significant investment of time and energy. Second, there is the fickle nature of the technology itself as discussed below. Nevertheless, the handful of laboratories initially able to target genes successfully has expanded quickly with the training of new young investigators, and this expansion is likely to continue much further with the recent publication of several excellent volumes containing detailed chapters on experimental protocols (Joyner, 1993; Wassarman and DePamphilis, 1993; Hogan et al., 1994).

## 6.4.2 Creating "gene knock-outs"

Once a particular gene has been cloned and characterized, the steps involved in obtaining a mouse with a *null* mutation in the corresponding locus can be outlined briefly as follows. First, one must design and construct an appropriate targeting vector in which the gene of interest has been disrupted with a positive selectable marker; in the most commonly used protocol, a negative selectable marker is also added at a position that flanks the gene sequence. The most commonly used positive selectable marker is the neomycin resistance (*neo*) gene, and the most commonly used negative selectable marker is the thymidine kinase (*tk*) gene.

The second step involves the introduction of the targeting vector into a culture of embryonic stem (ES) cells (usually derived from the 129 strain) followed by selection for those cells in which the internal positive selectable marker has become integrated into the genome without the flanking negative selectable marker. The third step involves screening for clones that have integrated the vector by homologous recombination rather than by the more common non-homologous recombination in random genomic sites.

Once "targeted clones" have been identified, the fourth step involves the production of chimeric embryos through the injection of the mutated ES cells into the inner cavity of a blastocyst (usually of the B6 strain), and the placement of these chimeric embryos back into foster mothers who bring them to term. A recently developed alternative approach to chimera formation through the aggregation and spontaneous incorpation of ES cells into cleavage stage embryos has the advantage of not requiring sophisticated microinjection equipment (Wood et al., 1993).

The experiment is deemed a success if the ES cells successfully enter the germ line of the chimeric animals as demonstrated by breeding. If the disrupted gene is indeed transmitted through the germ line, the first generation of offspring from the chimeric founder will include heterozygous animals that can be intercrossed to produce a second generation with individuals homozygous for the mutated gene. The nomenclature rules that are used to name all newly created mutations are described in Section 3.4.4.

### 6.4.3  Creating subtle changes

A second generation of homologous recombination strategies have been developed to allow the placement of specific small mutations into a locus without the concomitant presence of disrupting intragenic selectable markers. The ability to create subtle changes in a gene could provide an investigator with the tools required to dissect apart the function of a gene product one amino acid residue at a time. A number of different approaches toward this goal have been described. The most promising of these, called "hit and run," is based on the generation of ES cell lines that have undergone homologous recombination with a targeting vector, followed by selection for an intrachromosomal recombination event that eliminates the selectable markers and leaves behind just the mutated form of the gene (Joyner et al., 1989; Hasty et al., 1991; Valancius and Smithies, 1991; Fiering et al., 1993).

Unfortunately, at the time of this writing, the *hit and run* protocols are still extremely demanding and, with each experiment, an investigator will only obtain a single mutant allele at the locus of interest. An alternative strategy is to break the problem into two separate tasks: (1) knocking out the gene completely in one strain by standard homologous recombination; and (2) the independent production of one or more transgenic lines that contain subtly altered mutant versions of the gene. By breeding the knock-out line with one of the transgenic lines, it becomes possible to generate a new line of animals in which the original wild-type allele has been replaced (although not at the same site) with a specially designed transgene allele. There are several advantages to this approach. First, the methodology required for simply knocking out a gene is more straightforward and better developed at the time of this writing than the hit and run methodology. Second, gene targeting in ES cells requires much more time and effort than the production of transgenic mice by nuclear injection. Thus, when an investigator wishes to study a variety of alleles at a particular locus, it may be much easier to create a single line of mice by gene targeting and then breed it to different transgenic lines. The one potential disadvantage to this approach is that, in some cases, the transgene construct may not be regulated properly and accurate patterns of expression may not occur in the animal, even when the transgene is linked to its own promoter/enhancer.

### 6.4.4  Potential problems

Even when a laboratory has mastered all of the protocols required to perform gene targeting, the difference between success and failure can still be a matter of luck. Some DNA sites appear highly impervious to homologous recombination, whereas sites a few kilobases away may be much more open to integration. But even at the same site, the frequency of homologous versus non-homologous recombination events can vary by a factor of ten from one day to the next (Snouwaert et al., 1992).

At the time of writing, many of the factors responsible for success remain unknown. However, one critical factor that has recently become evident is the need to use source DNA for the targeting construct that has been cloned from the same strain of mice used for the derivation of the ES cell line into which the construct will be placed (van Deursen and Wieringa, 1992). In other words, the highest levels of gene replacement are obtained when the incoming DNA is *isogenic* with the target DNA. Apparently, the homologous recombination process is very sensitive to the infrequent nucleotide polymorphisms that are likely to distinguish different inbred strains from each other. In most cases today, ES cell lines have been derived from the 129/SvJ mouse strain (but see the subsection 6.4.5 below), and thus it is usually wise to build DNA constructs with clones obtained from 129 genomic libraries.

### 6.4.5 The "129 mouse"

The original 129/SvJ mouse, and the one still available from the Jackson Laboratory, has an off-white coat color caused by homozygosity for the pink-eye dilution ($p$) mutation, and forced heterozygosity for the chinchilla ($c^{ch}$) and albino ($c$) alleles at the linked albino locus ($c^{ch} p/c p$). In contrast, the "129 mouse" that serves as the source of most ES cell lines used for homologous recombination has a wild-type agouti coat color. What is the basis for this difference?

The answer is a historical one that centers on the work of Leroy Stevens, a cancer geneticist, now retired, who worked at the Jackson Laboratory. Stevens had observed that the 129/SvJ strain was unique in the occurrence of spontaneous testicular teratomas at an unusually high rate of 3–5% in male animals. As means to understand the genetic parameters responsible for tumor incidence better, Stevens set out to determine whether any of a variety of well-characterized single locus mutations that affect either tumor incidence or germ-cell differentiation would interact with the 129 genome in a manner to increase or decrease the natural frequency of tumor formation in this strain. One of the mutations that he tested was Steel ($Sl$), which plays an important role in the differentiation of germ cells as well as melanocytes and hematopoetic cells. The $Sl$ mutation expresses a dominant visible phenotype—the lightening of the normal wild-type black agouti coat color so that it has a "steely" appearance and a reduction of pigment in the distal half of the tail. Unfortunately, it is impossible to see this phenotypic alteration on the $c^{ch} p/c p$ coat of 129/SvJ mice, which already have a nearly complete loss of pigment production. Thus, to follow the backcrossing of $Sl$ onto 129/SvJ, it was necessary to replace the mutant alleles at the $c$ and $p$ loci with wild-type alleles. It is the triple congenic 129/Sv-$Sl/+$, $+^c$ $+^p$ line produced by Stevens that acted as the founder for all "129 mice" that have been used in ES cell work.[41]

## 6.5 FURTHER USES OF TRANSGENIC TECHNOLOGIES

### 6.5.1 Insertional mutagenesis and gene trapping

As indicated earlier in this chapter, one side product of many transgenic experiments is the generation of mice in which a transgene insertion has disrupted an endogenous gene with a consequent effect on phenotype. Unlike spontaneous or mutagen-induced mutations, "insertional mutations" of this type are directly amenable to molecular analysis because the disrupted locus is tagged with the transgene construct. Unexpected insertional mutations have provided molecular handles not only for interesting new loci but for classical loci, as well, that had not been cloned previously (Meisler, 1992).

When insertional mutagenesis, rather than the analysis of a particular transgene construct, is the goal of an experiment, one can use alternative experimental protocols that are geared directly toward gene disruption. The main strategies currently in use are based on the introduction into ES cells of beta-galactosidase reporter constructs that either lack a promoter (Gossler et al., 1989; Friedrich and Soriano, 1991) or are disrupted by an intron (Kramer and Erickson, 1981). The constructs can be introduced by DNA transfection or within the context of a retrovirus (Robertson, 1991). It is only when a construct integrates into a gene undergoing transcriptional activity that functional beta-galactosidase is produced and producing cells can be easily recognized by a color assay. Of course, the production of "beta-gal" will usually mean that the normal product of the disrupted gene can not be made and thus this protocol provides a means for the direct isolation of ES cells with tagged mutations in genes that function in embryonic cells. Mutant cells can be incorporated into chimeric embryos for the ultimate production of homozygous mutant animals that will display the phenotype caused by the absence of the disrupted locus. This entire technology, referred to as "gene trapping" (Joyner et al., 1992), is clearly superior to traditional chemical methods for the production of mutations at novel loci.

### 6.5.2 A database and a repository of genetically engineered mice

A computerized database (called TBASE) has been developed to help investigators catalog the strains that they produce and find potentially useful strains produced by others (Woychik et al., 1993). The database is available over the Internet through the Johns Hopkins Computational Biology Gopher and WWW Servers and is linked to the on-line mouse databases maintained by the Jackson Laboratory Informatics Group (see Appendix B).

Although the gene replacement technology has been employed with success by increasing numbers of laboratories, it is still the case, and likely to remain so, that an enormous amount of time and effort goes into the production of each newly engineered mouse strain. Clearly, it does not make sense to derive strains with the same gene replacement more than once. However, with the high costs of animal care and maintenance, it is often difficult for researchers to maintain strains that they are no longer actively using. Furthermore, many individual research colonies are contaminated by various viruses and, as such, virus-free facilities are reluctant to import mice from anywhere other than reputable dealers. The Jackson Laboratory has recently come to the rescue by setting up a clearing house to preserve what are likely to be the most useful of these strains for other members of the worldwide research community. For the first time, JAX will be importing mice from large numbers of individual researchers. Each strain will be rederived by Caesarean section into a germ-free barrier facility, and will be made available for a nominal cost, without experimental restriction, to all members of the research community.

### 6.5.3 The future

With the various technologies that have now been developed to manipulate the genomes of embryonic cells combined with ever more sophisticated molecular tools, it can be stated without exaggeration that the sky is the limit for what can be accomplished with the mouse as a model genetic system. It is always impossible to predict what the future holds, but one can imagine the use of both gene addition and gene replacement technologies as routine tools for assessing the functions of sequential segments of DNA obtained by walking along each mouse and human chromosome. With recent reports of success in the insertion of intact yeast artificial chromosomal (YAC)-sized DNA molecules of 250 kb or more in length into the germ line of transgenic animals, it becomes feasible to analyze even larger chunks of DNA for the presence of interesting genetic elements (Jakobovits et al., 1993; Schedl et al., 1993). In fact, it is only with experiments of this type that it will be possible to uncover completely all of the pathways through which a gene is regulated, and all of the pathways through which a gene product may function. Just as the study of neurons in isolation can not possibly provide a clue to human consciousness, the study of individual genes outside of the whole animal can not possibly provide a clue to the network of interactions required for the growth and development of a whole mouse or person.

# 7

# Mapping in the Mouse: An Overview

## 7.1 GENETIC MAPS COME IN VARIOUS FORMS

The remaining chapters in this book will be devoted to the process and practice of genetic mapping in the mouse. Although mapping was once viewed as a sleepy pastime performed simply for the satisfaction of knowing where a gene mapped as an end in itself, it is now viewed as a critical tool of importance to many different areas of biological and medical research. Mapping can provide a means for moving from important diseases to clones of the causative genes, which, in turn, can provide tools for diagnosis, understanding, and treatment. In the opposite direction, mapping can be used to uncover functions for newly derived DNA clones by demonstrating correlations with previously described variant phenotypes. Mapping can also be used to dissect out the heritable and non-heritable components of complex traits and the mechanisms by which they interact. The purpose of this chapter is to provide a primer on classical genetics and to give an overview of mapping in the mouse, with further details provided in subsequent, more focused chapters.

### 7.1.1 Definitions
#### 7.1.1.1 Genes and loci

In the prerecombinant DNA era, all genes were defined by the existence of alternative alleles that produced alternative phenotypes that segregated in genetic crosses. Today, with the use of molecular technologies, the ability to recognize genes has expanded tremendously. Monomorphic genes (those with only a single allele) can now be recognized through their transcriptional activity alone. Recognition of putative genes within larger genomic sequences can also be accomplished through the identification of open reading frames, flanking tissue-specific enhancers and other regulatory elements, internal splicing signals, and sequence conservation across evolutionary lines. Sequence-specific epigenetic phenomena such as imprinting, methylation, and DNase sensitivity can also be used to elucidate the existence of functional genomic elements.

Mouse geneticists use the term *locus* to describe any DNA segment that is distinguishable in some way by some form of genetic analysis. In the prerecombinant DNA era, only genes distinguished by phenotype could be recognized as loci. But today, with the use of molecular tools, it is possible to distinguish "loci" in the genome that have no discernible function at all. In fact, any change in the DNA sequence, no matter how small or large, whether in a gene or elsewhere, can be followed potentially as an alternative allele in genetic crosses. When alternative alleles exist in a genomic sequence that has no *known* function, the polymorphic site is called an *anonymous locus*.[42] With an average rate of polymorphism of one base difference in a thousand between individual chromosome homologs within a species, the pool of potential anonymous loci is enormous. Classes of anonymous loci and the methods by which they are detected and used as genetic markers will be the subject of Chapter 8.

### 7.1.1.2 Maps

A *genetic map* is simply a representation of the distribution of a set of *loci* within the genome. The loci included by an investigator in any one mapping project may bear no relation to each other at all, or they may be related according to any of a number of parameters including functional or structural homologies or a predetermined chromosomal assignment. Mapping of these loci can be accomplished at many different levels of resolution. At the lowest level, a locus is simply *assigned* to a particular chromosome without any further localization. At a step above, an assignment may be made to a particular subchromosomal region. At a still higher level of resolution, the relative order and approximate distances that separate individual loci within a linked set can be determined. With ever-increasing levels of resolution, the order and interlocus distances can be determined with greater and greater precision. Finally, the ultimate resolution is attained when loci are mapped onto the DNA sequence itself.

The simplest genetic maps can contain information on as few as two linked loci. At the opposite extreme will be complete physical maps that depict the precise physical location of all of the thousands of genes that exist along an entire chromosome. The first step toward the generation of these complete physical maps has recently been achieved with the establishment of single contigs of overlapping clones across the length of two complete human chromosome arms (Chumakov et al., 1992; Foote et al., 1992). By the time this book is actually read, it is likely that complete contigs across other human—as well as mouse—chromosomes will also be attained. However, it is still a long journey from simply having a set of clones to deciphering the genetic information within them.

There is actually not one, but three distinct types of genetic maps that can be derived for each chromosome in the genome (other than the Y). The three types of maps—*linkage, chromosomal,* and *physical*—are illustrated in Fig. 7.1, and are distinguished both by the methods used for their derivation and the metric used for measuring distances within them.

### 7.1.2 Linkage maps

The *linkage map*, also referred to as a *recombination map*, was the first to be developed soon after the rediscovery of Mendel's work at the beginning of the 20th century. Linkage maps can only be constructed for loci that occur in two or more heritable forms, or *alleles*. Thus, monomorphic loci—those with only a

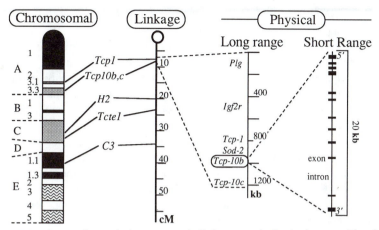

**Figure 7.1**  Comparison of chromosomal, linkage, and physical maps. The three types of maps are shown for loci on Chr 17. The chromosomal and linkage maps extend over the entire chromosome with lines connecting the relative positions of loci mapped in both. An example of a long-range physical map for a 1,200 kb genomic interval around the *Tcp10b* locus (Barlow et al., 1991) is shown along with a short-range physical map for the gene itself.

single allele—cannot be mapped in this fashion. Linkage maps are generated by counting the number of offspring that receive either parental or *recombinant* allele combinations from a parent that carries two different alleles at two or more loci. Analyses of this type of data allow one to determine whether loci are "linked" to each other and, if they are, their relative order and the relative distances that separate them (see Section 7.2).

A *chromosomal assignment* is accomplished whenever a new locus is found to be in linkage with a previously assigned locus. Distances are measured in centimorgans, with one centimorgan equivalent to a crossover rate of 1%. The linkage map is the only type based on classical breeding analysis. The term "genetic map" is sometimes used as a false synonym for "linkage map;" a genetic map is actually more broadly defined to include both chromosomal and physical maps as well.

### 7.1.3 Chromosome maps

The *chromosome map* (or *cytogenetic* map) is based on the karyotype of the mouse genome. All mouse chromosomes are *defined* at the cytogenetic level according to their size and banding pattern (see Fig. 5.1), and ultimately, all chromosomal assignments are made by direct cytogenetic analysis or by linkage to a locus that has been mapped previously in this way. Chromosomal map positions are indicated with the use of band names (Figs 5.2 and 7.1). Inherent in this naming scheme is a means for ordering loci along the chromosome (see Section 5.2).

Today, several different approaches, with different levels of resolution, can be used to generate chromosome maps. First, in some cases, indirect mapping can be accomplished with the use of one or more *somatic cell hybrid lines* that contain only portions of the mouse karyotype within the milieu of another species' genome. By correlating the presence or expression of a particular mouse gene with the presence of a mouse chromosome or subchromosomal region in these cells, one can obtain a chromosomal, or subchromosomal, assignment (see Section 10.2.3).

The second approach can be used in those special cases where karyotypic abnormalities appear in conjunction with particular mutant phenotypes. When the chromosomal lesion and the phenotype assort together, from one generation to the next, it is likely that the former causes the latter. When the lesion is a deletion, translocation, inversion, or duplication, one can assign the mutant locus to the chromosomal band that has been disrupted.

Finally, with the availability of a locus-specific DNA probe, it becomes possible to use the method of *in situ hybridization* to visualize the location of the corresponding sequence within a particular chromosomal band directly. This approach is not dependent on correlations or assumptions of any kind and, as such, it is the most direct mapping approach that exists. However, it is technically demanding and its resolution is not nearly as high as that obtained with linkage or physical approaches (see Section 10.2.2).

### 7.1.4 Physical maps

The third type of map is a *physical map*. All physical maps are based on the direct analysis of DNA. Physical distances between and within loci are measured in basepairs (bp), kilobasepairs (kb) or megabasepairs (mb). Physical maps are arbitrarily divided into short range and long range. Short-range mapping is commonly pursued over distances ranging up to 30 kb. In very approximate terms, this is the average size of a gene and it is also the average size of cloned inserts obtained from cosmid-based genomic libraries. Cloned regions of this size can be easily mapped to high resolution with restriction enzymes and, with advances in sequencing technology, it is becoming more common to sequence interesting regions of this length in their entirety.

Direct long-range physical mapping can be accomplished over megabase-sized regions with the use of rare-cutting restriction enzymes together with various methods of gel electrophoresis, referred to generically as *pulsed-field gel electrophoresis* or PFGE, which allow the separation and sizing of DNA fragments of 6 mb or more in length (Schwarz and Cantor, 1984; den Dunnen and van Ommen, 1991). PFGE mapping studies can be performed directly on genomic DNA followed by Southern blot analysis with probes for particular loci (see Section 10.3.2). It becomes possible to demonstrate physical linkage whenever probes for two loci detect the same set of large restriction fragments upon sequential hybridizations to the same blot.

Long-range mapping can also be performed with clones obtained from large insert genomic libraries such as those based on the yeast artificial chromosome (YAC) cloning vectors, since regions within these clones can be readily isolated for further analysis (see Section 10.3.3). In the future, long-range physical maps consisting of overlapping clones will cover each chromosome in the mouse genome. Short-range restriction maps of high resolution will be merged together along each chromosomal length, and ultimately, perhaps, the highest level of mapping resolution will be achieved with whole chromosome DNA sequences.

### 7.1.5 Connections between maps

In theory, linkage, chromosomal, and physical maps should all provide the same information on chromosomal assignment and the order of loci. However, the relative *distances* that are measured within each map can be quite different. Only the physical map can provide an accurate description of the actual length of DNA that separates loci from each other. This is not to say that the other two types of maps are inaccurate. Rather, each represents a version of the physical map that has been modulated according to a different parameter. Cytogenetic distances are modulated by the relative packing of the DNA molecule into different chromosomal regions. Linkage distances are modulated by the variable propensity of different DNA regions to take part in recombination events (see Section 7.2.3).

In practice, genetic maps of the mouse are often an amalgamation of chromosomal, linkage, and physical maps, but at the time of writing, it is still the case that classical recombination studies provide the great bulk of data incorporated into such integrated maps. Thus, the primary metric used to chart interlocus distances has been the centimorgan. However, it seems reasonable to predict that, within the next 10 years, the megabase will overtake the centimorgan as the unit for measurement along the chromosome.

## 7.2 MENDEL'S GENETICS, LINKAGE, AND THE MOUSE

### 7.2.1 Historical overview

By the time the chemical nature of the gene was uncovered, genetics was already a mature science. In fact, Mendel's formulation of the basic principles of heredity was not even dependent on an understanding of the fact that genes existed within chromosomes. Rather, the existence of genes was inferred solely from the expression in offspring of visible traits at predicted frequencies based on the traits present in the parental and grandparental generations. Today, of course, the field of genetics encompasses a broad spectrum of inquiry from molecular studies on gene regulation to analyses of allele frequencies in natural populations, with many subfields in between. To distinguish the original version of genetics—that of Mendel and his followers—from the various related fields that developed later, several terms have been coined including "formal" genetics, "transmission" genetics, or "classical" genetics. Transmission genetics is the most informative term, since it speaks directly to the feature that best characterizes the process by which Mendelian data are obtained—through an analysis of the transmission of genotypes and phenotypes from parents to offspring.

Mendel himself only formulated two of the three general features that underlie all studies in transmission genetics from sexually reproducing organisms. His formulations have been codified into two laws. The first law states, in modern terms, that each individual carries two copies of every gene and that only one of these two copies is transmitted to each child. At the other end of this equation, a child will receive one complete set of genes from each parent, leading to the restoration of a genotype that contains two copies of every gene. Individuals (and cells) that carry two copies of each gene are considered "diploid."

Mendel's first law comes into operation when diploid individuals produce "haploid" gametes—sperm or eggs—that each carry only a single complete set of genes. In animals, only a certain type of highly specialized cell—known as a "germ cell"—is capable of undergoing the transformation from the diploid to the haploid state through a process known as meiosis. At the cell division in which this transformation occurs, the two copies of each gene will separate or *segregate* from each other, and move into different daughter (or brother) cells. This event provides the name for Mendel's first law: "the law of segregation." Segregation can only be observed from loci that are heterozygous with two distinguishable alleles. As a result of segregation, half of an individual's gametes will contain one of these alleles and half will contain the other. Thus, a child can receive either allele with equal probability.[43]

While Mendel's first law is concerned with the transmission of individual genes in isolation from each other, his second law was formulated in an attempt to codify the manner in which different genes are transmitted relative to each other. In modern terms, Mendel's second law states that the segregation of alleles from any one locus will have no influence on the segregation of alleles from any other locus. In the language of probability, this means that each segregation event is independent of all others and this provides the name for Mendel's second law: "the law of independent assortment."

Independent assortment of alleles at two different loci—for example, *A* and *B*—can only be observed from an individual who is heterozygous at both with a genotype of the form *A/a, B/b* as illustrated in Fig. 7.2. Each gamete produced by such an individual will carry only one allele from the *A* locus and only one

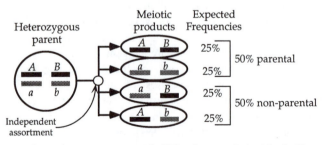

**Figure 7.2** Independent assortment of alleles from unlinked loci. Illustration of a parent heterozygous for alternative alleles at the unlinked *A* and *B* loci, and the four classes of meiotic products that it will generate with equal frequencies.

allele from the *B* locus. Since the two alleles are acquired independently of each other, it is possible to calculate the probability of any particular allelic combination by simply multiplying together the probability of occurrence of each alone. For example, the probability that a gamete will receive the *A* allele is 0.5 (from the law of segregation) and the probability that this same gamete will receive the *b* allele is similarly 0.5. Thus, the probability that a gamete will have a combined *A b* genotype is 0.5 × 0.5 = 0.25. The same probabilities are obtained for all four possible allelic combinations (*A B, a b, A b, a B*). Since the number of gametes produced by an individual is very large, these probabilities translate directly into the frequencies at which each gamete type is actually present and, in turn, the frequency with which each will be transmitted to offspring (Fig. 7.2).

As we all know today, Mendel's second law holds true only for genes that are not linked together on the same chromosome.[44] When genes *A* and *B* are linked, the numbers expected for each of the four allele sets becomes skewed from 25% (Fig. 7.3). Two allele combinations will represent the linkage arrangements on the parental chromosomes (for example, *A B* and *a b*) and these combinations will each be transmitted at a frequency of greater than 25%. The remaining two classes will represent recombinant arrangements that will be transmitted at a frequency below 25%. In the extreme case of absolute linkage, only the two parental classes will be transmitted, each at a frequency of 50%. At intermediate levels of linkage, transmission of the two parental classes together will be greater than 50% but less than 100%.

In 1905, when evidence for linkage was first encountered in the form of loci whose alleles did not assort independently, its significance was not appreciated (Bateson et al., 1905). The terms *coupling* and *repulsion* were coined to account for this unusual finding through some sort of underlying physical force. In a genetics book from 1911, Punnett imagined that alleles of different genes might "repel one another, refusing, as it were, to enter into the same zygote, or they may attract one another, and becoming linked, pass into the same gamete, as it were by preference" (Punnett, 1911). What this hypothesis failed to explain is why alleles found in repulsion to each other in one generation could become coupled

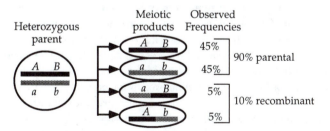

**Figure 7.3**   Non-independent assortment of alleles from linked loci. Illustration of a parent heterozygous for alternative alleles at the linked *A* and *B* loci, and the four classes of meiotic products that it will generate. As a consequence of linkage, the parental combinations of alleles will be transmitted to gametes at a higher frequency than the recombinant combinations. In this particular example, the *A* and *B* loci show recombination at a frequency of 10%.

to each other in the next generation. But even as Punnett's genetics text was published, an explanation was at hand. In 1912, Morgan and his colleagues proposed that coupling and repulsion were actually a consequence of co-localization of genes to the same chromosome: coupled alleles are those present on the same parental homolog and alleles in repulsion are those present on alternative homologs (Morgan and Cattell, 1912; Fig. 7.3). Through the process of crossing over, alleles that are in repulsion in one generation (for example, the A and b alleles in Fig. 7.3) can be brought together on the same homolog—and thus become coupled—in the next generation. In 1913, Sturtevant used the rates at which crossing over occurred between different pairs of loci to develop the first linkage map with six genes on the *Drosophila* X chromosome (Sturtevant, 1913). Although the original rationale for the terms coupling and repulsion was eliminated with this new understanding, the terms themselves have been retained in the language of geneticists (especially human geneticists). Whether alleles at two linked loci are coupled or in repulsion is referred to as the *phase of linkage*.

The purpose of this chapter is to develop the concepts of transmission genetics as they are applied to contemporary studies of the mouse. This discussion is not meant to be comprehensive. Rather, it will focus on the specific protocols and problems that are most germane to investigators who seek to place genes onto the mouse linkage map and those who want to determine the genetic basis for various traits that are expressed differently by different animals or strains.

### 7.2.2 Linkage and recombination

#### 7.2.2.1 The backcross

*Genetic* linkage is a direct consequence of the *physical* linkage of two or more loci within the same pair of DNA molecules that define a particular set of chromosome homologs within the diploid genome. Genetic linkage is demonstrated in mice through breeding experiments in which one or both parents are detectably heterozygous at each of the loci under investigation. In the simplest form of linkage analysis—referred to as a backcross—only one parent is heterozygous at each of two or more loci, and the other parent is homozygous at these same loci. As a result, segregation of alternative alleles occurs only in the gametes that derive from one parent, and the genotypes of the offspring provide a direct determination of the allelic constitution of these gametes. The backcross greatly simplifies the interpretation of genetic data because it allows one to jump directly from the genotypes of offspring to the frequencies with which different meiotic products are formed by the heterozygous parent.

For each locus under investigation in the backcross, one must choose appropriate heterozygous and homozygous genotypes so that the segregation of alleles from the heterozygous parent can be followed in each of the offspring. For loci that have not been cloned, the genotype of the offspring can only be determined through a phenotypic analysis. In this case, if the two alleles present in the heterozygous parent show a complete dominant/recessive relationship, then the other parent must be homozygous for the recessive allele. For example, the A allele at the agouti locus causes a mouse to have a banded "agouti" coat color, whereas the *a* allele determines a solid "non-agouti" coat color. Since the *A* allele is dominant to *a*, the homozygous parent must be *a/a*. In an *A/a* × *a/a* backcross, the occurrence of agouti offspring would indicate the transmission of the *A* allele from the heterozygous parent, and the occurrence of non-agouti offspring would indicate the transmission of the *a* allele.

In the case just described, the wild-type allele (*A*) is dominant and the mutant allele (*a*) is recessive. Thus, the homozygous parent must carry the mutant allele (*a/a*) and express a non-agouti coat color. In other cases, however, the situation is reversed with mutations that are dominant and wild-type alleles that are recessive. For example, the *T* mutation at the *T* locus causes a dominant shortening of the tail. Thus, if the *T* locus were to be included in a backcross, the heterozygous genotype would be *T/+* and the homozygous genotype would be wild-type ( *+/+* ) to allow one to distinguish the transmission of the *T* allele (within short-tailed offspring) from the *+* allele (within normal-tailed offspring).

As discussed in Chapter 8, most loci are now typed directly by DNA-based techniques. As long as both DNA alleles at a particular locus can be distinguished from each other,[45] it does not matter which is chosen for inclusion in the overall genotype of the homozygous parent. The same holds true for all phenotypically defined loci at which pairs of alleles act in a codominant or incompletely dominant manner. In all these cases, the heterozygote ($A^1/A^2$, for example) can be distinguished from both homozygotes ($A^1/A^1$ and $A^2/A^2$).

### 7.2.2.2 Map distances

In the example presented in Fig. 7.3, an animal is heterozygous at both of two linked loci, which results in two complementary sets of coupled alleles—*A B* and *a b*. The genotype of this animal would be written as follows: *AB/ab*.[46] In the absence of crossing over between homologs during meiosis, one or the other coupled set—either *A B* or *a b*—will be transmitted to each gamete. However, if a crossover event does occur between the *A* and *B* loci, a non-parental combination of alleles will be transmitted to each gamete. In the example shown in Fig. 7.3, the frequency of recombination between loci *A* and *B* can be calculated directly by determining the percentage of offspring formed from gametes that contain one of the two non-parental, or "recombinant," combinations of alleles. In this example, the recombination frequency is 10%.

To a first degree, crossing over occurs at random sites along all of the chromosomes in the genome. A direct consequence of this randomness is that the farther apart two linked loci are from each, the more likely it is that a crossover event will occur somewhere within the length of chromosome that lies between them. Thus, the frequency of recombination provides a relative estimate of genetic distance. Genetic distances are measured in centimorgans (cM) with 1 cM defined as the distance between two loci that recombine with a frequency of 1%. Thus, as a further example, if two loci recombine with a frequency of 2.5%, this would represent an approximate genetic distance of 2.5 cM. In the mouse, correlations between genetic and physical distances have demonstrated that 1 cM is, *on average*, equivalent to 2,000 kilobases. It is important to be aware, however, that the rate of equivalence can vary greatly due to numerous factors discussed in Section 7.2.5.

Although the frequency of recombination between two loci is roughly proportional to the length of DNA that separates them, when this length becomes too large, the frequency will approach 50%, which is indistinguishable from that expected with unlinked loci. The average size of a mouse chromosome is 75 cM. Thus, even when genes are located on the same chromosome, they are not necessarily *linked to each other* according to the formal definition of the term. However, a *linkage group* does include all genes that have been linked by association. Thus, if gene *A* is linked to gene *B*, and gene *B* is linked to gene *C*, the three genes together—*A B C*—form a linkage group even if the most distant members of the group do not exibit linkage to each other.

### 7.2.2.3 Genetic interference

A priori, one might assume that all recombination events within the same meiotic cell should be independent of each other. A direct consequence of this assumption is that the linear relationship between recombination frequency and genetic distance—apparent in the single digit centimorgan range—should degenerate with increasing distances. The reason for this degeneration is that, as the distance between two loci increases, so does the probability that multiple recombination events will occur between them. Unfortunately, if two, four, or any other even number of crossovers occur, the resulting gametes will still retain the parental combination of coupled alleles at the two loci under analysis as shown in Fig. 7.4. Double (as well as quadruple) recombinants will not be detectably different from non-recombinants. As a consequence, the observed recombination frequency will be less than the actual recombination frequency.

Consider, for example, two loci that are separated by a real genetic distance of 20 cM. According to simple probability theory, the chance that two independent recombination events will occur in this interval is the product of the predicted frequencies with which each will occur alone, which is 0.20 for a 20 cM distance. Thus, the probability of a double recombination event is $0.2 \times 0.2 = 0.04$. The failure to detect recombination in 4% of the gametes means that two loci

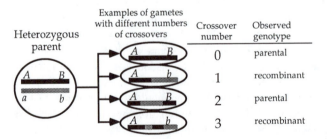

**Figure 7.4** Effects of multiple crossovers on the detection of recombinant chromosomes. Schematic representation of the transmission of a chromosomal interval containing two linked loci (*A* and *B*) to haploid gametes. When the number of crossovers occurring between the loci is odd, a recombinant chromosome can be detected. When the number of crossovers is even, no recombination is detected.

separated by 20 cM will only *show* recombination at a frequency of 0.16.[47] A similar calculation indicates that at 30 cM, the observed frequency of recombinant products will be even further removed at 0.21. In 1919, Haldane simplified this type of calculation by developing a general equation that could provide values for recombination fractions at all map distances based on the formulation just described. This equation is known as the "Haldane mapping function" and it relates the expected fraction of offspring with detectable recombinant chromosomes ($r$) to the actual map distance in morgans ($m$)[48] that separates the two loci (Haldane, 1919):

$$r = \frac{1}{2}(1 - e^{-2m})  \tag{7.1}$$

After working through this hypothetical adjustment to recombination rates, it is now time to state that multiple events of recombination on the same chromosome are not independent of each other. In particular, a recombination event at one position on a chromosome will act to interfere with the initiation of other recombination events in its vicinity. This phenomenon is known, appropriately, as "interference." Interference was first observed within the context of significantly lower numbers of double crossovers than expected in the data obtained from some of the earliest linkage studies conducted on *Drosophila* (Muller, 1916). Since that time, interference has been demonstrated in every higher eukaryotic organism for which sufficient genetic data have been generated.

Significant interference has been found to extend over very long distances in mammals. The most extensive quantitative analysis of interference has been conducted on human chromosome 9 markers that were typed in the products of 17,316 meiotic events (Kwiatkowski et al., 1993). Within 10 cM intervals, only two double-crossover events were found; this observed frequency of 0.0001 is 100-fold lower than expected in the absence of interference. Within 20 cM intervals, there were 10 double-crossover events (including the two above); this observed frequency of 0.0005 is still 80-fold lower than predicted without interference. As map distances increase beyond 20 cM, the strength of interference declines, but even at distances of up to 50 cM, its effects can still be observed (Povey et al., 1992).[49]

If one assumes that human chromosome 9 is not unique in its recombinational properties, the implication of this analysis is that for experiments in which fewer than 1,000 human meiotic events are typed, multiple crossovers within 10 cM intervals will be extremely unlikely, and within 25 cM intervals, they will still be quite rare. Data evaluating double crossovers in the mouse are not as extensive, but they suggest a similar degree of interference (King et al., 1989). Thus, for all practical purposes, it is appropriate to convert recombination fractions of 0.25, or less, directly into centimorgan distances through a simple multiplication by 100.

When it is necessary to work with recombination fractions that are larger than 0.25, it is helpful to use a mapping function that incorporates interference into an estimate of map distance. Since the effects of interference can only be determined empirically, one cannot derive such a mapping function from first principles.

Instead, equations have been developed that fit the results observed in various species (Crow, 1990). The best-known and most widely used mapping function is an early one developed by Kosambi (1944):

$$m_K = \frac{1}{4}\left[\ln(1 + 2r) - \ln(1 - 2r)\right] \tag{7.2}$$

By solving equation 7.2 for the observed recombination fraction, $r$, one obtains the "Kosambi estimate" of the map distance, $m_K$, which is converted into centimorgans through multiplication by 100. Later, Carter and Falconer (1951) developed a mapping function that assumes even greater levels of interference based on the results obtained with linkage studies in the mouse:[50]

$$m_{FC} = \frac{1}{4}\left\{\frac{1}{2}\left[\ln(1+2r) - \ln(1-2r)\right] + \tan^{-1}(2r)\right\} \tag{7.3}$$

Although it is clear that the Carter–Falconer mapping function is the most accurate for mouse data, the Kosambi equation was more easily solvable in the days before cheap, sophisticated hand-held calculators were available. Although the Carter–Falconer function is readily solvable today, it is not as well-known and not as widely used.

Interference works to the benefit of geneticists performing linkage studies for two reasons. First, the approximate linearity between recombination frequency and genetic distance is extended out much further than anticipated from strictly independent events.[51] Second, the very low probability of multiple recombination events can serve as a means for distinguishing the correct gene order in a three-locus cross, since any order that requires double recombinants among markers within a 20 cM interval is suspect. When all possible gene orders require a double or triple crossover event, it behooves the investigator to go back and reanalyze the sample or samples in which the event supposedly occurred. Finally, if the genotypings are shown to be correct, one must consider the possibility that an isolated gene conversion event has occurred at the single locus that differs from those flanking it.

### 7.2.3 Crossover sites are not randomly distributed

#### 7.2.3.1 Theoretical considerations in the ideal situation

Although genetic interference will restrict the randomness with which crossover events are distributed relative to each other within individual gametes, it will not affect the random distribution of crossover sites observed in large numbers of independent meiotic products. Thus, a priori, one would still expect the resolution of a linkage map to increase linearly with the number of offspring typed in a genetic cross. Assuming random sites of recombination, the average distance in centimorgans, between crossover events observed among the offspring from a cross can be calculated according to the simple formula (100/$N$), where $N$ is the number of meiotic events that are typed. For example, in an analysis of 200 meiotic events (200 backcross offspring or 100 intercross offspring), one will observe, on average, one recombination event every 0.5 cM. With 1,000 meiotic events, the average distance will be only 0.1 cM, which is equivalent to approximately 200 kb of DNA. Going further according to this formula, with 10,000 offspring, one would obtain a genetic resolution that approached 20 kb. This would be sufficient to separate and map the majority of average-size genes in the genome relative to each other.

Once again, however, the results obtained in actual experiments do not match the theoretical predications. In fact, the distribution of recombination sites can deviate significantly from randomness at several different levels. First, in general, the telomeric portions of all chromosomes are much more recombinogenic than are those regions closer to the centromere in both mice (de Boer and Groen, 1974) and humans (Laurie and Hulten, 1985). This effect is most pronounced in males and it leads to an effect like a rubber band when one tries to orient male and female linkage maps relative to each other (Donis-Keller et al., 1987). Second, different sites along the entire chromosome are more or less prone to undergo recombination. Third, even within the same genomic region, rates of recombination can vary greatly depending on the particular strains of mice used to produce the hybrid used for analysis (Seldin et al., 1989; Reeves et al., 1991; Watson et al., 1992). Finally, the sex of the hybrid can also have a dramatic effect on rates of recombination (Reeves et al., 1991).

### 7.2.3.2 Gender-specific differences in rates of recombination

Gender-specific differences in recombination rates are well known. In general, it can be stated that recombination occurs less frequently during male meiosis than during females meiosis. An extreme example of this general rule is seen in *Drosophila melanogaster* where recombination is eliminated completely in the male. In the mouse, the situation is not as extreme with males showing a rate of recombination that is, on average, 50–85% of that observed in females (Davisson et al., 1989). However, the ratio of male to female rates of recombination can vary greatly among different regions of the mouse genome. In a few regions, the recombination rates are indistinguishable between sexes, and in even fewer regions yet, the male rates of recombination exceed female rates. Nevertheless, the general rule of higher recombination rates in females can be used to maximize data generation by choosing gender appropriately for a heterozygous $F_1$ animal in a backcross. For example, to maximize chances of finding initial evidence for linkage, one could choose males as the $F_1$ animals, but to maximize the resolution of a genetic map in a defined region, it would be better to use females. These considerations are discussed further in Section 9.4.

### 7.2.3.3 Recombinational hotspots

The most serious blow to the unlimited power of linkage analysis has come from the results of crosses in which many thousands of offspring have been typed for recombination within small well-defined genomic regions. When the recombinant chromosomes generated in these crosses were examined at the DNA level, it was found that the distribution of crossover sites was far from random (Steinmetz et al., 1987). Instead, they tended to cluster in very small "recombinational hotspots" of a few kilobases or less in size (Zimmerer and Passmore, 1991; Bryda et al., 1992). The accumulated data suggest that these small hotspots may be distributed at average distances of several hundred kilobases apart from each other with 90% or more of all crossover events restricted to these sites.

The finding of recombinational hotspots in mice is surprising because it was not predicted from very high resolution mapping studies performed previously in *Drosophila*, which showed an excellent correspondence between linkage and physical distances down to the kilobase level of analysis (Kidd et al., 1983). Thus, this genetic phenomenon—like genomic imprinting (Section 5.5)—might be unique to mammals. Unlike imprinting, however, the locations of particular recombinational hotspots do not appear to be conserved among different subspecies or even among different strains of laboratory mice.

Figure 7.5 illustrates the consequences of hotspot-preferential crossing over on the relationship between linkage and physical maps. In this example, 2,000 offspring from a backcross were analyzed for recombination events between the fictitious A and F loci. These loci are separated by a physical distance of 1,500 kb and, in our example, 17 crossover events (indicated by short vertical lines on the linkage map) were observed among the 2,000 offspring. A recombination frequency of 17/2,000 translates into a linkage distance of 0.85 cM. This linkage distance is very close to the 0.75 cM predicted from the empirically determined equivalence of 2,000 kb to 1 cM. However, when one looks further at loci between A and F, the situation changes dramatically. The B and C loci are only 20 kb apart from each on the physical map but are 0.4 cM apart from each other

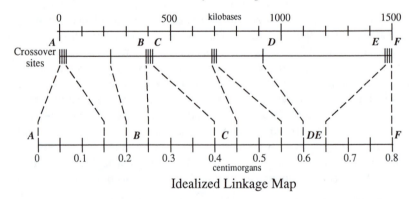

**Figure 7.5** The effects of recombination hotspots on the relationship between linkage and physical maps. Corresponding physical and linkage maps for a hypothetical genomic interval extending over six loci—A, B, C, D, E, and F. The linkage map is based on the analysis of 2,000 backcross offspring. Thus, the recombination fraction separating adjacent crossover events is 1/2,000 or 0.0005, which is equivalent to a linkage distance of 0.05 cM. Vertical lines on the linkage map represent the individual crossover events separated by this idealized distance. Vertical lines on the physical map represent the corresponding location of each crossover event within the DNA molecule. Hotspots are defined operationally by the clustering of crossover events—near locus A, between B and C, halfway between C and D, and between E and F.

on the linkage map because a hotspot occurs in the region between them. With random sites of crossing over, the linkage value of 0.4 cM would have predicted a physical distance of 800 kb. The reciprocal situation occurs for the loci $D$ and $E$, which are separated by a physical distance of 400 kb but which show no recombination in 2,000 offspring. In this case, random crossing over would have predicted a physical distance of less than 100 kb.

The existence and consequences of recombinational hotspots can be viewed in analogy to the quantized nature of matter. For experiments conducted at low levels of resolution—for example, in measurements of grams or centimorgans—the distribution of both matter and crossover sites will appear continuous. At very high levels of resolution, however, the discontinuous nature of both will become apparent. In practical terms, the negative consequences of hotspots on the resolution of a mouse linkage map will only begin to show up as one goes below the 0.2 cM level of analysis.

With the limited number of very large sample linkage studies performed to date, it is not possible to estimate the portion of the mouse genome that is dominated by hotspot-directed recombination. Furthermore, it is still possible that some genomic regions will allow unrestricted recombination as in *Drosophila*. Nevertheless, the available data suggest that for much of the genome, there will be an upper limit to the resolution that can be achieved in linkage studies based on a single cross. This limit will be reached at a point when the density of crossover sites passes the density of hotspots in the region under analysis. From the data currently available, it appears likely that this point will usually be crossed before one reaches 500 meiotic events corresponding to 0.2 cM or 400 kb. One strategy that can be used to overcome this limitation is to combine information obtained from several crosses with different unrelated inbred partners, each of which is likely to be associated with different hotspot locations. This approach is discussed more fully in Section 9.4.

### 7.2.3.4 Frequencies of recombination can vary greatly between different chromosomal regions

As mentioned previously, the telomeric portions of chromosomes show higher rates of recombination per DNA length than more centrally located chromosomal regions. However, there is still great variation in recombination rates even among different non-telomeric regions. Some 1 mb regions produce recombinants at a rate equivalent to 2 cM or greater, whereas other regions of equivalent size only recombine with a rate equivalent to 0.5 cM or less in animals of the same gender. This variation could be due to differences in the number and density of recombination hotspots. In addition, the "strength" of individual hotspots, in terms of recombinogenicity, may differ from one site to another. Such differences could be specified by the DNA sequences at individual hotspots or by the structure of the chromatin that encompass multiple hotspots in a larger interval. A final variable may be generalized differences in the rates at which recombination can occur in regions between hotspots. Many more empirical studies will be required to sort through these various explanations.

### 7.2.4 A history of mouse mapping

#### 7.2.4.1 The classical era

Although its significance was not immediately recognized, the first demonstration of linkage in the mouse was published in 1915 by the great 20th century geneticist J.B.S. Haldane (Haldane et al., 1915). What Haldane found was evidence for coupling between mutations at the albino (c) and pink-eyed dilution (p) loci, which we now know to lie 15 cM apart on chromosome 7. Since that time, the linkage map of the mouse has expanded steadily at a near-exponential pace. During the first 65 years of work on the mouse map, this expansion took place one locus at a time. First, each new mutation had to be bred into a strain with other phenotypic markers. Then further breeding was pursued to determine whether the new mutation showed linkage to any of these other markers. This process had to be repeated with different groups of phenotypic markers until linkage to one other previously mapped marker was established. At this point, further breeding studies could be conducted with additional phenotypic markers from the same linkage group to establish a more refined map position.

In the first compendium of mouse genetic data published in the *Biology of the Laboratory Mouse* in 1941 (Snell, 1941), a total of 24 independent loci were listed, of which 15 could be placed into seven linkage groups containing either two or three loci each; the remaining nine loci were found not to be linked to each other or to any of the seven confirmed linkage groups. By the time the second edition of the *Biology of the Laboratory Mouse* was published in 1966, the number of mapped loci had grown to 250, and the number of linkage groups had climbed to 19, although in four cases, these included only two or three loci (Green, 1966).

With the 1989 publication of the second edition of the *Genetic Variants and Strains of the Laboratory Mouse* (Lyon and Searle, 1989), 965 loci had been mapped on all 20 recombining chromosomes. However, even at the time that this map was actually prepared for publication (circa late 1987), it was still the case that the vast majority of mapped loci were defined by mutations that had been painstakingly incorporated into the whole genome map through extensive breeding studies.

#### 7.2.4.2 The middle ages: recombinant inbred strains

The first important conceptual breakthrough aimed at reducing the time, effort, and mice required to map single loci came with the conceptualization and establishment of recombinant inbred (abbreviated RI) strains by Donald Bailey and Benjamin Taylor at the Jackson Laboratory (Bailey, 1971, 1981; Taylor, 1978). As discussed in detail in Section 9.2, a set of RI strains provides a collection of samples in which recombination events between homologs from two different inbred strains are preserved within the context of new inbred strains. The power of the RI approach is that loci can be mapped relative to each other within the same "cross" even though the analyses themselves may be performed many years apart. Since the RI strains are essentially preformed and immortal, typing a newly defined locus requires only as much time as the typing assay itself.

Although the RI mapping approach was extremely powerful in theory, during the first two decades after its appearance, its use was rather limited because of two major problems. First, analysis was only possible with loci present as alternative alleles in the two inbred parental strains used to form each RI set. This ruled out nearly all of the many loci that were defined by gross phenotypic effects. Only a handful of such loci—primarily those that affect coat color—were polymorphic among different inbred strains. In fact, in the prerecombinant DNA era, the only other loci amenable to RI analysis were those that encoded: (1) polymorphic enzymes (called allozymes or isozymes) that were observed as differentially migrating bands on starch gels processed for the specific enzyme activity under analysis (Womack, 1979); (2) immunological polymorphisms detected at minor histocompatibility loci (Graff, 1978); and (3) other polymorphic cell surface antigens (called alloantigens or isoantigens) that could be distinguished with specially developed "allo-antisera" (Boyse et al., 1968). In retrospect, it is now clear that RI strains were developed ahead of their time; their power and utility in mouse genetics is only now—in the 1990s—being fully unleashed.

### 7.2.4.3 DNA markers and the mapping panel era

Two events that occurred during the 1980s allowed the initial development of a whole genome mouse map that was based entirely on DNA marker loci. The first event was the globalization of the technology for obtaining DNA clones from the mouse genome and all other organisms. Although the techniques of DNA cloning had been developed during the 1970s, stringent regulations in the US and other countries had prevented their widespread application to mammalian species like the mouse (Watson and Tooze, 1981). These regulations were greatly reduced in scope during the early years of the 1980s so that investigators at typical biological research facilities could begin to clone and characterize genes from mice. The globalization of the cloning technology was greatly hastened in 1982 by the publication of the first highly detailed cloning manual from Cold Spring Harbor Laboratory, officially entitled *Molecular Cloning: A Laboratory Manual,* but known unofficially as "The Bible" (Maniatis et al., 1982).[52]

Although DNA clones were being recovered at a rapid rate during the 1980s, from loci across the mouse genome, their general utilization in linkage mapping was not straightforward. The only feasible technique available at the time for mapping cloned loci was the typing of restriction fragment length polymorphisms (RFLPs). Unfortunately, as discussed earlier in this book (Sections 2.3 and 3.2), the common ancestry of the traditional inbred strains made it difficult, if not impossible, to identify RFLPs between them at most cloned loci.

The logjam in mapping was broken not through the development of a new molecular technique, but rather, through the development of a new genetic approach. This was the second significant event in terms of mouse mapping during the 1980s—the introduction of the interspecific backcross. Franαois Bonhomme and his French colleagues had discovered that two very distinct mouse species—*M. musculus* and *M. spretus*—could be bred together in the laboratory to form fertile $F_1$ hybrid females (Bonhomme et al., 1978). With the three million years that separate these two *Mus species* (Section 2.3), basepair substitutions have accumulated to the point where RFLPs can be rapidly identified for nearly every DNA probe that is tested. Thus, by backcrossing an interspecific super-heterozygous $F_1$ female to one of its parental strains, it becomes possible to follow the segregation of the great majority of loci that are identified by DNA clones through the use of RFLP analysis.

Although the "spretus backcross" could not be immortalized in the same manner as a set of RI strains, each of the backcross offspring could be converted into a quantity of DNA that was sufficient for RFLP analyses with hundreds of DNA probes. In essence, it became possible to move from a classical three-locus backcross to a several hundred locus backcross. Furthermore, the number of loci could continue to grow as new DNA probes were used to screen the members of the established "mapping panel" (until DNA samples were used up). The *spretus* backcross revolutionized the study of mouse genetics because it provided the first complete linkage map of the mouse genome based on DNA markers and because it provided mapping panels that could be used to map rapidly essentially any new locus that was defined at the DNA level.

### 7.2.4.4 Microsatellites

The most recent major advance in genetic analysis has come not from the development of new types of crosses, but from the discovery and utilization of PCR-based DNA markers that are extremely polymorphic and can be rapidly typed in large numbers of animals with minimal amounts of sample material. These powerful new markers—especially microsatellites—have greatly diminished the essential need for the *spretus* backcross and they have breathed new life into the usefulness of the venerable RI strains. Most importantly, it is now possible for individual investigators with limited resources to carry out independent, sophisticated mapping analyses of mutant genes or complex disease traits. As Philip Avner of the Institut Pasteur in Paris states: "If the 1980s were the decade of *Mus spretus*—whose use in conjunction with restriction fragment length polymorphisms revolutionized mouse linkage analysis, and made the mouse a formidably efficient system for genome mapping—the early 1990s look set to be the years of the microsatellite" (Avner, 1991). Microsatellites and other PCR-typable polymorphic loci are discussed at length in Section 8.3.

## 7.3  GENERAL STRATEGIES FOR MAPPING MOUSE LOCI

How should one go about performing a mapping project? The answer to this question will be determined by the nature of the problem at hand. Is there a particular locus, or loci, of interest that you wish to map? If so, at what level is the locus defined and at what resolution do you wish to map it? Is the locus associated with a DNA clone, a protein-based polymorphism, or a gross phenotype visible only in the context of the whole animal? Are you interested in mapping a transgene insertion site unique to a single line of animals? Do you have a new mutation found in the offspring from a mutagenesis experiment? Alternatively, are you isolating clones to be used as potential DNA markers for a specific chromosome or subchromosomal region with the need to know simply whether each clone maps to the correct chromosome or not? The answers to these questions will lead to the choice of a general mapping strategy.

### 7.3.1 Novel DNA clones

Gene cloning has become a standard tool for analysis by biologists of all types from those studying protein transport across cell organelles to those interested in the development of the nervous system. Genes are often cloned based on function or pattern of expression. With a cloned gene in hand, how does one determine its location in the genome? Today, the answer to this question is always through the use of an established mapping panel as described at length in Chapter 9. Mapping with established panels is relatively painless and very quick. Furthermore, it can provide the investigator with a highly accurate location within a single chromosome of the mouse genome. With these results in hand, it is always worthwhile to determine whether the newly mapped clone could correspond to a locus previously defined by a related trait or disease phenotype. This can be accomplished by consulting the most recent version of the genetic map for the region of interest. Maps and further genetic information for each mouse chromosome are prepared annually in reports by individual mouse chromosome committees. These reports are published together as a compendium in a special issue of *Mammalian Genome*. This information is also available electronically from the Jackson Laboratory (see Appendix B). If a relationship is suspected between a cloned locus and a phenotypically defined locus, further genetic studies of the type described in Chapter 9 can be pursued.

### 7.3.2 Transgene insertion sites

Transgene insertion sites are unique in that the inserted foreign sequence is present in its particular genomic location only in the founder of the transgenic line and those descendants to which the transgene has been transmitted. This uniqueness rules out the use of mapping panels for analysis when only the transgene itself is available as a probe. There are several general approaches to the mapping of transgene insertion sites, and each has advantages and disadvantages. The first approach is *in situ* hybridization (Section 10.2). The first advantage here is that the actual DNA used for embryo injection can now be used as a probe for mapping. Thus, one avoids the need to clone endogenous sequences that flank the insertion site in each and every founder line to be analyzed. A second advantage is that the analysis can be performed on a single animal and there is no need to carry out extensive crosses. The main disadvantage is the specialized nature of the *in situ* technique as mentioned previously.

A second approach is to clone genomic sequences that flank the inserted DNA from each founder line of interest. Once a flanking sequence is obtained, it can be analyzed like any other novel DNA sequence with the use of mapping panels as described in Section 9.3. The advantage to this approach is that it requires only standard molecular biology protocols. The disadvantage is that an additional cloning step is required for each founder line. Cloning endogenous sequences may be complicated by the chaotic nature of most transgene insertion events, which often have multiple copies of the transgene sequence intermingled with endogenous sequences.

A third approach is to follow the segregation of the transgene in relation to DNA markers that span the mouse genome in a standard backcross or intercross analysis as described in Section 9.4. The advantages to this approach are that only standard molecular biology protocols are required and there is no need for any cloning of endogenous sequences. The main disadvantage is the time and expense of generating and typing a novel mouse mapping panel for each transgenic line.

The choice of a mapping approach will be highly dependent on what is viewed as common practice in each investigator's laboratory. If one has access to the *in situ* hybridization technology, this will be the fastest and least expensive approach. If genomic library production and screening are commonly performed protocols, then the second approach would likely be the best one to follow. Finally, if an investigator has an active breeding program, and is facile at producing and analyzing large panels of mice, the third approach might be the easiest to follow.

### 7.3.3 Verification of region-specific DNA markers

When investigators are interested in the genetic analysis of a particular chromosome or subchromosomal region, they often begin by screening a specialized library that is enriched for clones from the region of interest (Section 8.4). In such cases, initial genetic mapping is limited to the question of whether a cloned sequence localizes to this region or not. The most efficient way to answer this question for a large number of clones is through the analysis of one or a few somatic cell hybrid lines that contain the chromosome of interest within the genetic background of another *host* species as described in Section 10.2. In the simplest cases, hybridization to a blot that contains gel-separated, restriction enzyme-digested DNA from three samples—mouse, the somatic cell hybrid line, and a cell line from the somatic cell host species—will provide the answer. Clones that are found to map to the region of interest can then be analyzed in more detail with mapping panels or other genetic tools developed for the particular project.

### 7.3.4 Loci defined by polypeptide products

In some cases, even today, the protein product of a locus may be identified before the locus itself is cloned. If the protein is truly of interest, it is likely that this state will be a temporary one, since numerous protocols have been devised to proceed backwards from a protein product to its coding sequence in the genome. Nevertheless, it is sometimes possible to map the gene that encodes a defined protein before a DNA clone becomes available. If the protein is associated with an enzymatic activity that is expressed constitutively—a so-called housekeeping function—it is often possible to assay for its expression among a panel of somatic cell hybrid lines, each of which contains a defined subset of mouse chromosomes as described in Section 10.2. As long as the mouse enzyme is generally expressed in somatic cells and is distinguishable from the homologous protein produced by the host species used to construct the somatic cell hybrid panel, a chromosomal assignment can be attained. Following along this line of analysis, subchromosomal mapping can be performed when somatic cell hybrid lines are available that contain defined segments of the chromosome in question. However, in most cases, the level of mapping resolution will still be quite low.

Linkage analysis can only be performed in those cases where different strains of mice are found to express distinguishable allelic forms of the protein. Protein polymorphisms are detectable in a number of different ways. In the earliest pre-recombinant DNA studies, assays were developed to detect specific enzymatic activities within mixtures of cellular proteins that had been separated by starch gel electrophoresis. Allelic differences involving charged amino acids caused enzyme molecules to migrate with different mobilities in a starch gel and the *in situ* detection system allowed the visualization of these alternative enzyme forms, which are known as "isozymes."

A more general approach to detecting allelic charge differences in proteins relies on the technique of isoelectric focusing, usually within the context of a two-dimensional polyacrylamide gel where the second dimension involves a molecular weight-based separation with sodium dodecyl sulphate (O'Farrell, 1975). High resolution two-dimensional gel electrophoresis can resolve up to 2,000 polypeptide spots from whole cell extracts (Garrels, 1983). Although this approach to mapping has been used with success in the past (Elliott, 1979; Silver et al., 1983), in most cases it is rather tedious since a separate two-stage gel must be run for each animal to be typed. However, when the sample size is small, for example, with two members of a congenic pair, a two-dimensional search for polypeptide polymorphisms becomes much more feasible (Silver et al., 1983).

A special class of polypeptide polymorphisms are those that are detected as antigenic differences through any of a variety of immunological assays. Most immunoassays are quick and easy to perform, and this allows the rapid mapping of genes that encode polymorphic antigens. A variety of other biochemical differences can result from alternative alleles at some loci, such as differences in enzyme kinetics. Any easily assayed difference can be exploited to map the underlying gene. Finally, in those cases where no polymorphism is detected, it makes sense to wait for a clone of the gene that can be used as a direct tool for mapping.

### 7.3.5 Mutant phenotypes

For loci defined by phenotype alone, rapid mapping is usually not possible. Interest in the new phenotype is likely to lie within its novelty and, as such, the parental strains used in all standard mapping panels are almost certain to be wild-type at the guilty locus. Thus, a broad-based recombinational analysis can be accomplished only by starting from scratch with a cross between mutant animals and a standard strain. Before one embarks on such a large-scale effort, it makes sense to consider whether the mutant phenotype, or the manner in which it was derived, can provide any clues to the location of the underlying mutation. Is the mutant phenotype similar to one that has been previously described in the literature? Does the nature of the phenotype provide insight into a possible biochemical or molecular lesion?

The most efficient way to begin a search for potentially related loci is to search through the detailed compilation of mouse loci and their effects in the mouse locus catalog (MLC) published in the *Genetic Variants and Strains of the Laboratory Mouse* (Lyon and Searle, 1989) and available on-line through the Internet at the Jackson Laboratory (see Appendix B). It is also worthwhile to consult the human equivalent of MLC called *Mendelian Inheritance in Man* and edited by Victor McKusick (1988). This database is also available on-line (and called OMIM) through Internet at the Genome Database maintained at Johns Hopkins University (see Appendix B). Phenotypically related loci can be uncovered by searching each of these electronic databases for the appearance of well-chosen keywords. Finally, one can carry out a computerized on-line search through the entire biomedical literature. Once again, this search need not be confined to the mouse since similarity to a human phenotype can be informative as well.

When a possible relationship with a previously characterized locus is uncovered, genetic studies can be directed at proving or disproving identity. This is most readily accomplished when the previously characterized locus—either human or mouse—has already been cloned. A clone can be used to investigate the possibility of aberrant expression from mice that express the new mutation, and with the strategies described in Section 9.4, one can follow the segregation of the cloned locus in animals that segregate the new mutation. Absolute linkage would provide evidence in support of an identity between the new mutation and the previously characterized locus.

Even if the previously characterized mutant locus has not yet been cloned, it may still be possible to test a relationship between it and the newly defined mutation. If the earlier mutation exists in a mouse strain that is still alive (or frozen), it becomes possible to carry out classical complementation analysis. This analysis is performed by breeding together animals that carry each mutation and examining the phenotype of offspring that receive both. If the two mutations— $m1$ and $m2$, for example—are at different loci, then the double mutant animals will have a genotype of $(+/m1, +/m2)$. If both mutations express a recessive phenotype, then this double mutant animal, with wild-type alleles at both loci, would appear wild-type; this would be an example of complementation. On the other hand, if the two mutations are at the same locus, then the double mutant animal would have a compound heterozygous genotype of $m1/m2$. Without any wild-type allele at this single locus, one would expect to see expression of a mutant phenotype; this would be an example of non-complementation.

Even if the previously characterized mutation is extinct, it may still be possible to use its previously determined map position as a test for the possibility that it did lie at the same locus as the newly uncovered mutation. This is accomplished by following the transmission to offspring of the newly uncovered mutation along with a polymorphic DNA marker that maps close to the previously determined mutant map position (methods for identifying appropriate DNA markers are discussed in Chapter 8). Close linkage between the new mutation and a DNA marker for the old mutation would suggest, although not prove, that the two mutations occurred at the same locus.

Finally, a similar approach can often be followed when the previously characterized mutation is uncloned but mapped in the human genome rather than the mouse. Most regions of the human genome have been associated with homologous regions in the mouse genome (Copeland et al., 1993; O'Brien et al., 1993). Thus, one can choose DNA markers from the region (or regions) of the mouse genome that is likely to carry the mouse gene showing homology to the mutant human locus. These markers can then be tested for linkage to the new mouse mutation. Again, the data would be only suggestive of an association.

In some cases, new mutations will be found to be associated with gross chromosomal aberrations. This is especially likely to be the case if the new mutation was first observed in the offspring from a specific mutagenesis study. Two mutagenic agents in particular—X-irradiation and the chemical chlorambucil—often cause chromosomal rearrangements (Section 6.1). Rearrangements can also occur spontaneously and when the mutant line is difficult to breed, this provides a hint that this might indeed be the case. In any case where the suspicion of a chromosomal abnormality exists, it is worthwhile analyzing the karyotype of the mutant animals. The observation of an aberrant chromosome—with a visible deletion, inversion, or translocation —should be followed up by a small breeding study to determine if the aberration shows complete linkage with the expression of the mutant phenotype. If it does, one can be almost certain that the mutation is associated with the aberration in some way. If the chromosomal aberration is a deletion, the mutant gene is likely to lie within the deleted region. With a translocation or inversion, the mutant phenotype is likely to be due to the disruption of a gene at a breakpoint. In all cases, the next step would be to perform linkage analysis with DNA markers that have been mapped close to the sites affected by the chromosomal aberration. The aberration itself may also be useful later as a tool for cloning the gene. This is especially true for translocations since the breakpoint will provide a distinct physical marker for the locus of interest.

Another possibility to consider is whether the mutation is sex-linked. This is easily demonstrated when the mutation is only transmitted to mice of one sex. Sex linkage almost always means X-chromosome linkage. If the mutation is recessive, a female carrier mated to a wild-type male will produce all normal females and 50% mutant males. If the mutation is dominant, a mutant male mated to a wild-type female will produce all normal males and all mutant females. Finally, if all efforts to map the novel phenotype by association fail, it will be necessary to set up a new mapping cross from scratch in which DNA markers from across the genome can be tested for linkage, as described in Section 9.4.

## 7.4 THE FINAL CHAPTER OF GENETICS

The fundamental goal of molecular genetics is to understand, at the molecular level, how genotype is translated into phenotype. To accomplish this goal, investigators everywhere are busy dissecting the genome into its component parts—the genes—and then tracking the pathway from the gene to its product to its role in the overall scheme of life. This contemporary approach to biological understanding can be divided into two parts: first, an investigator must obtain a clone of the gene, then second, he or she can use the clone in a large variety of experiments aimed at investigating the function of the gene.

## 7.4.1 From gene to function

There are two very different pathways to the analysis of gene function. One pathway begins with a mutant or variant phenotype, and follows this back to a clone of the guilty gene; this pathway will be discussed in the next Section. The other pathway begins with a clone of a transcription unit whose function is not understood, and proceeds to utilize this clone in various experiments aimed at uncovering gene function. With tens of thousands of uncharacterized transcription units sitting in every cDNA library, how does one go about choosing which ones to study. Often, clones have been chosen in a manner akin to a fishing expedition in which an investigator recovers clones from a cDNA library and selects a subset that shows a pattern of expression—among tissues or developmental stages—indicative of a potential role in a particular biological process. However, an ultimate goal of the human genome project is to characterize and understand the function of all genes in the genome (Hochgeschwender, 1992). In a more directed approach toward this goal, it is possible to walk down a cloned chromosomal region and pick up each transcription unit one by one for further analysis of function.

The first step in the analysis of a newly cloned gene is always to determine its sequence and compare it with all other sequences stored in databases such as GenBank and others. Sequence homologies in and of themselves can often be used to predict characteristics of the polypeptide encoded by the new gene under investigation. In some cases, the new product will contain a "domain" with homology to a specific "peptide motif" that is associated with a particular function in groups of previously characterized polypeptides. For example, one or more peptide motifs have been identified that are characteristic of DNA binding domains, membrane-associated domains, various enzymatic activities, receptor functions, and many others. Peptide motifs are almost always degenerate amino-acid sequences; they can vary in length from just three amino acids to over 100 residues.

Even when the new gene product does not contain any previously defined peptide motifs, standard search algorithms will sometimes allow the identification of previously cloned genes that are related by descent from a common ancestral sequence. Once again, if the function of the previously characterized gene has been determined, it can be used as a starting point for understanding the function of the new gene under investigation.

Although the sequence can sometimes provide clues to gene function, further experiments will always be required to demonstrate conclusively the role played by a particular gene product in the overall scheme of life. These further experiments can take two different forms: biochemical and genetic. A biochemical investigation is often begun by cloning the open reading frame into an expression vector, which is placed back into an appropriate host cell system for the "in vitro" synthesis of large quantities of the gene product. This can then be used to immunize rabbits or mice for the production of polyclonal antisera or monoclonal antibodies. These antibodies can then be used as a tool to investigate the expression and localization of the protein both among cells and within cells, and to purify the native protein from the mouse. The purified native protein can be analyzed for enzymatic activities and for its interactions with other molecules. Biochemical studies can often provide critical insight into the function of a particular polypeptide.

The genetic approach to understanding function flows from the ability to manipulate the expression of the selected gene within the mouse and then follow the phenotypic consequences of this manipulation. The two most powerful approaches to gene manipulation are based on targeted mutagenesis and the insertion of transgene constructs into the germ line of the mouse; both of these approaches are discussed in depth in Chapter 6. Targeted mutagenesis allows an investigator to produce a null mutation at the locus of interest, and determine how and where the absence of the corresponding gene product affects the animal, its tissues, and its cells. The transgenic technology can be used to produce animals that misexpress the gene and/or its product—in the wrong place, the wrong time, or the wrong form. The rationale for the use of both targeted mutagenesis and directed transgenesis is that by examining the perturbations in phenotype that occur in response to perturbations of the genotype, one can gain insight—by contrast—into the true function of the normal wild-type locus.

In some cases, the genetic approach will be the one that uncovers the function of a gene, and in other cases, it will be the biochemical approach. However, these two approaches are entirely complementary and thus together they are likely to provide more information than either one can alone.

### 7.4.2 From phenotype to gene

The second pathway to deciphering the relationship between genotype and phenotype is based on the initial observation of an interesting new variant that distinguishes one group of individuals from another. Variants may be observed in the context of either deleterious mutations or polymorphic differences in common traits such as growth, life span, disease resistance or various physiological parameters. In all of these cases, the phenotype will be available for analysis before the causative gene or genes. The process by which one moves from a phenotypic difference to the gene (or genes) responsible is referred to as positional cloning.

There are two stages in the process of positional cloning. The first stage is the focus of a major portion of this book: the use of formal linkage analysis and other genetic approaches—as tools—to find flanking DNA markers that must lie very close to the gene of interest. With these markers in hand, one can move to the second stage of this pathway: cloning across the region that must contain the gene responsible for the phenotype, and then identifying the gene itself apart from all other genes and non-genic sequences within this region. This second stage will be discussed in Chapter 10.

### 7.4.3 The molecular basis of complex traits

With all of the new approaches to mapping that have been developed over the last few years, it has become possible, for the first time, to follow the segregation of the whole genome from each parent to each offspring in a cross. This, in turn, has allowed investigators to consider the exciting possibility of approaching the genetic basis for quantitative, polygenic, and multifactorial traits. In fact, most common types of phenotypic differences that distinguish one individual from another are due to the interaction of alleles at more than one locus and expression is often modified by environmental factors as well. The available inbred strains provide a treasure chest of polygenic differences that control characteristics as diverse as size, life span, reproductive performance, aggression, and levels of susceptibility or resistance to particular diseases, both infectious and inherited. The golden age of mammalian genetics beckons: the genetic components of any and all traits that show variation between different mice are now amenable to dissection with classical genetic tools that can provide a means for obtaining clones of all of the genes involved, which can, in turn, be used as tools to understand each trait at the molecular level.

# Genetic Markers

## 8.1 GENOTYPIC AND PHENOTYPIC VARIATION

Linkage analysis can only be performed on loci that are polymorphic[53] with two or more distinguishable alleles. Naturally occurring polymorphic loci with clear single-gene effects are rarely observed in wild animal populations. In the laboratory, however, it is possible to identify and breed animals with mutations at many different loci (see Chapter 6). Over the last 90 years, thousands of independent mouse mutations have been characterized in various laboratories. In fact, as discussed previously, a primary reason for the initial choice of the mouse as an experimental genetic system was the collection of rare genetic variants present in the hands of the fancy mouse breeders (see Chapter 1). However, even this variation is restricted in its scope and usefulness for geneticists. This is because of the severe limitation in the number of phenotypic markers that can be incorporated into any one cross. Although over 50 independent loci have been characterized with effects on coat color (Silvers, 1979), it is impossible to follow more than a handful at any one time since mutant alleles at any one locus will act to obscure the expression of mutant alleles at other loci. With mutant alleles that affect viability in some way, the problem of sorting out overlapping phenotypes becomes even more severe.

Prerecombinant DNA geneticists were able to circumvent these problems by performing large numbers of different crosses, each of which tested overlapping subsets of phenotypic markers. Thus, as illustrated in Fig. 8.1, the loci *A*, *B*, and *C* were mapped in one cross; *C*, *D*, and *E* were mapped in a second cross, and *A*, *D*, and *F* were mapped in a third cross. If linkage was observed in each of these individual crosses among the three loci mapped therein as shown, it was then possible to develop a linkage map that encompassed all six loci even though, for example, the *B* locus was never mapped directly relative to *D*, *E* or *F*; and *A* was never mapped relative to *E*. By extension, it is possible to combine data obtained in hundreds of crosses to map hundreds of phenotypically defined loci to form linkage maps that extend across all 19 mouse autosomes and the *X* chromosome.

Mapping in the prerecombinant DNA era was tedious and was generally performed by investigators dedicated to this task alone. However, with the results of the first generation of cloning and sequencing studies, the scientific community became aware of the existence of a hidden level of enormous genetic

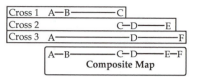

**Figure 8.1** Piecemeal construction of a linkage map. Illustration of three crosses that each provide linkage information for only three loci. By integrating linkage distances from each cross, it becomes possible to build a consensus map that contains all six loci.

variation that occurs naturally in all mammalian populations (Botstein et al., 1980). The frequency of DNA variation that exists between chromosome homologs from two unrelated individuals of the same species (including mice and humans) appears to be on the order of one nucleotide substitution or small length change in every 200–500 bp. Since the mammalian genome has a size of $3 \times 10^9$ bp, this frequency implies a total number of genetic differences between any two unrelated individuals of the order of *six million* per haploid genome set. In a comparison of individuals from separate species, such as *M. musculus* and *M. spretus*, the level of variation will be even higher.

In the prerecombinant DNA era, alleles could only be distinguished in terms of an altered phenotype; thus only genes could have alleles, and the demonstration of a genetic locus was dependent on the expression of alternative phenotypes. Today, every variant nucleotide in the genome is a potential locus. To say that DNA variation provides a larger reservoir for use in genetic studies than phenotypic variation is a vast understatement. Furthermore, and of most importance, there is essentially no limit to the number of these loci that can be mapped simultaneously within a single cross.

All simple forms of DNA variation fall into three classes: (1) base pair substitutions; (2) short regions of deletion or tandem duplication; and (3) insertions or translocations. Examples from each of these classes can be detected as RFLPs and/or by PCR-based protocols. These major tools for DNA allele detection will be discussed separately in the following two sections of this chapter.

## 8.2 RESTRICTION FRAGMENT LENGTH POLYMORPHISMS

### 8.2.1 The molecular basis for RFLPs

A restriction fragment length polymorphism is defined by the existence of alternative alleles associated with restriction fragments that differ in size from each other.[54] RFLPs are visualized by digesting DNA from different individuals with a restriction enzyme, followed by gel electrophoresis to separate fragments according to size, then blotting and hybridization to a labeled probe that identifies the locus under investigation. A RFLP is demonstrated whenever the Southern blot pattern obtained with one individual is different from the one

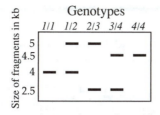

**Figure 8.2**    Demonstration of RFLPs by Southern blot analysis. A representation of five different Southern blot genotypes detected with a single probe is shown. The alleles present in each genotype are distinguished by numbers at the top of each lane. Each numbered allele in this Figure corresponds to the identically numbered genomic restriction map shown in Fig. 8.3.

obtained with another individual. An illustration of a result of this type is shown in Fig. 8.2. In this example, DNA samples from five individual mice were digested with the same enzyme, electrophoresed side by side on a gel and probed with the same clone of a single-copy DNA sequence. The five patterns detected are all different from each other and are representative of five different genotypes. The simplest interpretation of data of this type is that the first and last samples shown in the Figure are homozygous for a different restriction fragment while the middle samples are all heterozygous with different combinations of alleles. This simple interpretation can be tested and confirmed (or rejected) by simply breeding each of the animals to mates with different genotypes at this "RFLP locus" so that segregation of the two restriction fragment alleles can be demonstrated from putative heterozygotes and uniform transmission of the same restriction fragment allele can be demonstrated from putative homozygotes.[55]

RFLPs were the predominant form of DNA variation used for linkage analysis until the advent of PCR. Even now, in the PCR age, RFLPs provide a convenient means for turning an uncharacterized DNA clone into a reagent for the detection of a genetic marker. The main advantage of RFLP analysis over PCR-based protocols is that no prior sequence information, nor oligonucleotide synthesis, is required. Furthermore, in some cases, it may not be feasible to develop a PCR protocol to detect a particular form of allelic variation. Nevertheless, if and when a PCR assay for typing a particular locus is developed, it will almost certainly be preferable over RFLP analysis for the reasons to be described in Section 8.3.

The detection of a RFLP, in and of itself, does not provide information as to the mechanism by which it was created. Although the different-sized restriction fragments shown in Fig. 8.2 can be followed readily in a genetic cross, one cannot tell, from these data alone, how they differ from each other at the molecular level. In fact, RFLPs can be generated by all of the mechanisms through which DNA variation can occur. The simplest RFLPs are those caused by single base-pair substitutions. However, RFLPs can also be generated by the insertion of genetic material, such as transposable elements, or by tandem duplications, deletions, translocations, or other rearrangements.

Several different mechanisms of RFLP generation are illustrated in Fig. 8.3. In this set of hypothetical examples, the first chromosome represents the ancestral

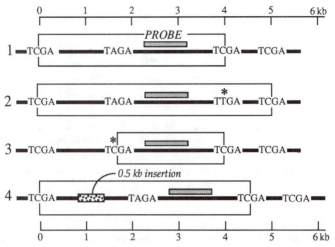

**Figure 8.3** RFLP generation by different molecular events. Chromosome 1 is the ancestral state, shown with three *Taq*I restriction sites (TCGA) and a fourth non-restriction site (TAGA). Chromosome 2 has undergone a C to T change (indicated by an asterisk) that destroys one of the restriction sites. Chromosome 3 has undergone an A to C change that creates a *Taq*I site. Chromosome 4 has had a 0.5 kb insertion. The boxed-in region on each chromosome is the restriction fragment that will be recognized on a Southern blot probed with the fragment indicated in the diagram.

state, and chromosomes 2, 3, and 4 represent different mutations from this state. In this example, DNA has been digested with the enzyme *Taq* I (with a recognition site of TCGA), fractionated and probed with a clone that recognizes the region shown in the Figure. The Southern blot results that would be obtained with animals that carry representative pairs of these chromosomes are shown in Fig. 8.2. The length of the restriction fragment that will be observed with each chromosome type is indicated by the boxed-in region. Chromosome 1 has *Taq*I sites that flank the probed region at a distance of 4 kb from each other. In chromosome 2, the right-flank *Taq*I site has been mutated (the base substitution is marked with a *); a previously more distal *Taq*I site now becomes the new flanking site, leading to the production of a 5 kb restriction fragment. In chromosome 3, a mutation has occurred with an opposite effect, causing the creation of a *Taq*I site where none existed before; this new *Taq*I site becomes the left-flank site, leading to the production of a smaller restriction fragment. Finally, in chromosome 4, an insertion has occurred within the region between the two flanking *Taq*I sites, leading to an actual increase in the length of the region between these same two sites. More complicated scenarios can be built upon these simple examples with restriction sites created or removed from within the probed region itself, or with new restriction sites brought in with inserted DNA elements. A final class of RFLPs is commonly generated through the expansion and contraction of families of tandemly repeated DNA elements as illustrated in Fig. 8.4. Loci having this type of organization are referred to as minisatellites, or VNTRs, and will be discussed separately in Section 8.2.3.

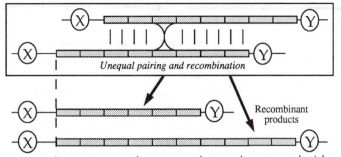

**Figure 8.4** Length variation can be generated at tandem repeat loci by unequal pairing and crossing over. This Figure illustrates the general principle behind the generation of DNA variation at tandem repeat loci such as minisatellites and microsatellites. Individual repeat units are represented by boxes. The flanking sites "X" and "Y" allow detection of length variation at the locus, and correspond to nearest flanking restriction sites in the case of minisatellites or targets for PCR primers in the case of microsatellites.

Attempts to identify RFLPs between different inbred strains of mice often meet with limited success even after testing with large numbers of enzymes. In one study, RFLPs were identified at only 30% of the single copy loci tested with 22 different restriction enzymes (Knight and Dyson, 1990). Furthermore, when RFLPs are identified, they are almost always di-allelic binary systems—the insertion, deletion, or restriction site change is either present or absent.[56] Unfortunately, di-allelic loci can only be mapped in crosses where the two parental chromosomes carry the two alternative alleles. Thus, even if a RFLP is identified between two inbred strains of mice, there is no guarantee that another pair of strains will also happen to carry alternative alleles. As a consequence, only a subset of the RFLP markers developed for analysis of one cross between traditional mouse strains will be of use for mapping in a cross between any other pair of inbred strains.

A major leap in mouse genetics came with the observation of an extremely high rate of RFLP detection between the common *M. musculus*-based inbred strains and the independent species *M. spretus*. As described in Section 2.3, these species can breed under laboratory conditions to produce interspecific $F_1$ hybrids. Although $F_1$ males are sterile, the $F_1$ females are fertile and they can be backcrossed to either parent to obtain offspring, which can be analyzed for linkage (Bonhomme et al., 1978; Avner et al., 1988). More recently, the observation of an increased rate of RFLPs has been extended to comparisons between the inbred strains and wild-derived samples of *M. m. castaneus*, which are more closely related to each other than either is to *M. spretus* (Fig. 2.2). The pros and cons of performing genetic studies with interspecific or intersubspecific crosses are discussed in Section 9.4.2.1.

### 8.2.2 Choice of restriction enzymes to use for RFLP detection

With so many restriction enzymes available, how does one decide which ones are the best to use in the search for RFLPs? Obviously, cost is an important consideration. Another consideration is whether the enzyme is optimally active with genomic DNA obtained from animal tissues.[57] However, a critical consideration is the rate at which RFLPs can be detected based on the enzyme that is chosen.

A systematic study of RFLP detection between B6 and *M. spretus* DNA subsequent to digestion with one of ten different enzymes has been reported by LeRoy et al. (1992). One hundred and ten anonymous DNA sequences of less than 4 kb in length were used as probes. The highest rate of RFLP detection— 63%—was observed with DNA digested with *Taq*I. The second highest rate— 56%—was observed with *Msp*I. In decreasing order of effectiveness were the enzymes *Bam*HI (50%), *Xba*I (47%), *Pst*I (44%), *Bgl*II (41%), *Hind*III (39%), *Pvu*II (38%) *Rsa*I (38%), and *Eco*RI (33%). It is ironic that, of the ten enzymes tested, the one most commonly used in molecular biological research—*Eco* RI—was the worst one, by a long shot, at detecting polymorphisms.

A theoretical explanation for the observation that *Taq*I and *Msp*I are more likely than other enzymes to detect RFLPs can be found in the dinucleotide CpG, which is at the center of both recognition sites. This dinucleotide is unusual in two respects. First, it is present in mammalian genomes at a frequency one-fifth of that expected from base composition alone. Second, when it is present, the cytosine within the dinucleotide is usually methylated.[58] As it turns out, the latter fact explains the former because methylated cytosine has a propensity to undergo spontaneous deamination to form thymidine. This complete transition is not recognized as abnormal by the repair machinery present in mammalian cells, and thus methylated-CpG dinucleotides serve as one-way hotspots for mutation (Barker et al., 1984). As a consequence, the CpG dinucleotide is relatively rare, and when it does occur in a methylated form, it is more likely to mutate than any other dinucleotide. Even an unmethylated CpG can undergo a spontaneous mutation from cytosine to uracil; however, this abnormal nucleotide is more likely to be recognized and repaired. Nevertheless, in those few cases where repair does not occur, the uracil will pair with an adenosine in the following round of DNA replication, leading to the same substitution as found with methylated CpGs.

Thus, *Taq*I and *Msp*I are the most useful enzymes for the identification of RFLPs. Both enzymes recognize four basepair sites, *Taq*I recognizes TCGA and *Msp*I recognizes CCGG. If nucleotides were randomly distributed across the genome, *Taq*I and *Msp*I sites would be distributed at average distances of 270 bp and 514 bp, respectively.[59] However, as a consequence of the paucity of CpG dinucleotides, these two restriction enzyme sites are actually found much less frequently in mammalian DNA. Empirical data indicate restriction fragment size distributions that average 2.9 and 3.5 kb for *Taq*I and *Msp*I respectively (Barker et al., 1984).

In practice, the enzyme *Taq*I is the better choice of the two for use in RFLP analysis. It is relatively cheap and it works well with animal DNA samples that other enzymes refuse to cut (presumably aided by the high temperature at which the digestion is carried out). *Msp*I is somewhat more sensitive to contaminants within animal tissue DNA samples, but is a good second choice. When the results obtained with *Taq*I and *Msp*I are combined, the Guénet group detected RFLPs at

74% of the loci tested for variation between *spretus* and *musculus* (LeRoy et al., 1992). When the results obtained with *Xba*I were added in, 79% of the loci were polymorphic. When the results obtained with the remaining seven enzymes were included, RFLPs were detected at 83% of the loci. The take-home lesson from this study is that it is most cost-effective to search for RFLPs on standard 1% agarose gels with just three enzymes—*Taq*I, *Msp*I, and *Xba*I. If the search is unsuccessful at this point, it would appear that the locus under analysis is not highly polymorphic at the DNA level, and in those cases where the locus is just "one more marker," it is probably not worth pursuing further. On the other hand, if the locus is of importance in and of itself, it makes sense to pursue more sensitive, PCR-based avenues of polymorphism detection, such as single-strand conformation polymorphism (Section 8.3.3) or linked microsatellites (Section 8.3.6).

### 8.2.3 Minisatellites: variable number tandem repeat loci

In contrast to traditional RFLPs caused by basepair changes in restriction sites, a special class of RFLP loci present in all mammalian genomes is highly polymorphic with very large numbers of alleles. These "hypervariable" loci were first exploited in a general way by Jeffreys and his colleagues (1985) for genetic mapping in humans (1985).

Hypervariable RFLP loci of this special class are known by a number of different names, including variable number tandem repeat (VNTR) loci and *minisatellites*, which is the more commonly used term today. Minisatellites are composed of unit sequences that range from 10 to 40 bp in length and are tandemly repeated from tens to thousands of times. Although various functions have been suggested for minisatellite loci as a class, none of these has withstood the test of further analysis (Jarman and Wells, 1989; Harding et al., 1992). Rather, it appears most likely that minisatellite loci (like microsatellite loci described in Section 8.3.6) evolve in a neutral manner through expansion and contraction caused by unequal crossing over between out of register repeat units as shown in Fig. 8.4 (Harding et al., 1992). Recombination events of this type will yield reciprocal products which both represent new alleles with a change in the *number* of repeat units.

The frequency with which new alleles are created at minisatellite loci—of the order of $10^{-3}$ per locus per gamete—is much greater than the classical mutation rate of $10^{-5}$–$10^{-6}$ (Jeffreys et al., 1988). This leads to a much higher level of polymorphism between unrelated individuals within a population. At the same time, one change in a thousand gametes is low enough so as not to interfere with the ability to follow minisatellite alleles in classical breeding studies.

Length polymorphisms at minisatellite loci are most simply detected by digestion of genomic DNA samples with a restriction enzyme that does not cut within the minisatellite itself but does cut within closely flanking sequences. As with all other RFLP analyses, the restriction digests are fractionated by gel electrophoresis, blotted and hybridized to probes derived from the polymorphic locus. However, unlike traditional point mutation RFLPs, minisatellites are caused by, and reflect, changes in the actual size of the locus itself.

The best restriction enzymes to use for minisatellite analysis are those with 4 bp recognition sites such as *Hae*III, *Hin*fI or *Sau*3A; it is likely that one of these enzymes will not cut within the relatively short minisatellite unit sequence, but will cut within several hundred basepairs of flanking sequence on both sides. Standard 1% agarose gels with maximal separation in the 1–4 kb range are usually best for the resolution of minisatellite bands; however, conditions can be optimized for each minisatellite system under analysis.

There is nothing special about the unit sequence present within minisatellites, which are defined only by their repeated nature and their repeat unit size. Thus, it is not possible to develop a general protocol for identifying all minisatellite sequences within the genome, and there is no way of knowing how many loci of this type are actually present. However, significant homology (indicative of evolutionary relatedness) often exists among unlinked minisatellite loci that are scattered throughout the genome. Homologies that allow cross-hybridization define minisatellite families that can have as few as two and as many as 50 members (Nakamura et al., 1987). It is often possible to take advantage of these cross-homologies to map ten or more minisatellite loci as independent RFLPs within single Southern blot hybridization patterns.

The simultaneous detection of 10–40 unlinked and highly polymorphic loci provides a whole genome "fingerprint" pattern, which is very likely to show differences between any two unrelated individuals (Jeffreys et al., 1985). These DNA fingerprints provide a powerful tool in human forensic analysis in the absence of any knowledge as to the map location of any of the individual loci that are being detected (Armour and Jeffreys, 1992). DNA fingerprinting per se is of much less use in the analysis of laboratory animals, who do not bring paternity suits or stand trial for rape or murder. However, fingerprinting can allow field biologists to follow individual animals in wild populations subjected to repeated capture and release sampling. It can also be used to monitor the integrity of inbred strains of mice, and for the characterization and comparison of different breeds of domesticated animals that have commercial importance.

New minisatellite families are uncovered by chance, by cross-hybridization with probes defined in other species (Jeffreys et al., 1987), or by the use of "synthetic tandem repeats" of arbitrary 14–20 mer oligonucleotides (Mariat and Vergnaud, 1992). The first analysis of minisatellites in the mouse was performed with the use of several human minisatellite sequences as probes (Jeffreys et al., 1987). The results obtained in the analysis of a set of recombinant inbred strains (described at length in Chapter 9) demonstrated the expected high level of polymorphism as well as a high level of stability over time, both of which are critical properties for a useful mapping tool. Julier and his colleagues have performed more detailed mapping studies with a larger panel of human minisatellite probes (Julier et al., 1990) and, in collaboration with Mariat and colleagues, they have also performed minisatellite mapping with the use of arbitrary oligonucleotides of 14–16 bases in length (Mariat et al., 1993). With the 29 human-derived minisatellite probes tested, these authors found that 48% gave well-resolved complex fingerprint patterns upon hybridization to the mouse genome. With a set of 24 arbitrary oligonucleotides that were preselected for detection of minisatellites in humans, 23 were found to detect polymorphic loci in the mouse as well.

In an initial analysis with just 11 of the human minisatellite probes, a total of 115–234 restriction fragment differences were detected in pairwise comparisons among a series of seven *M. musculus*-derived inbred strains. The least number of polymorphic loci was observed in a comparison of C3H/He and DBA/2J; the highest number were observed between SJL/J and 129/Sv. Approximately twice as many polymorphisms were observed in pairwise comparisons between *M. musculus*-derived strains and a *M. spretus* inbred line.

The 11 characterized probes were used to follow the segregation of minisatellite alleles in a higher resolution analysis of the BXD set of RI strains as described in Chapter 9 (Julier et al., 1990). The 346 polymorphic bands followed in this study sorted into 166 independent loci, approximately half of which were represented by a single restriction fragment, with the remaining represented by two or more fragments. As expected, in several cases, new fragments were detected in particular RI strains that were not present in either of the parental inbred strains from which they were generated, attesting to the rapid rate at which minisatellite loci mutate to new alleles.

Mapping with multilocus minisatellite probes is most effective for whole genome studies rather than for single chromosomes analyses. Thus, like the two-dimensional RFLP and RAPD technologies described below, minisatellite mapping is actually of greatest use for the initial development of whole genome "framework maps" of relatively uncharacterized species, of which the mouse is not one.

### 8.2.4 Dispersed multilocus analysis with cross-hybridizing probes

Minisatellite families are just one example of dispersed, cross-hybridizing loci that can be mapped simultaneously by Southern blot analysis. Another class of this type includes those gene families that have multiple members dispersed to unlinked chromosomal locations. In general, protein-encoding genes will be much less polymorphic than minisatellite loci; thus, simultaneous mapping of multiple members of gene families through RFLP analysis is best accomplished with interspecific backcrosses of the *spretus—domesticus* type. In one such study, probes for just two gene families—ornithine decarboxylase and triose phosphate isomerase—were combined with a probe for the highly polymorphic mouse mammary tumor virus (MMTV) elements (described below) in traditional Southern blot studies to detect and map a total of 28 loci to 16 of the 19 mouse autosomes (Siracusa et al., 1991).

A third broad class of cross-hybridizing loci is represented by the endogenous retroviral and retroviral-like elements that have been dispersed to random positions throughout the genome. A number of different families and subfamilies of this class have been identified (see Section 5.4.1). The best characterized of these (with average copy number per haploid genome in parentheses) are MMTVs (4–12), ecotropic MuLVs (0–10), non-ecotropic MuLVs (40–60), VL30s (~200), and IAPs (~2,000). In all of these cases, polymorphisms are a consequence of the recent integration of proviral elements so that particular elements are present in the genomes of some strains but not others; thus, each polymorphism is represented by a binary plus/minus system.

Both the MMTVs and ecotropic MuLVs are present at copy numbers that are suitable for mapping by standard agarose gel electrophoresis. By combining data from various crosses, Jenkins and colleagues (1982) mapped a total of 18 ecotropic MuLV integration sites that were named *Emv*-1 through *Emv*-18. In similar studies, 26 MMTV integration sites have been mapped among various inbred strains (Kozak et al., 1987); these have been named *Mtv*-1 through *Mtv*-26.

The non-ecotropic MuLV elements are present at a copy number, which is somewhat too high for complete resolution of all elements on standard agarose gels. To overcome this problem, and to obtain maximal mapping information, it is possible to take advantage of the subfamily structure of this class of elements. Oligonucleotides that recognize different subsets of 10–30 loci per genome have been used as Southern blot probes with excellent resolving power (Frankel et al., 1990, 1992). In general, 30–50% of the non-ecotropic viral elements are shared in any one pairwise comparison of inbred strains. By combining data from different sets of recombinant inbred lines, Frankel and colleagues were able to map over 100 non-ecotropic integration sites; these have been named with the prefixes polytropic murine virus (*Pmv*-), modified polytropic murine virus (*Mpmv*-), or xenotropic murine virus (*Xmv*-) according to the particular oligonucleotide that cross-hybridized to each element. With the MMTV and various MuLV families, it is still possible to use the same probes to map even more integration sites through the examination of strains that were not previously studied.

The retroviral-like families IAP (Lueders and Kuff, 1977) and VL30 (Courtney et al., 1982; Keshet and Itin, 1982) are present in 200 and 2,000 copies, respectively, per haploid genome. These families and others of the same class contain a large potential reservoir of useful genetic markers. However, their copy number is much too high to allow the resolution of individual family members in traditional Southern blot studies with restriction digested DNA samples. It is typically difficult to resolve more than 20 bands in a traditional one-dimensional hybridization pattern. Furthermore, as the copy number of cross-hybridizing bands increases, the resolution of individual bands actually decreases as more and more merge into each other to eventually form a continuous smear.

In theory, this problem could be alleviated in two different ways. The first approach would be the same as that used for the non-ecotropic loci, which is to reduce the complexity of the Southern blot pattern with the use of oligonucleotide probes that detect small subsets of the whole family. The validity of this approach has been demonstrated for the IAP family of elements (Meitz and Kuff, 1992).

A second, very different approach is based on increasing resolving power, rather than decreasing complexity, by fractionating genomic DNA in two sequential dimensions. This can be accomplished as illustrated in Fig. 8.5. DNA samples are first subjected to digestion with a restriction enzyme that cuts relatively infrequently (step 1 in the Figure) followed by fractionation on

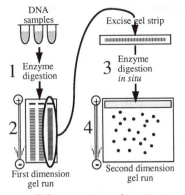

**Figure 8.5** Two-dimensional fractionation of restriction fragments. Illustration of the general steps behind various methods of two-dimensional fractionation of DNA based on the distribution of two or more classes of restriction sites.

an agarose gel (step 2). At the completion of electrophoresis, each sample-containing gel lane is excised and incubated directly with a second restriction enzyme that cuts more frequently (step 3). Finally, the gel slice itself is used as the sample for a second round of electrophoresis in a direction perpendicular to the first round (step 4). At the completion of this second dimension run, the gel is blotted, hybridized to the high copy-number probe, and autoradiographed.

Separation of DNA fragments in two dimensions, rather than one, should theoretically provide a "power of two" increase in resolution, from approximately 20 bands to 400. In fact, over 130 restriction-fragment "spots" have been observed in individual two-dimensional patterns obtained with probes for the IAP and VL30 families (Sheppard et al., 1991; Sheppard and Silver, 1993). In general, each spot represents a single retroviral-like locus of the type defined by the probe used for hybridization. The $X$ coordinate of the spot measures the distance between flanking restriction sites produced in the first digestion. The $Y$ coordinate provides a measure of the distance between the two closest restriction sites (of either type) that flank the locus after the double digestion.

The main advantage of a two-dimensional mapping approach is that large numbers of loci from each animal can be mapped simultaneously. There are two main disadvantages to the general use of this approach for analyzing a large cross. First, only one animal can be analyzed within each gel. Second, from start to finish, each gel run can take 5 days to complete, and there is very little tolerance for mistakes of any kind throughout the protocol.

### 8.2.5 Restriction landmark genomic scanning

A significant variation on two-dimensional RFLP analysis has been developed by Hayashizaki and his colleagues (Hatada et al., 1991). With this novel protocol, restriction sites are scanned directly without the intervention of probes for specific loci. This can be accomplished through the direct end-labeling of a class of restriction sites that are generated by a rare-cutting enzyme followed by additional rounds of restriction digestion and gel separation. Briefly, the first restriction digestion is carried out with a rare-cutting enzyme like *Not*I which has an 8 bp recognition site that is present, on average, only once per megabase in the mouse genome (a first component of step 1 in Fig. 8.2). Digestion with *Not*I will produce a total of only ~3,000 fragments from each haploid genome. Labeling of *Not*I sites is accomplished by filling-in the single strand restriction site overhangs with radioactive nucleotides. Subsequently, the *Not*I fragments are reduced in size by digestion with a second enzyme having a 6 bp recognition site that produces fragments with an average size of 4–6 kb (a second component of step 1). Although the total number of restriction fragments per genome is increased 200-fold by this second digestion, only those fragments that have an original *Not*I site at one end will be labeled. This total mixture is fractionated by agarose gel electrophoresis (step 2) and then digested in situ with a third enzyme that has a 4 bp recognition site and thus cuts very frequently in the genome (step 3). The average size of restriction fragments has now been reduced to several hundred basepairs. The gel strip containing each sample is now placed on top of a polyacrylamide gel and a second orthogonal dimension of electrophoresis is carried out (step 4).

This RFLP protocol differs from all those described previously in that the rare restriction sites are visualized directly without the use of probes that light up particular loci or locus families. Thus, the complete set of fragments that flank both sides of every *Not*I site in the genome of an individual will be displayed in the pattern that is obtained. The $X$ coordinate of each labeled spot will be a measure of the distance between the first labeled *Not*I restriction site and the nearest neighbor second restriction site. The $Y$ coordinate of each spot will be a measure of the distance between the first labeled restriction site and the nearest neighbor third restriction site. Polymorphisms can arise from changes that affect any of the three restriction sites that define each spot.

Since the rare restriction sites themselves are labeled, blotting and hybridization steps are eliminated and autoradiographs are obtained by direct exposure of gels to film. The elimination of two lengthy steps significantly reduces the overall time required to process each sample. In addition, without blotting and hybridization, spots are much more sharp and well delineated from each other. Resolution is also improved with the use of a polyacrylamide, rather than agarose, medium in the second dimension of separation. Hayashizaki, Hatada and colleagues have reported the detection of several thousand spots on two-dimensional gels derived from individual mice (Hatada et al., 1991). Analysis of the BXD set of RI strains with this protocol has allowed the mapping of 473 polymorphic loci.

The advantages in resolution notwithstanding, the restriction landmark genomic scanning (RLGS) protocol is still technically demanding and it still allows the processing of only one sample per gel. Like other multiplex whole genome scanning methods, it is actually of greatest utility for the initial development of whole genome maps of relatively uncharacterized species.

## 8.3 POLYMORPHISMS DETECTED BY PCR

Without a doubt, the polymerase chain reaction (PCR) represents the single most important technique in the field of molecular biology today. What PCR accomplishes in technical terms can be described very simply: it allows the rapid and unlimited amplification of specific nucleic acid sequences that may be present at very low concentrations in very complex mixtures. Within less than a decade after its initial development, it has become a critical tool for all practicing molecular biologists, and it has served to bring molecular biology into the practice of many other fields in the biomedical sciences and beyond. The reasons are several fold. First, PCR provides the ultimate in sensitivity—single DNA molecules can be detected and analyzed for sequence content (Li et al., 1988; Arnheim et al., 1991). Second, it provides the ultimate in resolution—all polymorphisms, from single base changes to large rearrangements, can be distinguished by an appropriate PCR-based assay. Third, it is extremely rapid— for many applications, it is possible to go from crude tissue samples to results within the confines of a single work day. Finally, the technique is an agent of democracy—once the sequences of the pair of oligonucleotides that define a particular PCR reaction are published, anyone anywhere with the funds to buy the oligonucleotides can reproduce the same reaction on samples of his or her choosing; this stands in contrast to RFLP analyses in which investigators are often dependent upon the generosity of others to provide clones to be used as probes. Numerous books and thousands of journal articles have been published on the principles and applications of the technique [Erlich (1989) and Innis et al. (1990) are two early examples].

Although the applications of PCR are as varied as the laboratories in which the technique is practiced, this section will focus entirely on six general applications that are relevant to the detection and typing of genetic variation in the mouse. Four of these applications are based on the PCR amplification of particular loci that have been previously characterized at the sequence level. In these cases, primer pairs must be chosen to be as specific as possible for the locus in question in order to avoid artifactual PCR products. Computer programs are available to assist in primer design (Lowe et al., 1990; Dietrich et al., 1992) but manual inspection is usually adequate. One must be careful to avoid self-complementarity within any one primer and the presence of complementary sequences between the two primers. Also, potential primers should be screened with use of a sequence comparison program to avoid homology with the highly repeated elements B1, B2, and L1 (see Section 5.4). The primer length should be at least 20 bases, the G:C content should be at least 50%, and the melting temperature should be at least 60°C.

Even when all of these conditions are adhered to, it is still possible to find that a particular pair of primers will not work properly to amplify a specific locus into a reproducible product that can be clearly distinguished from artifactual background bands. There are a variety of approaches that one can take to eliminate such problems (Erlich, 1989; Innis et al., 1990), but if all else fails, one should replace one or both primers with alternatives derived from other nearby flanking sequences that also fit the rules listed above.

### 8.3.1 Restriction site polymorphisms

#### 8.3.1.1 Overview

Rapid, highly efficient PCR-based assays can be designed to detect all RFLPs—previously defined by Southern blot analysis— as long as the nature of the RFLP is understood and sequence information flanking the actual polymorphic site is available. A pair of PCR primers that flank this site can then be synthesized according to the rules just described and tested for their ability to amplify a specific product that can be readily identified as an ethidium bromide-stained band by gel electrophoresis.

With the simplest and most common type of RFLP illustrated in Fig. 8.6, the polymorphism results from a single nucleotide difference that provides a recognition site for a restriction enzyme in one allelic form and not the other. A polymorphism of this type can be rapidly detected by: (1) amplifying the region around the polymorphic site from each sample; (2) subjecting the amplified material to the appropriate restriction enzyme for a brief period of digestion; and (3) distinguishing the undigested PCR product from the smaller digested fragments by gel electrophoresis. By choosing primers that are relatively equidistant to and sufficiently far from the polymorphic site, one can easily resolve allelic forms on agarose or polyacrylamide gels as illustrated in Fig. 8.6.

This PCR-based protocol provides results much more rapidly and is much easier to carry out than the Southern blot alternative, which requires blotting, probe labeling, hybridization, and autoradiography. Since the major expense involved with PCR is in the initial sequencing of the locus and the synthesis of PCR primers, it is also less costly in all cases where one expects to type large numbers of samples for the particular locus in question.

Even RFLPs caused by more complex mutational events can be analyzed by PCR. Figure 8.7 illustrates the logic behind devising PCR strategies for detecting deletions, insertions, inversions, and translocations. The only requirement is a knowledge of the sequences that surround the breakpoints associated with each particular genetic event.

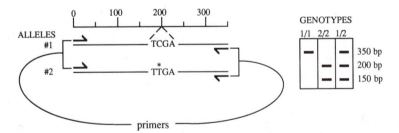

**Figure 8.6**   Restriction site polymorphism detection by PCR. In this example, primers have been chosen that are 200 bp upstream and 150 bp downstream of a polymorphic *Taq*I site. The initial PCR product amplified from both restriction site alleles will have the same length of 350 bp. However, the two alleles can be distinguished by subjecting the PCR products to digestion with *Taq*I and then fractionating the DNA by gel electrophoresis. At the right is an illustration of the ethidium bromide patterns that will be observed with each of the three possible genotypes.

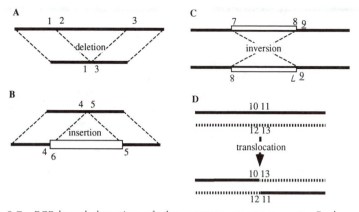

**Figure 8.7** PCR-based detection of chromosome rearrangements. Each number represents the position of an oligonucleotide sequence that has been used to generate a PCR primer. In panel A, primers 1 and 2 will amplify a product only from the non-deleted chromosome, and primers 1 and 3 will amplify a product only from the deleted chromosome. The specificity of this last primer pair is based on the empirical observation of a maximum distance of ~1 kb beyond which PCR amplification is extremely inefficient under normal reaction conditions. In the three other panels, different combinations of primers can be used similarly to distinguish wild-type and mutant alleles.

### 8.3.1.2 3′-Untranslated regions as a mapping resource

An important resource for the identification of new restriction site polymorphisms that can be typed by PCR is the 3′-untranslated (3′-UT) regions of transcripts. These regions are not under the same selective constraints as coding sequences and are frequently just as polymorphic as random non-transcribed genomic regions. However, 3′-UT regions are direct markers for the 3′ ends of genes. They are usually not interrupted by introns and are often sufficiently divergent between different members of a gene family to allow locus-specific analysis.

Most cDNA libraries are constructed from cDNA molecules that have been initiated by priming from the poly(A) tail present at the 3' end of the mRNA. For all clones recovered from these libraries, it is straightforward to obtain sequence information for a few hundred basepairs of the 3′-UT region directly adjacent to the poly(A) tail. This sequence information can be used to design a pair of PCR primers that can be used, in turn, to amplify and sequence the same region from a different strain or species of mice such as *M. spretus* or *M. m. castaneus*. In a comparison of 2,312 bp present in 3′-UT regions derived from 22 random mouse cDNA clones, Takahashi and Ko (1993) found an overall polymorphism rate of one change in every 92 bp. These single base changes translated into restriction site polymorphisms within nine of the 22 clones analyzed. With primers already in-hand, these newly identified polymorphisms provide PCR markers for the direct mapping of corresponding genes that are indistinguishable in their coding regions.

### 8.3.2 Detection of allelic changes defined by single basepairs

#### 8.3.2.1 Hybridization and single basepair changes

Although PCR detection of RFLPs is an improvement over Southern blot detection, the real advantage of PCR lies within its nearly universal ability to discriminate alleles differing by single base changes even when they do not create or destroy any known restriction site. In fact, most random basepair changes will be of the non-RFLP type, and before the advent of PCR, there was no efficient means by which these alleles could be easily followed in large numbers of samples. It was this limitation that led originally to the development of the PCR protocol (Saiki et al., 1986).

The inability to detect single base changes on Southern blots was a consequence of both theoretical limitations inherent in the process of hybridization as well as practical limitations in the sensitivity of nucleic acid probes and the elimination of background noise. With Southern blot analysis, the sensitivity at which target sequences can be detected within a defined sample is directly proportional to the length of the probe. For example, a 1 kb probe will hybridize to ten times the amount of target sequence as a 100 bp probe (having the same specific activity), and this will lead to a signal which is ten times stronger. It is for this reason that it is always best to use the longest probes possible for traditional Southern blot studies as well as for other protocols such as in situ hybridization. Signal strength is important not simply to reduce the amount of time required for autoradiographic exposure, but also to allow detection over the background "noise" inherent in any hybridization experiment. If conditions are at all less than optimal, the signal to noise ratio will drop below 1.0 as the probe size is reduced below 100–200 bp.

The only forces holding the two strands of a DNA double helix together are the double or triple hydrogen bonds that exist within each basepair. Individual hydrogen bonds are very weak, and it is only when they are added together in large numbers that the double helix has sufficient stability to avoid being split apart by normal thermal fluctuations. Thus, for DNA molecules having a size in the range from a few basepairs up until a critical value of ~ 50 bp, the length itself plays a critical role in the determination of whether the helix will remain intact or fall apart. However, once this upper boundary is crossed, length is no longer a factor in thermal stability. In effect, there is only a small window— ~ 10 to 40 bp—over which it is possible to obtain differential hybridization of the probe to the target based on differences in hybrid length.

But how could length make a difference in allele detection when both the target and probe lengths are held constant? The answer is that the *effective* length of a hybrid is determined by the *longest* stretch of DNA that does not contain any mismatches. Thus, when a probe of 21 bases in length hybridizes to a target that differs at a single base directly in the middle of the sequence, the effective length of the hybrids that are formed is only 10 bp. Since a 10 bp hybrid is significantly less stable than a 21 bp hybrid, it is becomes easy to devise hybridization conditions (essentially by choosing the right temperature) such that the perfect hybrid will remain intact while the imperfect hybrid will not. In contrast, the thermal stability of a 50 bp hybrid is not sufficiently different from the thermal stability of a 100 bp hybrid (of equivalent sequence composition) to allow detection by differential hybridization.

Thus, in 1985, the detection of single base differences through differential Southern blot hybridization was not possible because of two counteracting problems. First, it was only with very short probes—oligonucleotides of less than 50 bases—that single base changes provided a large enough difference to be readily detected. However, it was only with much longer probes—of several hundred bases or greater—that signal strength and signal to noise ratio were sufficient to allow specific detection of the target sequence in any allelic form within the high complexity mouse genome. How could one break this impasse?

The answer, of course, was to focus on the target sequences rather than the probe or hybridization conditions. PCR provided a means to increase the absolute amount of target sequence, as well as the target to non target ratio, by virtually-unlimited orders of magnitude. This, in turn, results in a proportional increase in potential signal strength, which, in turn, allows one to use short oligonucleotides for hybridization and which, in turn, allows for the detection of single base differences in a simple plus/minus assay.

### 8.3.2.2 Allele-specific oligonucleotides

Once alternative alleles have been sequenced and a single basepair change between the two has been identified, it becomes possible to design a PCR protocol that allows one to follow their segregation (Farr, 1991). First, PCR primers are identified that allow specific amplification of a region that encompasses the variant nucleotide site (for this application, the length of the product is not critical and can be anywhere from 150 to 400 bp in length). Next, two allele-specific oligonucleotides (ASOs) are produced in which the variant nucleotide is as close to the center as possible considering other factors described at the beginning of Section 8.3. The ideal ASO length is 19–21 bases —short enough to allow differential hybridization based on a single base change and long enough to provide a high probability of locus specificity. The two ASOs are used with defined samples to determine a temperature at which positive hybridization is obtained with target DNA containing the correct allele but not with target DNA containing only the alternative allele.

Typically, a sample of genomic DNA is subjected to PCR amplification with the locus-specific primers, aliquots of the amplified material are spotted into two "dots," and each is probed with labeled forms of each of the two ASOs. Hybridization at one dot but not the other is indicative of a homozygote for that allele, while hybridization at both dots is indicative of a heterozygote that carries both alleles.

The power of this protocol for allele detection is its simplicity. The elimination of gel running saves both time and allows for easy automation. With large amounts of target sequence, it becomes possible to use non-radioactive labeling protocols that are safer and allow for long-term storage of labeled probes (Helmuth, 1990;

Levenson and Chang, 1990). However, there are pitfalls that are important to keep in mind. First, some variants may be refractory to reproducible PCR analysis because of problems inherent in the sequence that surrounds the site of the base change. Second, plus/minus assays of any kind are subject to the problem of false negatives. (One can always insert a gel-running step prior to hybridization to be certain that amplified material is actually present in the aliquot under analysis.) Finally, in the analysis of mice derived from anything other than a defined cross, there is always the risk that a third novel allele will exist that cannot be detected by either of the two ASOs developed for the analysis. An animal heterozygous for such a novel allele (along with one of the two known alleles) could be falsely characterized as homozygote for the one known allele present, since the protocol is not quantitative. Nevertheless, even with these pitfalls, the PCR/ASO protocol remains a useful tool for genetic analysis.

### 8.3.2.3 The oligonucleotide ligation assay

An alternative protocol for the detection of well-defined alleles that differ by single base changes has been developed by Hood and his colleagues (Landegren et al., 1988, 1990). This method, called the oligonucleotide ligation assay (OLA) or ligase-mediated gene detection, is predicated on the requirement for proper base pairing at the 3'-end of one oligonucleotide as well as the 5'-end of an adjacent oligonucleotide before ligase can work to form a covalent phosphodiester bond. The conceptual framework behind the protocol is illustrated in Fig. 8.8. First, the potential target sequence is amplified by PCR. Next hybridization is carried out simultaneously with two oligonucleotides complementary to

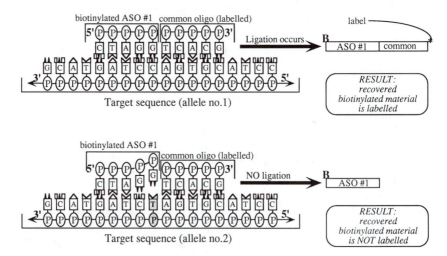

**Figure 8.8**  Ligase-mediated gene detection. Illustration of the molecular basis for this technique. The top illustration shows a target sequence that has hybridized without mismatch to two oligonucleotides, so ligation is possible. The bottom illustration shows a target sequence with a single base change located at the 3'-end of the allele-specific oligonucleotide. The inability to base pair at this position prevents ligation.

sequences that are adjacent to each other and directly flank the variant nucleotide. The variant base itself will be either complementary or non-complementary to the most 3'-base of the first allele-specific oligonucleotide. This ASO is modified ahead of time with an attached biotin moiety but is not otherwise labeled. The second oligonucleotide, which is labeled radioactively or non-isotopically, extends across an adjacent sequence that is common to both alleles under analysis. Ligase is also present in the reaction, and if both oligonucleotides are perfectly matched with the target sequence, the ligase will create a covalent bond between them. If a mismatch occurs at the junction site, the two oligonucleotides will *not* become ligated. Biotinylated material can be easily and absolutely separated from non-biotinylated material with the use of a streptavidin matrix and the resulting sample can then be tested for the presence of the label associated with the second oligonucleotide.

There are two main advantages to using the ligase-mediated detection protocol as a substitute for the PCR/hybridization protocol described in the previous section. First, the chance of a false positive arising from the OLA protocol is essentially zero. Second, the OLA protocol is highly amenable to automation. However, as in all plus/minus assays, proper controls are critical to rule out the possibility of false negatives.

### 8.3.2.4 The ligase chain reaction

By combining OLA together with the exponential amplification strategy of PCR, a new technique has been developed that is referred to as the ligase chain reaction LCR (Barany, 1991; Weiss, 1991). Like OLA, detection of nucleotide differences with the LCR protocol is based upon a requirement for perfect pairing at the two sites that flank the break between two oligonucleotides in order for ligase to form a phosphodiester bond between them as illustrated in Fig. 8.8. The difference is that in the LCR protocol, four oligonucleotides are used corresponding to the regions that flank the polymorphic site on *both* strands of the target DNA molecule. Therein lies the mechanism of amplification. If the target sequence provides a match, both sets of flanking oligonucleotides will become ligated. After denaturation, each of the newly created double-length oligonucleotides can now act as a template for a new set of oligonucleotides to basepair and ligate. Thus, LCR proceeds by rounds of annealing in the presence of a heat-stable ligase followed by denaturation and then annealing again. The only difference in the thermocycler pattern from that used for PCR is the elimination of the elongation step. At the end of the process, the products of LCR can be detected easily in the same manner used for OLA as shown in Fig. 8.8. In contrast, detection of polymorphic sites with PCR requires a follow-up protocol—either hybridization to an allele-specific oligonucleotide or restriction digestion followed by gel electrophoresis.

The essential difference between LCR and the original OLA protocol is sensitivity: LCR requires far less starting material since the product is amplified during the protocol. The advantages of the LCR protocol over PCR are several. First, like OLA, LCR will not produce false positives. Second, because the product of LCR is directly assayable without further detection schemes, the process is much more amenable to automation and is much more likely to be quantitative.

The disadvantage of LCR is that it can only be used to detect single base substitutions that have been previously characterized by sequence analysis.

### 8.3.3 Single strand conformation polymorphism

#### 8.3.3.1 Historical background

There are many circumstances where it is most useful to be able to follow genes—in contrast to anonymous sequences—directly within experimental crosses. A number of different approaches have already been described but they all have limitations. Gene-associated RFLPs are detected between different species—*M. spretus* and traditional inbred *M. musculus* strains, for example—at a reasonable frequency, but are much more difficult to find among the inbred *M. musculus* strains themselves. To use any of the approaches dependent on ASOs, it is first necessary to sequence the locus in question from different strains of mice, identify basepair variants, synthesize the ASOs as well as other locus-specific primer(s), test their specificity and optimize the reaction conditions for each locus. And at this point, one still only has a protocol for distinguishing two allelic states.

A geneticist's perfect protocol for detection and analysis of polymorphisms at any locus would satisfy the following criteria. First, it would allow the detection of any and all basepair variants in a DNA region as multiple alleles. Second, it would not require prior sequence information from each allele. Third, it would not require the synthesis of ASOs. Fourth, the assay protocol itself would require no special equipment or special skills above and beyond that found in a standard molecular biology laboratory. Finally, the assay would be rapid and the results would be readily reproducible.

All of these criteria have been satisfied, to a good degree, with a simple protocol that takes advantage of the fact that even single nucleotide changes can alter the three-dimensional equilibrium conformation that single strands will assume at low temperatures (Orita et al., 1989a). If a sample of DNA is denatured at high temperature and then quickly placed onto ice, reformation of DNA hybrids will be inhibited. Instead, each single strand will collapse onto itself in what is often called a random coil. In fact, it is now clear that each single strand will assume a most-favored conformation based on the lowest free energy state. Presumably, the most favored state is one in which a large number of bases can form hydrogen bonds with each other. Even a single nucleotide change could conceivably disrupt the previous most favored state and promote a different one, which, if different enough, would run with an altered mobility on a gel. Different allelic states of a locus that are detected with this protocol are called *single-strand conformation polymorphisms* (SSCPs) (Beier et al., 1992; Beier, 1993).

#### 8.3.3.2 Denaturing gradient gel electrophoresis

The development of the SSCP protocol was an outgrowth of an earlier technique that allowed the detection of single base changes in genomic DNA upon electrophoresis through an increasing gradient of denaturant (Fischer and Lerman, 1983). This technique is called denaturing gradient gel electrophoresis (DGGE). Small changes in sequence have dramatic effects on the point in the denaturing gradient at which particular double-stranded genomic restriction fragments would split into single strands. With the attachment of a "GC-clamp"—composed of a stretch of tightly bonding G:C basepairs—the DNA fragment could be held together with a double helix in the clamped region attached to the open single strands present in the melted region. This two-phase molecule would be very resistant to further migration in the gel and would essentially *stop in its track*. Two allelic forms of a genomic fragment that differed by even a single basepair would undergo this transition at different denaturant points and this would be observed as different migration distances in the denaturing gradient gel. In the original protocol, different allelic forms were detected directly within total genomic DNA upon Southern blotting and hybridization to a locus-specific probe. At the time DGGE was developed, there was no other means available for detecting basepair changes that did not alter restriction sites, and thus DGGE expanded the polymorphism horizon. Unfortunately, DGGE requires the use of custom-made equipment and is tedious to perform on a routine basis.

In recent years, the DGGE protocol has been modified for use in conjunction with PCR as a means for the initial detection of allelic variants among samples recovered from different individuals within a population (Sheffield et al., 1989). The differential migration of PCR products can be detected directly in gels with ethidium bromide staining and variant alleles can be excised from the gel for sequence analysis. The main advantage to the use of DGGE is that nearly all single base substitutions can be detected (Myers et al., 1985). Thus, it is ideal for the situation where one wants to search for rare variant alleles among individuals within a population without the need for sequencing through all of the wild-type alleles that will be present in most samples. Nevertheless, DGGE does not scale up easily and, thus, it is not the method of choice when another less labor-intensive protocol can also be used for the detection of allelic variants. In many cases, the better protocol will be SSCP.

### 8.3.3.3 The SSCP protocol and its sensitivity

One of the main virtues of the new SSCP detection protocol developed by Orita et al. (1989a, 1989b) is its simplicity. In its basic form, PCR products are simply denatured at 94°, cooled to ice temperature to prevent hybrid formation, and then electrophoresed on standard non-denaturing polyacrylamide gels.

The two strands of the PCR product will usually run with very different mobilities and base changes can act to further alter the mobility of each strand. Thus, what appears to be a single PCR product by standard analysis can split into four different bands on an SSCP gel, if the original DNA sample was heterozygous for base changes that altered the mobility of both strands.

Various studies have analyzed pairs of PCR products known to differ by single base substitutions to obtain an estimate of the fraction that can be distinguished by the SSCP protocol. In one such study, 80% of 228 variant PCR products were distinguishable (Sheffield et al., 1993). In another study, the rate of detection was 80–90% (Michaud et al., 1992). However, when this last set of samples was analyzed under three different electrophoretic conditions, the detection rate was an astonishing 100%.

### 8.3.3.4 A powerful tool for the detection of polymorphisms among classical inbred strains

When SSCP analysis was performed on a set of PCR products amplified from either the 3′-untranslated or intronic regions of 30 random mouse genes, 43% showed polymorphism in a comparison of the classical inbred strains B6 and DBA, and 86% showed polymorphism between B6 and *M. spretus* (Beier et al., 1992). The rate at which SSCP polymorphisms are detected between B6 and DBA is much greater than that observed with RFLP analysis, and in the same ballpark as the polymorphism frequencies observed for microsatellites described in Section 8.3.6.

There are a number of advantages to the SSCP approach over other systems for detecting polymorphisms: (1) SSCP has potential applicability to all unique sequences, both within genes and non-genic regions; (2) PCR primers can be designed directly from cDNA sequences; and (3) if one amplified region does not show polymorphism, one can always move upstream or downstream to another. However, it is still likely to be the case that microsatellites—described in Section 8.3.6—will have larger numbers of different alleles and this makes them more useful in straightforward fingerprinting approaches. Together, SSCP and microsatellite analysis provide a powerful pair of PCR-based tools for classical linkage analysis with recombinant panels derived from both intra- and interspecific crosses.

### 8.3.4 Random amplification of polymorphic DNA

With all of the PCR protocols described so far, there is an absolute requirement for pre-existing sequence information to design the primers upon which specific amplification depends. In 1990, two groups demonstrated that single short random oligonucleotides of arbitrary sequence could be used to prime the amplification of genomic sequences in a reproducible and polymorphic fashion (Welsh and McClelland, 1990; Welsh et al., 1991; Williams et al., 1990). This protocol is called random amplification of polymorphic DNA (RAPD). The principle behind the protocol is as follows. Short oligonucleotides of random sequence will, just by chance, be complementary to numerous sequences within the genome. If two complementary sequences are present on opposite strands of a genomic region in the correct orientation and within a close enough distance from each other, the DNA between them can become amplified by PCR. Each amplified fragment will be independent of all others and, by chance, will likely be of different length as well; if few enough bands are amplified, all will be resolvable from each other by gel electrophoresis. Different oligonucleotides will amplify completely different sets of loci.

RAPD polymorphisms result from the fact that a primer hybridization site in one genome that is altered at a *single* nucleotide in a second genome can lead to the elimination of a specific amplification product from that second genome as illustrated in Fig. 8.9. If, for example, the random primer being used has a length of 10 bases, then each PCR product will be defined by 20 bases (10 in the primer

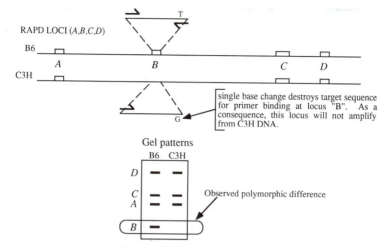

**Figure 8.9**   Detection of polymorphic loci with the RAPD technique. Corresponding chromosomes from B6 and C3H are indicated schematically as horizontal lines. Boxes on each line represent genomic fragments (RAPD loci) that can be amplified with a particular primer. Locus B is polymorphic. An illustration of the gel pattern that would be obtained with amplifiable products from B6 and C3H is shown.

target at each end) that are all susceptible to polymorphic changes.[60] The resulting polymorphism will be detected as a di-allelic +/− system.

If one starts with the assumption that complete complementarity between primer and target is required for efficient amplification, it becomes possible to derive a general equation to predict the approximate number, $A$, of amplified bands expected as PCR products from a genome of complexity $C$ that is primed with a single oligomer of length $N$.[61] For amplified fragments of 2 kb or smaller in size[62] the equation is:

$$A = \left( \frac{4{,}000C}{16^N} \right) \tag{8.1}$$

Let us use $2 \times 10^9$ as an estimate for the complexity of the single-copy portion of the mouse genome (see Section 5.1.2) and solve equation 8.1 for primer lengths that vary from eight to 11. With $N = 8$, the predicted number of PCR products is 18,626—far too many to resolve by any type of gel analysis. With $N = 9$, the equation predicts 116 PCR products, which is still too high a number. With $N = 10$, the prediction is 7.2 products, and with an 11-mer, the prediction is 0.45 products. Thus, the use of random 10-mers would be most appropriate for obtaining a maximal number of easily resolvable bands from the mouse genome.

Optimizations of the complete RAPD protocol, from the parameters upon which primer sequences are chosen to the conditions used for PCR, have been published (Williams et al., 1990; Nadeau et al., 1992). It is actually possible to obtain multiple PCR products with primers longer than 10-mers when relaxed reaction conditions are used to allow amplification from mismatched target sequences. In fact, one group has suggested that 12–14-mer primers are optimal (Nadeau et al., 1992). It is also possible to increase the predicted number of products by a simple factor of three by including two unrelated random primers of the same length in each PCR reaction. However, in any case where the number of visible PCR products goes above 12–20, it would become necessary to use polyacrylamide gels, rather than agarose gels, in order to clearly resolve each band; thus, the trade-off for the detection of more loci is a more time-consuming analysis. In the end, the protocol that requires the least amount of time for typing *per locus* is the one that should be chosen. Since different laboratories often excel at different techniques, the optimal conditions for RAPD analysis should be determined independently in each laboratory.

A comprehensive RAPD analysis of the two most well-characterized inbred strains—B6 and DBA—has been performed with 481 independent 10-mers used singly in PCR reactions (Woodward et al., 1992). An average of 5.8 PCR products per reaction were observed, which is not very different from that predicted from equation 8.1. In a direct strain comparison, 95 reproducible differences were observed between B6 and DBA among the complete set of 2,900 discrete bands detected. Assuming that each polymorphism results from a single nucleotide change in one of the two primer targets and all such changes are detectable, it becomes possible to calculate the average sequence difference between these two strains at 1.6 changes per 1,000 nucleotides. This low level of polymorphism is not unexpected given the high degree of relatedness known to exist among all of the classical inbred strains (see Section 2.3.4).

Using RAPD as a method for detecting polymorphisms between B6 and DBA would appear to be rather inefficient—on average, only one polymorphism was detected among every five primer reactions that were run. A second negative factor is that all RAPD polymorphisms are binary +/− systems. Thus, as discussed above for RFLPs, a polymorphism detected between one pair of strains may not translate into use for another pair of strains. Furthermore, it is not possible to distinguish animals that are heterozygous at any locus from those that are homozygous for the "+" allele. Thus, on average, only half of the RAPD polymorphisms detected between two strains would be mappable among offspring from a backcross to one parent, and with an intercross mapping system, the RAPD approach is even more limited (see Section 9.4.3).

Nevertheless, there are many features that speak to the utility of the RAPD approach. Foremost among these is the relative speed and ease with which results can be obtained—there is no need for blotting or radioactive hybridization, and a complete analysis from start to finish can be performed within a single working day, unlike RFLP or minisatellite studies. Unlike all other PCR-based protocols, RAPD primers are not dependent on the results of costly cloning and sequencing studies, and once they are obtained, the main cost per sample is the DNA polymerase used for PCR. Thus, even in comparisons of inbred strains, the RAPD protocol may be more efficient in the long run relative to other techniques for generating random DNA markers. Additionally, cloning of RAPD fragments can be rapidly accomplished after the simple recovery of ethidium bromide-detected bands. Cloned RAPD loci will have an advantage over minisatellites and microsatellites in that RAPD loci need not, and most will not, contain repetitive sequences.

As is the case with traditional RFLP loci, the interspecies level of RAPD polymorphism is much greater than that observed among the traditional inbred strains. The data of Serikawa et al. (1992) indicate a five-fold increase in the number of polymorphic bands observed in comparisons between *M. spretus* and traditional laboratory strains; this increase parallels the known increase in genetic diversity. Thus, the RAPD technology will be even more efficient for marker development in crosses that incorporate one parent that is not derived from one of the traditional inbred strains.

Like minisatellite analysis, the RAPD protocol can provide genomic fingerprints that simultaneously scan loci dispersed throughout the genome. In an analysis of 32 representative inbred strains maintained at the Jackson Laboratory, Nadeau and colleagues (1992) defined 29 unique strain fingerprints with the use of only six primers. Thus, RAPD would appear to provide an efficient and easy means by which to monitor the genetic purity of inbred lines on an ongoing basis.

Finally, it should be mentioned that while the RAPD protocol is a useful, and important, addition to the arsenal of tools available for genetic analysis of the mouse, it is of vastly greater importance for genetic studies of other species, including both animals and plants, that are not well-characterized at the DNA level. For these other species, the RAPD technology can provide a unique method for the rapid development of genetic markers and maps even before DNA libraries and clones are available.

### 8.3.5 Interspersed repetitive sequence PCR

The principle behind the RAPD approach is that oligonucleotides having essentially a random sequence will be present at random positions in the genome (of the mouse and every other species) at a frequency that can be predetermined mathematically. Thus, by choosing oligonucleotides of an appropriate size and by running PCR amplification reactions under the appropriate conditions, one can control the number of independent genomic fragments that are amplified such that they can be optimally resolved by a chosen system of gel electrophoresis. Although useful for linkage analysis, the RAPD approach does not allow the discrimination of mouse sequences from those of other species and, thus, it cannot be used as a means for recovering mouse genomic fragments from cells that contain a defined portion of the mouse genome within the context of a heterologous genetic background.

An alternative approach that, like RAPD, also allows the simultaneous PCR amplification of multiple genomic fragments is based on the natural occurrence of highly repeated DNA elements that are dispersed throughout the genome. Three families of mouse repeat elements—B1, B2, and L1—are each present in approximately 100,000 copies (see Section 5.4). Amplification of these elements in and of themselves with repeat-specific primers would not be useful, first, because their copy numbers are too great to be resolved by any available gel system and, second, because there would be no way of distinguishing most individual elements from each other since most would be of the same consensus size. However, if instead one used a repeat-specific primer that "faced-out" from the element, one would amplify only regions of DNA present between two elements that were sufficiently close to each and in the correct orientation to allow the PCR reaction to proceed. The number of instances in which two elements would satisfy these conditions will be much lower than the total copy number since, on average, these elements will be spaced apart at distances of approximately 30 kb[63] and, for all practical purposes, PCR amplification does not occur over distances greater than 1–2 kb. By working with (1) one or a combination of two or more primers together that (2) hybridize to whole families or subsets of elements within a family from (3) two ends of the same element or from different elements, one can adjust the number of PCR products that will be generated to obtain the maximal number that can be resolved by a chosen system of gel electrophoresis (Herman et al., 1992).

This general protocol is referred to as *Interspersed Repetitive Sequence (IRS) PCR* (IRS-PCR). It was first developed for use with the highly repeated *Alu* family of elements in the human genome (Nelson et al., 1989) and was subsequently applied to the mouse genome (Cox et al., 1991). Cox and co-workers used individual primers representing each of the major classes of highly dispersed repetitive elements—B1, B2, and L1—to amplify genomic fragments from inbred B6 mice (*M. musculus*) and *M. spretus*. Although the IRS-PCR patterns obtained upon agarose gel electrophoresis were extremely complex, it was possible to see clear evidence of species specificity. To simplify the patterns, these workers blotted and sequentially hybridized the IRS-PCR products to simple sequence oligonucleotides (12-mers containing three tandem copies of a tetramer) present frequently in the genome but only, by chance, in a subset of the amplified inter-repeat regions. The simplified patterns obtained allowed the identification and mapping of 13 new polymorphic loci.

Clearly, the utility of IRS-PCR as a general mapping tool is no greater than that of the RAPD technique or any other protocol that allows the random amplification of PCR fragments from around the genome. However, the real power of IRS-PCR is not in general mapping but in the identification and recovery of mouse-specific sequences from interspecific cell hybrids as discussed in the Section 8.4.

## 8.3.6 Microsatellites: simple sequence length polymorphisms

### 8.3.6.1 The magic bullet has arrived

Although the ability to identify and type simple basepair substitutions changed the face of genetics, it has not been a panacea. Finding RFLPs within a cloned region is often not easy; when they are found, their polymorphic content is often limited and di-allelic; finally, typing large numbers of RFLP loci by Southern blot analysis is relatively labor-intensive. Non-RFLP base changes can also be difficult to find, although this task has become easier with the development of the SSCP protocol. However, most SSCPs still show a limited polymorphic content with just two distinguishable alleles. Minisatellites are much more polymorphic than loci defined by nucleotide substitutions and minisatellite probes often allow one to simultaneously type multiple loci dispersed throughout the genome. However, minisatellite elements as a class have no unique sequence characteristics and are recognized only by the Southern blot patterns they produce when they are used as probes. The number of minisatellite loci uncovered to date numbers less than 1,000. Thus, in general, minisatellites cannot provide specific handles for typing newly cloned genes or genomic regions. Other methods of multilocus analysis described previously suffer from the same limitations.

In the next chapter, it will be seen that one very important use of DNA markers is not to follow particular genes of interest in a segregation analysis but rather to provide "anchors" that are spaced at uniform distances along each chromosome in the genome. Together, these anchor loci can be used to establish "framework maps" for new crosses, which, in turn, can be used for the rapid mapping of any new locus or mutation that is of real interest. If the number of anchors is sufficient, it will only take a single cross to provide a map position for the new locus. There is no need for anchor loci to represent actual genes. Their only purpose is to mark particular points along the DNA molecule in each of the chromosomes in a genome.

There are three criteria that define perfect anchor loci. First, they should be extremely polymorphic so that there is good chance that any two chromosome homologs in a species will carry different alleles. Second, they should be easy to identify so that one can develop an appropriate set of anchors for the analysis of any complex species that a geneticist wishes to study. Finally, they should be easy to type rapidly in large numbers of individuals.

Now with the dawn of the 1990s has come what may indeed be the magic bullet that geneticists (who study the mouse as well as all other mammals) have been waiting for—a genomic element with unusually high polymorphic content, that is present at high density throughout all mammalian genomes examined, is easily uncovered and quickly typed: the microsatellite. A microsatellite—also known as a simple sequence repeat (or SSR)—is a genomic element that consists of a mono-, di-, tri- or tetrameric sequence repeated multiple times in a tandem array.

Unlike other families of dispersed cross-hybridizing elements—such as B1, B2, and L1—in which individual loci are derived by retrotransposition from common ancestral sequences (see Section 5.4), individual microsatellite loci are almost certainly derived de novo, through the chance occurrence of short simple sequence repeats that provide a template for unequal crossover events (as illustrated in Fig. 8.4) that can lead to an increase in the number of repeats through stochastic processes. In general, microsatellite loci are not conserved across distant species lines, for example, from mice to humans, and it seems unlikely that these elements—which are practically devoid in information content—have any functionality either to the benefit of the host genome[64] or in and of themselves. Microsatellites do not appear to be selfish elements (discussed in Section 5.4). Rather, microsatellites, like minisatellites, are simply genomic quirks that result from errors in recombination or replication.

Microsatellites containing all nucleotide combinations have been identified. However, the class found most often in the mouse genome contains a $(CA)_n \cdot (GT)_n$ dimer, and is often referred to as a CA repeat. The existence of CA repeats, their presence at high copy number and their dispersion throughout the genomes of a variety of higher eukaryotic species was first demonstrated a decade ago by several different laboratories (Miesfeld et al., 1981; Hamada et al., 1982; Jeang and Hayward, 1983). Although independent examples of CA-repeat polymorphisms surfaced over the following decade, it was not until 1989 that three groups working independently uncovered sufficient evidence to suggest that microsatellites as a class were intrinsically extremely polymorphic (Pickford, 1989; Weber and May, 1989; Pedersen et al., 1993). Further systematic studies have confirmed the high level of polymorphism associated with many microsatellite loci in all higher eukaryotes that have been looked at.

### 8.3.6.2 Typing by PCR

Without PCR, most microsatellites would be useless as genetic markers. Allelic variation[65] is based entirely on differences in the number of repeats present in a tandem array rather than specific basepair changes. Thus, the only way in which alleles can be distinguished is by measuring the total length of the microsatellite. This is most readily accomplished through PCR amplification of the micro-satellite itself along with a small amount of defined flanking sequence on each side followed by gel electrophoresis to determine the relative size of the product as illustrated in Fig. 8.10.

Microsatellite loci can be identified in two ways—by searching through DNA sequence databases or by hybridization to libraries or clones with an appropriate oligonucleotide such as $(CA)_{15}$. In the former case, flanking sequence information is obtained directly from the database. In the latter case, it is first necessary to sequence across the repeat region to derive flanking sequence information. A unique oligonucleotide on each side of the repeat is chosen for the production of a primer according to the criteria described at the beginning of Section 8.3. It is best to choose two primers that are as close to the repeat sequence as possible—the smaller the PCR product, the easier it is to detect any absolute difference in size. Variations in the length of PCR products can be detected by separation on NuSieve™ agarose (FMC Corp.) gels (Love et al., 1990; Cornall et al., 1991) or polyacrylamide gels (Weber and May, 1989; Love et al., 1990). Agarose gels are easier to handle, but polyacrylamide gels provide

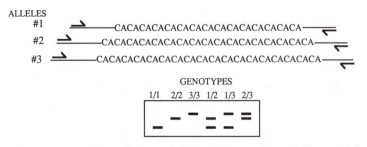

**Figure 8.10** Microsatellite alleles and their detection. Three different alleles at a microsatellite locus composed of a CA repeat. The half-arrows represent the locus-specific primers used for amplification. Illustrations of the various gel patterns that would be observed with different allele combinations are indicated.

higher resolution. When alleles are difficult to resolve with native gels, it is often possible to improve the level of resolution by running denaturing gels. Bands are detected by ethidium bromide or silver staining of gels, or by autoradiography of PCR products formed with labeled primers.

An even higher level of cost efficiency can be achieved by combining two or more loci for simultaneous analysis through multiplex PCR. Samples can be combined before the PCR reaction —if the different primer pairs have been shown not to cause combinatorial artifacts—or after the PCR reaction but before the gel run. In all cases, the entire process is amenable to automation.

### 8.3.6.3 Classification and frequency of microsatellites

Microsatellites can be classified first according to the number of nucleotides in the repeat unit. Mononucleotide and dinucleotide repeat elements are quite common; with each subsequent increment in nucleotide length—from trinucleotide to tetranucleotide to pentanucleotide—the frequency of occurrence drops quickly. *Perfect* microsatellites are those that contain a single uninterrupted repeat element flanked on both sides by non-repeated sequences (Weber, 1990). A large proportion of microsatellite loci are *imperfect* with two or more runs of the same repeat unit interrupted by short stretches of other sequences. The polymorphic properties of imperfect microsatellites are determined by the longest stretch of perfect repeat within the locus. Not infrequently, microsatellites are of an *imperfect* and *compound* nature, with a mingling of two or more distinct runs of different repeat units.

The most common microsatellites in the mouse genome are members of the dinucleotide class. With complementarity and frame-shift symmetry, there are only four unrelated types of dinucleotide repeats that can be formed— $(CA)_n:(GT)_n$, $(GA)_n:(CT)_n$, $(CG)_n:(GC)_n$, and $(TA)_n:(AT)_n$. Of these four, two are not useful as microsatellite markers for different reasons: $(CG)_n:(GC)_n$ is present only infrequently within all mammalian genome (as discussed in Sections 8.2.2 and 10.3.4.4), and long $(TA)_n:(AT)_n$ stretches do not allow for stable hybrid formation at the temperature normally used for PCR strand elongation.[66] Of the remaining two classes, CA repeats are found most often in the mouse genome. Furthermore, although CA repeats have been found in all eukaryotes examined, they are absent from prokaryotes. This fact greatly simplifies the task of screening for their presence in traditional *E. coli*-based libraries.

Based on a quantitative dot blot analysis, Hamada and his colleagues (1982a, 1982b) estimated the number of CA repeat loci in the mouse genome at ~ 100,000, equivalent to an average of one locus every 30 kb. Another estimate of CA-repeat copy number was obtained by scanning 287 kb of mouse genomic sequences entered in GenBank for $(CA)_n$, where $n$ was six or greater (Stallings et al., 1991). This analysis found CA repeats once every 18 kb on average. The difference between these two estimates can be accounted for entirely by sequences having only 6–9 repeats, which are too short to be detected by the hybridization-based dot blot analysis (Weber, 1990).

The second most frequent microsatellite class in the mouse genome is the GA repeat, which occurs at a frequency of approximately half that observed for CA repeats (Cornall et al., 1991). GA-repeat loci are just as likely to be polymorphic as CA-repeat loci. Thus by screening for both simultaneously, one can increase the chances of finding a useful microsatellite by 50%.

The mononucleotide repeat poly(A·T) is found in the mouse genome at a frequency similar to, if not greater than, the CA repeats. However, it is often contained within the highly dispersed B1, B2, and L1 repeats, which are themselves present in ~ 100,000 copies per haploid genome. Thus, random screens for poly(A·T) tracts will frequently land investigators in these more extensive repetitive regions where it will be difficult to derive locus-specific primer pairs for PCR analysis. Nevertheless, if one is aware of this pitfall, it becomes possible to use computer programs to assist one in this task, and it is often possible to type microsatellite-containing B1 (or B2 or L1) elements (Aitman et al., 1991). Mononucleotide repeats, both within and apart from the more complex repeat elements, are just as likely to be polymorphic as dinucleotide repeats (Aitman et al., 1991). However, a second potential pitfall with long poly(A·T) tracts is that, as is the case with long $(TA)_n \cdot (AT)_n$ dinucleotide tracts, there is a reduced melting temperature which necessitates the use of PCR elongation steps under conditions of reduced specificity, leading to an increased incidence of artifactual products. As a consequence of these pitfalls, poly(A·T) tracts have been used much less frequently as a source of polymorphic microsatellite markers. The microsatellite poly(C·G) is not associated with either of these pitfalls, but it is much less frequently observed—by an order of magnitude—in the mouse genome (Aitman et al., 1991).

Tri- and tetranucleotide repeat unit microsatellites are also present in the genome, but at a frequency ten-fold below that of the dinucleotide $(CA)_n$ and $(GA)_n$ loci (Hearne et al., 1992).[67] As such, they will be represented much less often in genomic libraries and individual clones. However, once uncovered, these higher-order microsatellites are much better to work with than the dinucleotide loci. The level of polymorphism observed with the tri- and tetranucleotide loci appears similar to that observed with CA and GA repeats, but alleles are much more readily resolved with 3–4 bp mobility shifts for each repeat unit difference. Furthermore, ladders of artifactual PCR products commonly seen with dinucleotide repeats do not appear as often or as intensely with higher-order repeat unit loci (Hearne et àl., 1992).

### 8.3.6.4 Polymorphism levels and mutation rates

As is the case with minisatellite loci, the generation of new microsatellite alleles is not due to classical mechanisms of mutagenesis. Rather, the number of tandem repeats is altered as a consequence of mispairing, or slippage, during recombination or replication within the tandem repeat sequence. As illustrated in Fig. 8.4, events of this type will create new alleles by expanding or contracting the size of the locus. The frequency with which these events occur is a function of the number of repeats in the locus with a sigmoidal distribution. CA-microsatellites with 10 or fewer repeat units are unlikely to show polymorphism; with 11–14 repeat units, there is an intermediate and climbing probability of detecting polymorphism; with 15 repeat units or more, there is a maximal probability of detecting polymorphism (Weber, 1990; Dietrich et al., 1992). Thus, to maximize the probability of detecting polymorphism, one should focus analyses on CA-repeat loci having $n \geq 15$. Hybridization screens can be set up to accomplish this task by probing blots with a $(CA)_{15}$ oligonucleotide under high stringency conditions of 65°C with $0.1 \times SSC$ (Dietrich et al., 1992).

A large number of laboratories have now reported the results of investigations into the frequencies at which microsatellite polymorphisms are detected in comparisons of two or more inbred strains or mouse species. The actual results would be expected to vary depending on the method used to recover microsatellites (because this will determine the lower boundary for repeat number) and the method used to type the PCR products (because agarose gels are less resolving than polyacrylamide gels). In an analysis of over 300 CA-repeat microsatellites that are predominantly of the $n \geq 15$ class, an average polymorphism rate of approximately 50% was observed in pairwise comparisons among nine classical *M. musculus* inbred strains; the lowest level of polymorphism observed was 35% between DBA/2J and C3H/HeJ, and the highest was 57% between B6 and LP/J (Dietrich et al., 1992). Not unexpectedly, even higher levels of polymorphism were observed in pairwise comparisons between classical inbred strains and other *Mus* species or subspecies. The rate of polymorphism between B6 and *M. m. castaneus* was 77%, and between B6 and *M. spretus*, it was ~90% (Love et al., 1990; Dietrich et al., 1992). For a small but significant number of loci, the primers designed to amplify an inbred strain locus failed to amplify an allelic product from the *M. spretus* genome (Love et al., 1990); this is almost certainly due to an interspecific polymorphism in a target sequence recognized by one of the flanking primers.

A number of investigators have attempted to measure the rate at which new microsatellite alleles are created. This can be readily accomplished in the mouse where the relationships among a large number of different inbred strains have been well documented and it is possible to count the generations that separate various strains from each other (Bailey, 1978). The results of these studies indicate that the rate of *mutation* is highly variable—over at least an order of magnitude. This variability could be a consequence of genomic position effects but the mechanism of allele generation must be clarified before one can say for sure. In the most comprehensive analysis to date, Dietrich and colleagues (1992) analyzed the average rate of mutation at 300 loci within the BXD set of recombinant inbred strains. The average mutation rate was calculated at one in 22,000 per locus per generation, which is 5–50-fold greater than that normally attributed to mutagenesis at classical loci. This average microsatellite "mutation" rate is high enough to allow the generation of a large amount of polymorphism among individuals within a species, but low enough to allow one to follow the segregation of two or more alleles accurately from one generation to the next within a typical genetic cross.

### 8.3.6.5 The awesome power of microsatellites

The high level of polymorphism associated with microsatellites (as a class) represents just one component of their rapid rise to become the "genetic tool of choice" for mappers working with all animal species. Their uniqueness and power also lies within the ease with which they can be uncovered, the ease with which they can be typed, and the ease with which they can be disseminated. To develop a panel of microsatellite loci for analysis of the mouse genome, Todd and his colleagues simply searched through the EMBL and GenBank databases for entries that contained $(CA)_{10}$, $(GA)_{10}$, or their complements (Love et al., 1990). To increase the size of this panel for higher resolution mapping analysis, genomic libraries constructed to contain short inserts were screened with CA-repeat probes, and positive clones were isolated and sequenced (Cornall et al., 1991; Dietrich et al., 1992).

Using a combined panel of 317 microsatellite loci, the Whitehead/MIT Genome Center developed a first-generation whole mouse genome linkage map with an average spacing of 4.3 cM (Dietrich et al., 1992). With the publication of the oligonucleotide sequences that define and allow the typing of each locus, the markers became available to everyone in a democratic fashion. As of January 1994, the Whitehead group had defined and mapped over 3,000 microsatellite loci.[68] Up to date mapping, strain distribution and sequence information on all of these loci can be obtained electronically as described in Appendix B. Furthermore, the commercial concern Research Genetics, Inc. has made life even easier for the mouse genetics community by offering each primer pair in this panel at a greatly reduced cost relative to custom DNA synthesis.

Since microsatellite typing is PCR based, and there is usually no need for blotting or probing, results can be obtained rapidly with a minimal expenditure of often precious material and always precious man- and woman-hours. Dietrich and colleagues (1992) reported that two scientists can "genotype new crosses for the entire genome in a few weeks per cross," which represents an order of magnitude improvement over RFLP-based approaches.

Microsatellites can serve not only as tags for anonymous loci but for functional genes as well. Stallings and his colleagues (1991) found that 78% of the clones from a mouse cosmid library have CA repeats. If one also searched for GA repeats, the percentage of microsatellite-positive cosmid clones would be even greater. An even higher probability of identifying microsatellite loci—close to 100%—can be achieved with gene-containing clones recovered from larger insert libraries constructed with yeast artificial chromosomes (YACs) or special prokaryotic vectors (see Section 10.3.3). Small fragments that contain the microsatellite can be subcloned and sequenced to identify a unique set of flanking primers for genetic analysis. Microsatellites can truly be viewed as universal genetic mapping reagents.

During the 1980s, the difficulties encountered in the search for RFLPs among the classical inbred strains led to the emergence of the interspecific cross—between a M. musculus-derived inbred strain and *M. spretus*—which became a critical tool for the development of the first high-resolution DNA-locus-based maps of the mouse genome (Avner et al., 1988; Copeland and Jenkins, 1991; Section 9.3). Interspecific backcross panels still represent a powerful tool for mapping newly characterized DNA clones. However, with microsatellites, it is now possible to go back to classical crosses among *M. musculus* strains to map interesting phenotypic variants as discussed in Section 9.4.

## 8.4  REGION-SPECIFIC PANELS OF DNA MARKERS

A large fraction of the gene mapping studies performed today have as an ultimate goal the cloning of a phenotypically defined locus based on its chromosomal position. This process of *positional cloning* (discussed in detail in section 10.3) is still rather tedious, and it is usually dependent on two experimental tools that exist in the form of panels. The first panel consists of DNA samples obtained from the offspring of a cross set up to uncover recombination events between and among the phenotypically defined locus and nearby marker loci. The types of crosses that can be used and the number of offspring to be analyzed are topics of the following chapter. In all cases, analysis of a large number of offspring is required to have a reasonable chance at identifying recombination breakpoints that are close to the locus of interest.

Identification of the recombination breakpoints that lie closest to the locus of interest is dependent on the availability of a sufficient number of region-specific polymorphic DNA markers. This is the second panel of tools. Ideally, one would like to have at hand a set of markers, such as microsatellites, distributed at average distances of a few hundred kilobases apart. This would provide sufficient resolution for the mapping of recombination sites (Section 7.2.3 and Fig. 7.5) as well for the recovery of overlapping YAC clones (Section 10.3.3).

Before 1994, most regions of the genome were not covered to this degree, and it was nearly always necessary for investigators to pursue special strategies to increase the size of the region-specific marker panel. However, as this section is being written, the average whole genome density of mapped microsatellite markers has reached one per megabase, and within a year's time, it should be one per 500 kb. Furthermore, contigs of overlapping YAC clones have been developed for two complete human chromosome arms—21q and the Y (Chumakov et al., 1992; Foote et al., 1992), and it is only a matter of time before additional human chromosomes and mouse chromosomes are added to this list. If an ordered, whole chromosome library is available, one can go directly to the clones that span the region of interest to derive polymorphic marker loci. This could be readily accomplished, for example, by screening for microsatellites within these clones.

Thus, what follows will soon be of historical interest only for mouse geneticists: approaches that investigators have used in the past to generate region-specific panels of DNA markers. These approaches have been included here for two reasons. First, to enable all readers to appreciate earlier work in this area of mouse molecular genetics. Second, to describe tools that may still be critical for geneticists working on organisms whose genomes are less well characterized than that of the mouse.

All rational approaches to region-specific cloning are based on fractionating the genome such that only a single chromosome or defined subchromosomal region from the species of interest is accessible *prior* to the recovery of clones that can be tested for use as DNA markers. Genome fractionation protocols fall into several classes with certain advantages and disadvantages. The major classes of genome fractionation methods are described in the following subsections.

### 8.4.1 Chromosome microdissection

The most direct means for genome fractionation is based on "microscopic dissection" (or *microdissection* as it is commonly called) of the region of interest from spreads of metaphase chromosomes on glass slides. This technique was first developed for the isolation of polytene chromosome bands from *Drosophila* salivary gland chromosomes (Scalenghe et al., 1981), and was later modified for use with mammalian chromosomes (Röhme et al., 1984). To aid in the identification of the correct chromosome, one can start with cells from mice in which the chromosome is marked karyotypically within the context of a single Robertsonian chromosome (Röhme et al., 1984, see Section 5.2). Micro-dissection is an extremely tedious protocol that is difficult to master and it is this difficulty that is its main drawback. However, the most skilled practitioners can circumscribe the region of dissection to a few chromosomal bands. This can represent a 100-fold enrichment from the whole genome, with almost no contamination from unlinked chromosomal regions. Although chromosome microdissection was developed prior to PCR, it is when the two techniques are combined that the power of this approach becomes apparent with the potential for generating thousands of markers from a very well-defined subchromosomal interval (Ludecke et al., 1989; Bohlander et al., 1992). Detailed protocols for performing chromosome microdissection followed by cloning have been described in a monograph by Hagag and Viola (1993).

### 8.4.2 Chromosome sorting by fluorescent-activated cell sorter

A less tedious protocol for direct genomic fractionation is based on the utilization of a fluorescent-activated cell sorter (FACS) to separate the metaphase chromosome of interest away from all other chromosomes (Gray et al., 1990). The starting material for this protocol must come from a cell line in which this chromosome is physically distinguishable from all others. Sources of such chromosomes include cells from animals with an appropriate Robertsonian translocation (Bahary et al., 1992) or interspecific somatic cell hybrid lines that contain only the foreign chromosome or subregion of interest (Section 10.2.3). The material obtained from a typical FACS sort is likely to be 50–70% pure, equivalent to an enrichment factor of about ten-fold, with the remaining material due to contaminants from other chromosomes. The resolution of the FACS chromosome fractionation protocol is clearly much less than that possible with microdissection and this is its main drawback. The main advantage of this protocol is that a greater amount of material can be recovered and used directly to construct chromosome-specific large-insert genomic libraries (Bahary et al., 1992).

### 8.4.3 Somatic cell hybrid lines as a source of fractionated material

A variety of somatic cell hybrid lines have been generated that contain only one or a few mouse chromosomes on the genetic background of a different species as described in Section 10.2.2. The host genomes used most often to create somatic cell hybrid lines of use to mouse geneticists are Chinese hamster and human. The main advantage of well-characterized somatic cell hybrid lines is the ease with which they can be used, and the unlimited amount of high-quality material that they can provide. The main disadvantage is that mouse genomic material is not alone, but mixed together with the whole genome of another species. Thus, to derive mouse-specific clones for use as markers, one must choose a protocol that allows the discrimination of mouse sequences from these other sequences, be they hamster or human. This can be accomplished by enlisting the highly repetitive element families B1, B2 and L1, that are unique to the mouse genome. The earlier approaches along this line were based on the construction of whole genome libraries from the cell line and then screening for mouse-containing clones with one or more repeat sequences (Kasahara et al., 1987). Of course, once such repeat element clones were obtained, it was imperative to subclone unique flanking sequences for use as DNA markers. More recently, the IRS-PCR technique described in Section 8.3.5 has been used with great success in the rapid recovery of mouse-specific sequences from somatic cell hybrid lines (Simmler et al., 1991; Herman et al., 1992). With IRS-PCR, there is no need to first prepare a whole genome library.

An obvious limitation to the recovery of region-specific probes with IRS-PCR is that amplification will only occur between repetitive elements that are relatively close to each other and in the correct orientation. With the use of just the B2 primer, Herman and her colleagues (1991) were able to amplify approximately one PCR product for each megabase of mouse DNA present in somatic cell lines containing portions of the mouse X chromosome. With the use of other repeat element primers, alone or in combination, additional loci could be amplified (Simmler et al., 1991; Herman et al., 1992). PCR fragments can be readily excised from gels for cloning or for direct use as probes for linkage analysis.

### 8.4.4 Miscellaneous approaches

Under special circumstances, other approaches can be considered for obtaining an enrichment of sequences from particular subregions of the genome. For example, if the region is contained within a defined NotI restriction fragment (or one derived from another infrequent cutter) that is sufficiently larger than the 1 mb average, it would be possible to excise the portion of a pulsed field gel that contained this fragment followed by amplification (IRS-PCR or random sequence) and cloning. This procedure could provide as much as a ten-fold enrichment for sequences within a multiple-megabase region (Michiels et al., 1987).

In another approach, Hardies and colleagues (Rikke et al., 1991; Rikke and Hardies, 1991; Herman et al., 1992) have taken advantage of the concerted evolution of L1 sequences that occurs within a species  (see Section 5.3.3.3)to develop specific oligonucleotides that recognize L1 subfamilies that are relatively unique to the genomes of either *M. spretus* or *M. musculus* (see Section 5.4.2). These oligonucleotides can be used to probe whole genome libraries made from animals congenic for a chromosomal region of interest from one species within the genetic background of the other species. This protocol has been validated in another laboratory (Himmelbauer and Silver, 1993) and could serve to provide a small number of new markers from those limited cases where the appropriate congenic lines have been constructed. In genetic terms, congenic strains are far superior to somatic cell hybrids because the region of interest can be more greatly circumscribed. As indicated in Fig. 3.6, after ten generations of backcrossing, the differential region will have an average length of 20 cM, and after 20 generations of backcrossing, the average differential length will be reduced to 10 cM.

Finally, in theory, one should be able to enrich for a region deleted in one genome, but not another, by subtractive hybridization. This approach has been tried in various formats that are all dependent on the use of a large excess of DNA from the deleted genome to drive hybrid formation with sequences that are also present in the non-deleted or "tester genome" (Kunkel et al., 1985). If the driver sequences are tagged in some way, they can be removed from the completed reaction mixture along with the tester sequences to which they hybridized. "Target sequences" unique to the tester genome—in other words, those that have been deleted from the driver genome—will all be left behind in the solution ready for analysis or cloning.

In practice, this approach has never worked as well as one would like because the high complexity of the mammalian genome prevents the hybridization reaction from going to completion. Even when subtractive steps are reiterated, the target sequences have only been enriched by a factor of 100–1000 at the very most. Thus, in its original form, this approach has lost favor. More recently, Wigler and his colleagues have built upon the subtractive hybridization approach to develop a PCR-based technique that is much more sensitive and highly resolving (Lisitsyn et al., 1993). This new technique, called representational difference analysis (RDA), can be used to purify to *completion* sequences that are deleted from one genome but not another that is otherwise identical. In theory, this same technique could also be used in a manner analogous to that described for the L1 sequences above, for the identification and cloning of new RFLPs that are present in the differential DNA segment that distinguishes two members of a congenic pair.

# 9

# Classical Linkage Analysis and Mapping Panels

## 9.1 DEMONSTRATION OF LINKAGE AND STATISTICAL ANALYSIS

### 9.1.1 Mapping new DNA loci with established mapping panels

When a new mouse locus has been defined at the DNA level, it can be mapped by three different approaches: somatic cell hybrid analysis, *in situ* hybridization, or formal linkage analysis. The first of these approaches is not applicable generally to the mouse because single chromosome hybrids have not been gathered together in a systematic way for the whole mouse genome. However, even in those cases where such hybrids exist, this type of analysis provides only a chromosomal assignment. The second approach—*in situ* hybridization—is more highly resolving than somatic cell hybrid analysis, but this protocol requires special expertise and the resolution is still less than that obtained routinely with linkage analysis. Both of these non-sexual mapping protocols have two advantages over all forms of linkage analysis. First, they do not require any prior knowledge of map positions for other loci. Second, they allow the mapping of non-polymorphic loci. Thus, in the early days of mouse molecular genetics, before many DNA markers had been placed onto the map, and before new methods for uncovering polymorphisms had been developed, both of these protocols served useful functions in the arsenal of general mapping tools.

Today, the method of choice for mapping a new locus defined at the DNA level will always be formal linkage analysis. There are two interrelated reasons for this. First, a whole genome mouse linkage map of very high density has been developed with thousands of polymorphic DNA markers already in place and new ones being added each month (Copeland et al., 1993). The second reason lies within the existence of various mouse "mapping panels" that have been established by a number of investigators at different institutions around the world.

A mapping panel is a set of DNA samples obtained from animals that carry random recombinant chromosomes produced within the context of a specific breeding scheme. The most widely used mouse mapping panels are of two specific types. One consists of representative DNA samples derived from each strain of a recombinant inbred (RI) set or group of RI sets. The approach to mapping with RI strains will be detailed in Section 9.2. The second type of widely used mapping panel contains samples derived from the offspring of an interspecific backcross between the two species, *M. musculus* and *M. spretus*. This approach will be discussed in Section 9.3. It is also possible to design mapping panels that are based on an intercross between two $F_1$ hybrid parents obtained in an interspecific or intersubspecific outcross between two different inbred strains (Dietrich et al., 1992).

The power of mapping panels lies within the database of information that is already available for a large number of previously typed loci in members of the same defined cohort of animals. The most useful panels have been typed for at least 200 independent DNA markers and, in fact, the most well-established panels have been typed for many more. In classical genetic terminology, this can be viewed as a multihundred point cross that provides linkage maps across the complete spans of all chromosomes in the genome.

Thus, the mapping of a new locus can be accomplished simply by genotyping each of the samples in the same cohort (or a subset thereof) for *just* the new locus of interest. It is never necessary to type more than 100 animals in the initial analysis and, as discussed in Sections 9.2 and 9.4, with a well-characterized panel, one can usually obtain a map position with the typing of 50 or fewer animals. A single investigator can easily carry out such an analysis in less than a week's time with the use of either a PCR analysis or Southern blotting. The results obtained are entered into the database containing all prior mapping information on the panel and a computational algorithm is used to determine the location of the new locus within the already-established linkage map. Essentially, this is accomplished by searching for concordant segregation between alleles at the new locus and those at one or more loci that have been previously typed on the same panel. With a well-established mapping panel, a first-order map position will always be obtained. A discussion of the two most important classes of mapping panels—recombinant inbred strains and the interspecific backcross—will be presented in Sections 9.2 and 9.3 of this Chapter.

### 9.1.2 Anchoring centromeres and telomeres onto the map

As discussed in Section 5.2, all 21 chromosomes in the standard mouse karyotype (19 autosomes and the X and Y) are extremely acrocentric. Even with very high-resolution light microscopy of extended prophase chromosomes, the centromere appears to lie at one end of each chromosome. Although there must be a segment of DNA containing at least a telomeric sequence that precedes the centromere, no unique sequence loci have ever been localized to this hypothetical segment. Thus, for all intents and purposes, one can view the genetic map of each chromosome as beginning with a centromere and ending with a telomere.

In the absence of centromere and telomere mapping information, a linkage map will be unanchored. As a result, the length of genetic material that lies beyond the furthermost marker at each end of the map will not be known. However, since both centromeres and telomeres are composed of repeated simple sequences that are shared among all chromosomes, their direct mapping requires special approaches.

### 9.1.2.1 Telomere mapping

All mammalian telomeres are composed of thousands of tandem copies of the same basic repeat unit TTAGGG (Moyzis et al., 1988; Elliott and Yen, 1991). Early sequence comparisons indicated that while the basic repeat unit was highly conserved, occasional nucleotide changes could arise anywhere within the large telomeric sequence present at the end of any chromosome. Elliott and Yen (1991) realized that one particular nucleotide change, from a G to a C in the sixth position of this repeat unit, would create a *Dde*I restriction site (CTNAG) that overlapped two adjacent repeats—[TTAGGC][TTAGGG]. In the absence of such a change, the enzyme *Dde*I would not cut anywhere inside a particular telomeric region which would remain intact within a restriction fragment of 20 kb or more in size. In contrast, one or more substitutions of the type described would allow *Dde*I to reduce a telomeric region into smaller restriction fragments that could be detected by probing a Southern blot with a labeled oligonucleotide (called TELO) consisting of five tandem copies of the consensus telomere hexamer (Elliott and Yen, 1991). To date, strain-specific telomeric *Dde*I RFLPs have allowed the inclusion of telomeres from six mouse chromosomes as segregating markers in linkage studies (Eicher and Shown, 1993; Ceci et al., 1994). More recently, another repeat sequence has been identified with a subtelomeric position in all mouse chromosomes (Broccoli et al., 1992). In the future, it may be possible to develop analogous strategies for mapping telomeres with this subtelomeric repeat as well.

### 9.1.2.2 Centromere mapping with Robertsonian chromosomes

Unfortunately, the satellite sequences present within all mouse centromeres are not amenable to the same type of mapping strategy just described. The problem is that each centromere contains about eight megabases of satellite sequences (Section 5.3.4), which is about 400 times larger than a telomere.

Consequently, base substitutions away from the consensus satellite sequence will be much more numerous; this will lead to whole genome Southern blot patterns, with any restriction enzyme, that are unresolvable smears.

So, how does one go about placing centromeres onto a linkage map? One approach is to mark the centromeres of individual homologs with a Robertsonian fusion (see Section 5.2). If a test animal is heterozygous for a particular Robertsonian chromosome, the segregation of the fused centromere can be followed in each offspring through karyotypic analysis. If the Robertsonian chromosome carries distinguishable alleles at linked loci, the recombination distance between the centromere and these linked loci can be determined by DNA marker typing.

Unfortunately, this approach is complicated by the finding that local recombination is suppressed in animals heterozygous for many Robertsonian chromosomes due to minor structural differences that interfere with meiotic pairing (Davisson and Akeson, 1993). Thus, the distance between the centromere and the nearest genetic locus is likely to be underestimated by this method.

### 9.1.2.3 Centromere mapping through secondary oocytes

A second approach to determining distances between centromeres and linked markers is based on the genetic analysis of large numbers of individual "secondary oocytes," which are the products of the first meiotic division. As shown in Fig. 9.1, sister chromatids remain together in the same nucleus after the first meiotic division. Thus, in the absence of crossing over, the secondary oocyte will receive one complete parental homolog or the other, and would appear "homozygous" for all markers upon genetic analysis. However, if crossing over does occur, the oocyte will receive both parental alleles at all loci on the telomeric side of the crossover event. Thus, all telomeric-side loci that were heterozygous in the parent will also appear heterozygous in the oocyte, but all centromeric-side loci will remain homozygous. The fraction of individual oocytes that are heterozygous for a particular genetic marker will be twice the linkage distance that separates that marker from the centromere, since only half of the haploid gametes generated from a double allele oocyte will actually carry the recombinant chromatid.

How does one go about determining the individual genotypes of large numbers of secondary oocytes? There are two basic protocols. The first to be developed was based on the clonal amplification of secondary oocytes within the form of ovarian teratomas (Eicher, 1978). Ovarian teratomas result from the parthenogenetic development of secondary oocytes into disorganized tumors that contain many different cell types. The inbred LT/Sv strain of mice undergoes spontaneous ovarian teratoma formation at a very high rate. This inbred strain in and of itself is not useful for oocyte-based linkage analysis, since it is homozygous at all loci, but it is possible to construct congenic

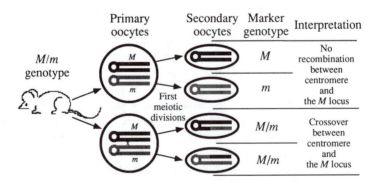

**Figure 9.1**   Illustration of centromere mapping by secondary oocyte analysis. The different meiotic outcomes possible for a single pair of homologs in females heterozygous for a marker locus *M* (with *M* and *m* alleles) are shown from the primary oocyte to the secondary oocyte stage. Sister chromatids are shown attached at their centromeres. Each single crossover event between the centromere and the marker *M* results in the formation of two secondary oocytes that contain both alleles (*Mm*). Thus, the total fraction of individual oocytes with two alleles is equal to twice the recombination distance between the centromere and the marker.

animals that are heterozygous for particular marker loci within an overall LT/ Sv genetic background. In the cases reported, these congenic animals retain the high rate of teratoma formation associated with the parental LT/Sv strain (Eppig and Eicher, 1983, 1988; Artzt et al., 1987). This approach is tedious in that a different congenic line has to be developed to map centromeres on each chromosome, but there is every reason to believe that the results obtained are an accurate measure of centromere-marker linkage distances in female mice.

An alternative protocol for genotyping oocytes is based on DNA amplification (by PCR) rather than cellular amplification. The main advantage to this approach is that genotyping can be performed on oocytes derived from any heterozygous female (Cui et al., 1992). Thus, in theory, this approach could be used to position the centromere relative to any marker on any chromosome. However, in practice, PCR amplification from single cells is difficult, and there is a high potential for artifactual results—such as amplification from one DNA molecule but not its homolog.

### 9.1.2.4 Centromere mapping by in situ hybridization or Southern blots

A third approach to positioning centromeres on linkage maps is based on direct cytological analysis. This approach is possible because of the divergence in centromeric satellite DNA sequences that has occurred since the separation of *M. musculus* and *M. spretus* from a common ancestor ~ 3 million years ago (see Section 5.3 and Fig. 2.2). In particular, the major satellite sequence in *M. musculus* is composed of a 234 bp repeat unit that is present in 700,000 copies distributed among all the centromeres. This same 234 bp repeat unit is only present in 25,000 copies spread among the centromeres in *M. spretus* (Matsuda and Chapman, 1991). The 28-fold differential in copy number can be exploited with the technique of *in situ* hybridization to readily distinguish the segregation of *M. musculus* centromeres from *M. spretus* centromeres in the offspring of an interspecific backcross. This approach has now been used to anchor all of the mouse chromosomes at their centromeric ends (Ceci et al., 1994). The only caveat to mention is the possibility that interspecific hybrids have a distorted recombination frequency in the vicinity of their centromeres.

A final possibility, that has yet to be validated, is the mapping of centromeres as RFLPs observed on Southern blots in the same manner as described for telomeres in Section 9.1.2.1. This approach may be possible with the use of a newly described repeat sequence that appears to be present in reasonable copy numbers adjacent to the centromeres of nearly every mouse chromosome (Broccoli et al., 1992).

### 9.1.3 Statistical treatment of linkage data

### 9.1.3.1 Testing the null hypothesis

Let us assume that two inbred strains of mice (B6 and C3H, for example) carry distinguishable alleles (symbolized by $b$ and $c$, respectively) at each of two fictitious loci *Xy1* and *Gh3* as shown in Fig. 9.2. An $F_1$ hybrid between B6 and C3H will be heterozygous at each locus with a genotype of $Xy1^c/Xy1^b$, $Gh3^c/ Gh3^b$. If these two loci are linked on a single chromosome, the $F_1$ hybrid will

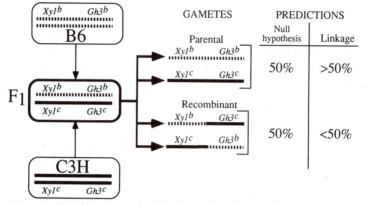

**Figure 9.2**   Predicted outcomes for two loci with linkage and no linkage. Illustration of the gametic products expected from $F_1$ hybrid animals heterozygous for two fictitious loci—*Xy1* and *Gh3*. The relative frequencies of parental and recombinant products will be a function of the linkage relationship that exists between the two loci.

have one homolog with the $Xy1^c$ and $Gh3^c$ alleles, and the other homolog with the $Xy1^b$ and $Gh3^b$ alleles. By definition, linkage means that the $F_1$ hybrid will produce a greater number of gametes carrying a parental set of alleles (either $Xy1^b\ Gh3^b$ or $Xy1^c\ Gh3^c$) than a recombinant set of alleles (either $Xy1^b\ Gh3^c$ or $Xy1^c\ Gh3^b$). As discussed at length in Section 7.2, the actual distance that separates the two loci will determine the strength of their linkage in terms of the fraction of recombinant gametes.

If one could determine the haploid genotype (or haplotype) of each and every sperm produced by a C3H × B6 hybrid male, one would know for sure whether the two loci in question are linked. But with the typing of a finite number of progeny in an experimental cross, the answer is often not as clear. Let us say that 100 offspring from the $F_1$ hybrid have been typed to test for linkage between *Xy1* and *Gh3* with the result that 62 carry parental allele combinations and 38 carry recombinant allele combinations. Do these data provide evidence in favor of the hypothesis: "*Xy1* and *Gh3* are linked"?

Unfortunately, there is a problem with a general hypothesis that states "genes *A* and *B* are linked" in that there is no precise prediction of what to expect in terms of data from a breeding experiment. This is because linkage can be very tight so that recombination would be expected rarely, or linkage can be rather loose so that recombination would be expected frequently. Of course, the strength of linkage, if indeed the genes under analysis are linked, is unknown at the outset of the experiment. In contrast, there is a precise prediction of what to expect from the so-called "null hypothesis" of *no* linkage between genes *A* and *B*. The prediction of this null hypothesis is that alleles at different genes will assort independently leading to a 50:50 ratio of gametes with parental or recombinant combinations of alleles.

Thus, whenever geneticists wish to determine whether their data provide evidence for linkage (of any degree), what they actually do is ask the following question: are these data significantly different from what one would expect if the two loci were *not* linked? With this well-defined "null hypothesis", it becomes possible to apply a statistical test to determine whether the data actually observed are significantly different from the expected outcome for no linkage. In the example above, with the analysis of 100 offspring, the null hypothesis would lead to a prediction of 50 animals with a parental allele combination and 50 animals with a recombinant allele combination in comparison with the observed results of 62 and 38, respectively. Are these two sets of numbers significantly different from each other? If the answer is yes, this would suggest that the null hypothesis is false and that the two loci are indeed linked. On the other hand, if the observed data are not significantly different from those expected from the null hypothesis, the question of linkage will remain unresolved—the two loci may be unlinked, but it may also be possible that the loci are linked and there are simply not enough data to detect it.

### 9.1.3.2 A comparison of mouse linkage data and human pedigree data

Before launching into a discussion of the statistical treatment of linkage data, it is important to illuminate a critical difference between linkage analysis in the mouse and in humans. In nearly all cases of linkage analysis in the mouse, the parental combinations of alleles—the so-called *phase of linkage*—will be known with absolute certainty. In the example above, if we assume that the two loci in question are linked, we know that the $Xyl^c$ and $Gh3^c$ alleles will be present on one homolog, and the $Xyl^b$ and $Gh1^b$ alleles will be present on the other homolog in the $F_1$ hybrid, as illustrated in Fig. 9.2. With this information, we can tell immediately upon typing whether an offspring carries a parental or recombinant combination of alleles.

More often than not, the phase of linkage is not known with certainty in the analysis of human pedigrees. As a consequence, human geneticists are forced to employ more sophisticated statistical tools that evaluate results in light of the probabilities associated with each possible phase relationship for each parent in a pedigree (Elston and Stewart, 1971). These *maximum likelihood estimation* (MLE) analyses are always performed by computer and they lead to the determination of *LOD* score graphs, which show the likelihood of linkage between two loci over a range of map distances (Morton, 1955). With most human pedigrees, it is impossible to count the actual number of recombination events that have occurred between two loci, and, as a consequence, it is impossible to determine even a *most likely* genetic distance separating two loci without the use of a computer. In contrast, all single recombination events can be clearly detected in two of the three most common types of mouse breeding protocols—the backcross and RI strains—and with the intercross, all but a small percentage of recombination events can also be distinguished unambiguously (see Fig. 9.3). With backcross and RI data in particular, linkage distance estimates can be easily determined by hand or with a simple calculator, and confidence limits around these estimates can be extrapolated from sets of tables (such as those in Appendix D).

### 9.1.3.3 The $\chi^2$ test for backcross data

The standard method for evaluating whether non-Mendelian recombination results are statistically significant is the "method of $\chi^2$." Upon calculating a value for $\chi^2$, one can use a look-up table to determine the likelihood that an observed set of data represents a chance deviation from the values predicted by a particular hypothesis. This determination can lead one to reject or accept the hypothesis that is being tested.

In its most general form, the $\chi^2$ statistic is defined as follows:

$$\chi^2 = \sum_{i=1}^{n} \frac{(obs_i - exp_i)^2}{exp_i} \tag{9.1}$$

where there are $n$ potential outcome classes, each of which is associated with an observed number $(obs_i)$ that is experimentally determined and an expected number $(exp_i)$ that is calculated from the hypothesis being tested. It is obvious from a quick examination of equation 9.1 that, as the differences between observed and expected values become larger, the calculated value of $\chi^2$ will also become larger. Thus the $\chi^2$ value is inversely related to the *goodness of fit* between the experimental results and the null hypothesis being tested, with a $\chi^2$ value of zero indicating a perfect fit. As the value of $\chi^2$ grows larger and larger, the likelihood that the experimental data can be explained by the null hypothesis becomes smaller and smaller.

Consider the case of a backcross with the (B6 × C3H) $F_1$ hybrid described above to analyze the possibility of linkage between the fictitious loci *Xyl* and *Gh3*. In terms of these two loci, the $F_1$ hybrid can produce four types of meiotic products, which will engender four experimental outcome classes (Fig. 9.2). If one makes the a priori assumption that the two parental classes represent different manifestations of the same outcome of no recombination, and the two other classes represent, for all practical purposes, reciprocal products of the same recombination event, then the data can be reduced in complexity to a set of just two outcomes—parental or recombinant.[69] In this case, the $\chi^2$ statistic becomes:

$$\chi^2 = \frac{(obs_r - exp_r)^2}{exp_r} + \frac{(obs_p - exp_p)^2}{exp_p} \tag{9.2}$$

where the $r$ subscript indicates recombinant and the $p$ subscript indicates parental.

Whenever the $\chi^2$ test is used to analyze data obtained from a backcross and the null hypothesis is one of no linkage, a further simplification of equation 9.2 can be accomplished. In this case, the *expected* values for parental and recombinant classes will both be equivalent to half the total number $(N)$ of offspring typed (which is the sum of the two observed values). Furthermore, the two observed values will both differ from the expected value by the same absolute number, and the square of each difference will yield the same positive value. Thus, the two terms in equation 9.2 can be combined to form:

$$\chi^2 = 2\frac{(obs_r - exp_r)^2}{exp_r} \qquad (9.3)$$

Equation 9.3 can be simplified even further by substituting each appearance of $exp_r$ with the equivalent expression $(obs_p + obs_r)/2$. The form of the equation that is so derived contains only the two experimentally obtained values as variables:

$$\chi^2 = \frac{(obs_p - obs_r)^2}{(obs_p + obs_r)} \qquad (9.4)$$

In plain English, equation 9.4 can be read as "square the difference, divide by the sum," and this simple calculation can often be performed through mental calculations alone.[70]

Now we can return to our example from above with 62 parental and 38 recombinant allele combinations and use equation 9.4 to determine the appropriate $\chi^2$ value. The difference between the numbers in the two observation classes (62-38 = 24) is squared to yield 576, and this value is divided by the total size of the sampled population (100) to yield 5.76.

One more piece of information is required before it is possible to translate a $\chi^2$ value into a measurement of significance—the number of "degrees of freedom" ($df$) associated with the particular experimental design. The "degrees of freedom" is always one less than the total number of potential outcome classes ($df = n -1$). The rationale for this definition is that it is always possible to determine the number of events that have occurred in the any one class by subtracting the sum of the events in all other classes from the total size of the sample set. In the backcross example under discussion, we have defined two potential outcome classes: recombinant and parental. Knowing the number in either class, along with the total sample size, provides the number in the other class. Thus, the number of degrees of freedom in this case is one.

With a $\chi^2$ value and the number of degrees of freedom in hand, one can proceed to a $\chi^2$ probability look-up table such as the one presented in Table 9.1. This table shows the $\chi^2$ values that are associated with different "$P$ values." A $P$ value is a measure of the probability with which a particular data set, or one even more extreme, would have occurred just by chance if the null hypothesis were indeed true. To obtain a $P$ value for the data set in the example under discussion, we would look across the row associated with one degree of freedom to find the largest $\chi^2$ value that is still less than the one obtained experimentally. In this case, this procedure yields the $\chi^2$ value of 3.8. Looking up the column from this $\chi^2$ value, we obtain a $P$ value of 0.05.[71] We have now reached the final goal of our statistical test for significance.

In this hypothetical example, our statistical analysis indicates that the data obtained would be expected to occur with a frequency of less than 5% if the two loci were not linked. However, is this result significant enough to prove linkage? To answer this question, it is very important to understand exactly what it is that the $\chi^2$ test and its associated $P$ value do and what they do not do. The outcome of a $\chi^2$ test cannot *prove* linkage or the absence thereof. It just provides one with a quantitative measure of significance. What is a significant result? Traditionally,

**Table 9.1** $\chi^2$ values associated with selected $P$ values for one to eight degrees of freedom ($df$)

|  | $P$ values[a] | | | |
| --- | --- | --- | --- | --- |
| $df$ | 0.05 | 0.01 | 0.001 | 0.0001 |
| 1 | 3.8 | 6.6 | 10.8 | 15.0 |
| 2 | 6.0 | 9.2 | 13.8 | 18.5 |
| 3 | 7.8 | 11.4 | 16.3 | 21.0 |
| 4 | 9.5 | 13.3 | 18.5 | 23.4 |
| 5 | 11.1 | 15.1 | 20.5 | 26.1 |
| 6 | 12.6 | 16.8 | 22.5 | 28.0 |
| 7 | 14.1 | 18.5 | 24.3 | 30.0 |
| 8 | 15.5 | 20.1 | 26.1 | 32.0 |

[a] $P$ values are determined as follows. First, the number of degrees of freedom associated with the experimental design is determined. For backcross and RI strain data, $df = 1$, and for intercross data, $df = 3$, as discussed in the text. For the evaluation of data from any one experiment, it is only necessary to look across the row of the table associated with the appropriate $df$ value. Next, the experimentally obtained value of $\chi^2$ should be compared to the values shown along this row in the table. The table values that flank the experimental $\chi^2$ value provide the limits of the interval within which the $P$ value lies. For example, if a $\chi^2$ value of 5.7 is obtained from an analysis of backcross data, the $P$ value will be less than 0.05 (but greater than 0.01). More accurate $P$ values can be obtained from graphical representations of $\chi^2$ provided in most statistics books.

scientists have chosen a $P$ value of 0.05 as an arbitrary cutoff. But with this choice, one will conclude *falsely* that linkage exists in one of every 20 experiments conducted on loci that are, in fact, not linked! As discussed below in Section 9.1.3.6, the interpretation of a $\chi^2$ value in modern genetic experiments that look simultaneously for linkage between a test locus and large numbers of genetic markers is subject to further restrictions that result from the application of Bayes' theorem.

### 9.1.3.4 The $\chi^2$ test for intercross data

It is instructive to consider the application of the $\chi^2$ test to a breeding protocol with more than two potential outcome classes. The most relevant example of this type is the intercross between two $F_1$ hybrid animals that are identically heterozygous at two loci with a genotype of *A/a, B/b*. Figure 9.3 illustrates the different types of $F_2$ offspring genotypes that are possible in the form of a Punnett square. In the absence of linkage between $A$ and $B$, one would expect each of the 16 squares shown to be represented in equal proportions among the $F_2$ progeny. If one compares the actual genotypes present in each square, one finds that there is some redundancy with only nine different genotypes in total. These are as follows (with their relative occurrences in the Punnett square shown in parenthesis): *A/A, B/B* (1), *a/a, b/b* (1), *A/a, B/b* (4), *A/A, B/b* (2), *A/a, B/B* (2), *A/a, b/b* (2), *a/a, B/b* (2), *A/A, b/b* (1), and *a/a, B/B* (1). However, as in the case of the backcross, these classes are not really independent of each other. In particular, two classes (*A/A, B/B* and *a/a, b/b*) result from the transmission of parental allele combinations from both $F_1$ parents (zero recombinants in Fig. 9.3), four classes (*A/A, B/b, A/a, b/b, a/a, B/b,* and *A/a, B/B*) result from a single recombination event in one parent or the other, two classes result from two

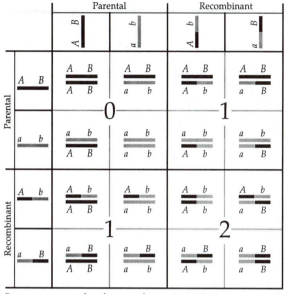

**Figure 9.3** Punnett square for the two locus $F_1 \times F_1$ intercross. The four possible gametes that can be produced by an $F_1$ hybrid mother or father heterozygous at the $A$ and $B$ loci ($A/a$, $B/b$) are shown along the left side and top of the overall square. The $F_2$ genotypes that result from the combination of each pair of gametes are shown within each smaller square. The large numbers represent the number of recombination events represented within each $F_2$ genotype. The four upper left squares have genotypes with no recombination events, the four lower left and four upper right squares have genotypes with one recombination event, and the four lower right squares have genotypes with two recombination events. However, the double heterozygous ($A/a$, $B/b$) genotype is non-informative, since it appears in both the zero and double recombinant groups.

recombination events ($A/A$, $b/b$ and $a/a$, $B/B$), and the final class ($A/a$, $B/b$) is ambiguous and could result from either no recombination in either parent or recombination events in both parents. Thus, the the nine genotypic outcomes from the double heterozygous intercross can be combined into four truly independent genotypic classes. By adding up the number of squares included within any one class and dividing by the total (16), one obtains the fraction of offspring expected in each in the case of the null hypothesis: zero recombinants—1/8; single recombinants—1/2; double recombinants—1/8; ambiguous (zero or two recombinants)—1/4. As discussed above, four outcome classes yield three degrees of freedom.

With this information, it becomes possible to set up a $\chi^2$ test to evaluate the evidence for linkage between two segregating loci typed among the progeny of an ($F_1 \times F_1$) intercross. The special $\chi^2$ equation for the intercross takes the following form:

$$\chi^2 = \frac{(obs_0 - N/8)^2}{N/8} + \frac{(obs_1 - N/2)^2}{N/2} + \frac{(obs_2 - N/8)^2}{N/8} + \frac{(obs_a - N/4)^2}{N/4} \quad (9.5)$$

where the subscript in each observational class indicates the number of recombination events ($obs_a$ is the ambiguous class) and $N$ is the total number of $F_2$ progeny typed. A comparison of the experimentally determined $\chi^2$ value with the critical values shown in the $df = 3$ row of Table 9.1 will allow a determination of a corresponding $P$ value.

### 9.1.3.5 Limitations and corrections in the use of the $\chi^2$ test

The $\chi^2$ test does have some limitations in usage. First, it cannot be applied to very small data sets, which are defined as those in which 20% or more of the outcome classes have expected values that are less than five (Cochran, 1954). With this rule, it is possible to set minimum sample sizes required for the analysis of backcross data at 10, and $F_1 \times F_1$ intercross data at 40. In actuality, a backcross or RI data set must include at least 13 samples to show significance in the case of no recombinants (based on the Bayesian correction described below). Furthermore, when the sample size is below 40 (in cases of one degree of freedom only), a more accurate $P$ value is obtained if one includes the *Yates correction* for small numbers. This is accomplished by subtracting 0.5 from the absolute difference between observed and expected values in the numerator of equation 9.3.

A final point is that $\chi^2$ analysis provides a general statistical test for significance that can be used with many different experimental designs and with null hypotheses other than the complete absence of linkage. As long as a null hypothesis can be proposed that leads to a predicted set of values for a defined set of data classes, then one can readily determine the goodness of fit between the null hypothesis and the data that are actually collected.

### 9.1.3.6 A Bayesian correction for whole genome linkage testing

If one has reason to believe *from other results* that two loci are just as likely to be linked as not, then the $P$ value obtained with the $\chi^2$ test can be used directly as an estimate of the probability with which the null hypothesis is likely to be true, and subtracting the $P$ value from the integer one provides a direct estimate of the probability of linkage. However, when a previously unmapped locus is being tested for linkage to a large number of markers across the genome, there is usually no a priori reason to expect linkage between the new locus and any one particular marker locus. If we assume a particular experimental design such that linkage is detectable out to 25 cM on both sides of an unmapped locus[72] and a total genome length of 1,500 cM, then the fraction of the genome in linkage with the novel locus will be $(25 + 25)/1,500 \approx 0.033$. In other words, out of 100 markers distributed randomly across the genome, one would expect only 3.3 actually to be in linkage with any particular test locus. But, if one accepts a $P$ value of 0.05 as providing evidence for linkage, then 5% of the unlinked 97 loci—or an additional ~5 loci—will be falsely considered linked according to this statistical test. As a consequence, the expected number of false positives—five—is larger than the expected number of truly linked loci—3.3. Thus, of the 8.3 positive markers expected, only 3.3 would be linked, and this means that a $P$ value of 0.05 has only provided a probability of linkage of 40%. This situation is clearly unacceptable.

**Table 9.2** Probabilities of linkage associated with critical $P$ values and swept radii

| P value | Swept radius [a] | | | | | |
|---------|-------|-------|-------|-------|-------|-------|
|         | 10 cM | 15 cM | 20 cM | 25 cM | 30 cM | 35 cM |
| 0.0500 | 21% | 29% | 35% | 40% | 44% | 48% |
| 0.0100 | 57% | 67% | 73% | 77% | 80% | 82% |
| 0.0050 | 73% | 80% | 84% | 87% | 89% | 90% |
| 0.0020 | 87% | 91% | 93% | 94% | 95% | 96% |
| 0.0015 | 90% | 93% | 95% | 96% | 96% | 97% |
| 0.0010 | 93% | 95% | 96% | 97% | 98% | 98% |
| 0.0005 | 96% | 98% | 98% | 99% | 99% | 99% |

[a] The swept radius is defined as the maximum distance over which linkage can be detected between two loci with 95% probability based on the power of the genetic approach used. Swept radii associated with data obtained from RI and backcross experiments are shown graphically in Figs 9.5 and 9.13, respectively.

The logical approach just discussed is referred to as Bayesian analysis after the statistician who first suggested that prior information on the likelihood of outcomes be included in calculations of probabilities. One can generalize from the example given to obtain a Bayesian equation for converting any $P$ value obtained by $\chi^2$ analysis of recombination data into an actual estimate of the probability of linkage:[73]

$$\text{Probability of linkage} = 1 - \left( \frac{P}{P + f_{\text{swept}}} \right) \tag{9.6}$$

where $P$ is the $P$ value obtained by $\chi^2$ analysis and $f_{\text{swept}}$ is the fraction of the genome over which linkage can be detected around each marker based on the power of the genetic approach used.[74] Solutions to equation 9.6 for some critical $P$ values and genomic distances are given in Table 9.2. Of interest are the $P$ values required to provide evidence for linkage with 95% probability. So long as the experimental design allows detection of linkage out to 15 cM, one can use a cutoff $P$ value of 0.001 as evidence for linkage between any two loci. In accepting linkage at $P < 0.001$, one is actually setting a limit for accepting less than one false positive result for every 20 true positive results. Later in this chapter, the Bayesian approach is used to calculate cutoff values for the demonstration of linkage with 95% probability in the case of RI strain data (Fig. 9.5) and backcross data (Fig. 9.13).

## 9.2 RECOMBINANT INBRED STRAINS

### 9.2.1 Overview

#### 9.2.1.1 Conceptualization and history of RI strains

Although a handful of isolated recombinant inbred (RI) mouse strains were developed during the 1940s and 1950s, the importance of these genetic

**Table 9.3**   Established sets of recombinant inbred strains

| Set | Extant strains | Female progenitor | Male progenitor | Number of loci typed[a] | Availability |
|---|---|---|---|---|---|
| AKXD | 22 | AKR/J | DBA/2J | 239 | JAX |
| AKXL | 16 | AKR/J | C57L/J | 287 | JAX |
| AXB/BXA | 30[b] | A/J | C57BL/6J | 209 | JAX |
| BXD | 26 | C57BL/6J | DBA/2J | 818 | JAX |
| BXH | 12 | C57BL/6J | C3H/HeJ | 329 | JAX |
| CXB | 13 | BALB/cBy | C57BL/6By | 281 | JAX |
| CXJ | 7 | BALB/cBy | SJL/J | 32 | JAX |
| CXS | 14 | BALB/cHeA | STS/A | 104 | J. Hilgers[b] |
| LXB | 3 | C57L/J | C57BL/6J | ? | JAX |
| LXPL[c] | 5 | C57L/J | PL/J | 28 | JAX |
| NX129[c] | 5 | NZB/B1NJ | 129/J | 36 | JAX |
| NX8 | 13 | NZB/ICR | C58/J | 37 | R. Riblet[d] |
| NXSM | 15 | NZB/B1NJ | SM/J | 145 | JAX |
| OXA | 14 | O20/A | AKR/FuRdA | 71 | J. Hilgers[e] |
| SWXJ | 14 | SWR/J | SJL/J | 11 | JAX |
| SWXL | 3 | SWR/J | C57L/J | 111 | JAX |

[a] This information was obtained from the Map Manager database (Manly, 1993; Appendix B) and is accurate as of August 1993 (R. W. Elliott, personal communication).
[b] Hilgers and Arends (1985), Hilgers and Poort-Keeson (1986).
[c] LXPL is called LXP, and NX129 is called NX9 by Manly (1993).
[d] Datta et al., (1982).
[e] ten Berg (1989).

constructs was not appreciated by any scientist prior to Dr Donald Bailey who first conceptualized their potential utility for linkage analysis while working at the National Institutes of Health in 1959 (Bailey, 1981). Bailey realized that established sets of RI strains could provide an efficient alternative to the lengthy process of classical breeding analysis that was required at that time to map any newly uncovered locus. Bailey moved with his first set of eight RI strains—the original CXB set—to the Jackson Laboratory in 1967. In 1969, Dr Benjamin Taylor, also at the Jackson Laboratory, took up the cause of the RI approach and set about developing, from several different pairs of progenitor parents, large sets of RI strains that are now standards for the mouse genetics community including BXD, BXH, AKXL, and AKXD (Taylor, 1978; Table 9.3). These standard RI sets, and others listed in Table 9.3 under JAX availability, can be purchased directly—in the form of either animals or DNA—from the Jackson Laboratory. Today, RI strains represent a critical tool in the arsenal used to map mouse genes and the RI approach has made its way into the study of other experimental organisms as well such as the plant *Arabidopsis* (Lister and Dean, 1993).

### 9.2.1.2 Construction and naming of RI strains

The construction of a set of RI strains is quite simple in theory and is illustrated in Fig. 9.4. One begins with an outcross between two well-established highly inbred strains of mice, such as B6 and DBA in the example

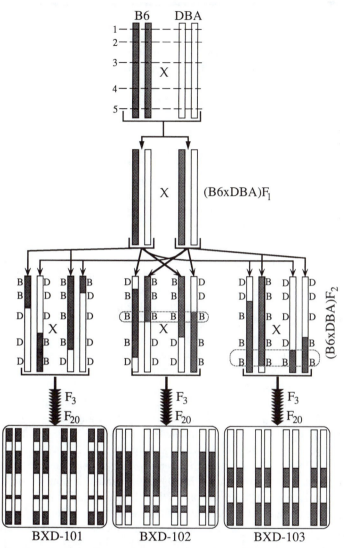

**Figure 9.4** Construction of a set of RI strains from B6 and DBA progenitors. A single pair of chromosome homologs are followed through the process of RI strain development. Genomic intervals derived from B6 are indicated by shading and intervals derived from DBA are indicated without shading for all genotypes in all generations. The map positions of five fictitious loci—numbered 1–5—are shown in the progenitor generation. In the $F_2$ generation, the genotypes at these five loci are indicated for six different animals with "B" signifying the B6 allele and "D" signifying the DBA allele. The two circled regions represent segments of the genome that are already fixed for one parental type within the two $F_2$ progenitors of particular incipient RI strains. At the $F_{20}$ generation, four separate animals with the same homozygous genotype are shown for each of the three newly constructed RI strains (fictitiously called BXD-101, BXD-102, and BXD-103).

shown. These are considered the progenitor strains. The $F_1$ progeny from this cross are all identical and thus, in genetic terms, they are all interchangeable. $F_1$ hybrid animals are bred to each other to produce a large set of $F_2$ animals. At this generation, siblings and cousins are no longer identical because of the segregation of B6 and DBA alleles from the heterozygous $F_1$ parents. As illustrated in Fig. 9.4 for a single pair of homologs, each of the $F_2$ animals will have a unique genotype with some loci homozygous for the B6 allele, some homozygous for the DBA allele, and some heterozygous with both alleles. At this stage, pairs of $F_2$ animals are chosen at random to serve as the founders for new inbred strains of mice. The offspring from each $F_2$ founder pair are maintained separately from all other offspring, and just two are chosen randomly for brother–sister mating to produce the next generation. The same process is repeated at each subsequent generation until at least 20 sequential rounds of strict brother–sister matings have been completed and a new inbred strain with special properties is established.[75]

Each of the new inbred strains produced according to this breeding scheme is called a "recombinant inbred" or "RI" strain. Different RI strains produced from the same pair of progenitor strains are considered members of the same "RI set." Each RI strain is named with an uninterrupted string of upper case letters and numbers that can be broken down into the following components: (1) a shortened version of the maternal progenitor strain name; (2) the upper case letter "X," (3) a shortened version of the paternal progenitor strain name; and (4) a hyphen and a number to distinguish each strain from all other strains in the same RI set.[76] The first three components of an RI strain name are also used alone as the name for the corresponding RI set.

Table 9.3 lists the 17 RI sets that have been used to a significant degree by members of the mouse genetics community (R.W. Elliott, personal communication). The most readily available of these are the ones that can be purchased directly from the Jackson Laboratory (JAX) in the form of live animals or DNA—14 RI sets with a total of 187 strains. The best-known and most well-characterized RI set is BXD with 26 extant strains formed from an initial cross between B6 females and DBA/2J males. Individual strains within this set are called BXD-1, BXD-2, BXD-5, and so on. (The BXD-3, BXD-4, and other missing strains are now extinct and their numbers have been retired.) By August of 1993, 818 loci had been typed on the BXD set. Unfortunately, as discussed earlier in this book (Sections 2.3.4 and 3.2.2), the two progenitors of the BXD set—B6 and DBA—are related to each other through the common group of founder animals used to develop most of the classical inbred strains at the beginning of the 20th century. As a consequence, a significant number of loci will show no variation between the two progenitors even with highly sensitive PCR protocols that detect SSCPs and microsatellites (see Section 8.3).

More recently, two progenitor mouse strains—A/J and C57BL/6J—known to differ in susceptibility to over 30 infectious or chronic diseases were used to establish a new pair of RI sets—AXB and BXA—based on reciprocal crosses (Marshall et al., 1992). Genetic surveys suggest that these progenitors are much less related to each other than B6 is to DBA, and with 31 genetically validated extant strains[77] available from the Jackson Laboratory, this paired set is the largest, and likely to be the most useful for both linkage analysis and the study of complex traits that differ between the two progenitors.

At any time, it is possible to add to the numbers of strains in a particular RI set by starting from scratch with a new cross between the original progenitor strains and following it through 20 generations of inbreeding. In this manner, J. Hilgers at the Netherlands Cancer Institute in Amsterdam and L. Mobraaten at the Jackson Laboratory have increased the size of the original CXB RI set with five and six new strains, respectively.

### 9.2.1.3 The special properties of RI strains

Like all inbred strains, RI strains are fixed to homozygosity at essentially all loci.[78] Thus, the genomic constitution of each strain can be maintained indefinitely by continued brother–sister matings, and the strain can be expanded into as many animals as required at any time. However, unlike the classical inbred strains, the genotype of an RI strain is greatly circumscribed. First, there are only two choices for the allele that can be present at each locus: thus, for every locus in every strain of the BXD RI set, only the B6 or the DBA allele can be present. Second, because there is only a limited number of opportunities for recombination to occur between the two sets of progenitor chromosomes before homozygosity sets in, complete homogenization of the genome can not take place. An illustration of this principle can be seen in Fig. 9.4. In this hypothetical situation, one can see genomic regions that are already frozen at the outset of the $F_2$ intercross in two of the three incipient RI strains. The two fixed regions are circled and (in this illustrative example) are both homozygous for B6 genomic material in the two $F_2$ parents that will act as founders for the new BXD-102 and BXD-103 strains. At each subsequent generation, additional regions will become frozen in either a B6 or a DBA state. After 20 generations of inbreeding, each RI strain will be represented by a group of animals that will all carry the same genomic tapestry with random patches from each of the two progenitors, as illustrated in the final row of Fig. 9.4.

It is this homozygous patchwork genomic structure that is the key to the power of the RI strains. This is because the boundaries that separate all of these genomic patches represent individual recombination events that are also permanently frozen. Every RI strain has different recombination sites distributed randomly throughout its genome. Thus, a set of RI strains can be used in essentially the same manner as the offspring from a mapping cross to obtain information on linkage and map distances. Like other mapping panels, it is only necessary to type each RI genotype once for a particular locus and the information obtained from typing different loci is cumulative. The major difference, of course, is that the particular genotype present within each RI strain can be propagated indefinitely whereas different offspring from a mapping cross will all have unique genotypes and finite life spans. Thus, there is no limit to the number of loci that can be typed within an RI strain. Furthermore, in the case of complex phenotypic variation, one can actually sample the same genotype multiple times to demonstrate instances of incomplete penetrance or variable expressivity as discussed in Section 9.2.5.

### 9.2.2 Using RI strains to determine linkage

#### 9.2.2.1 Strain distribution patterns

The major use of RI strains by mouse geneticists today is as a tool to determine linkage and map positions for newly derived DNA clones. As the first step in this process, an investigator should survey the progenitor strains for all the most useful RI sets to determine which of these sets can be typed for the presence of alternative alleles. The strains to be surveyed should include AKR/J, DBA/2J, C57L/J, A/J, C57BL/6J, C3H/HeJ, BALB/cJ, NZB/B1NJ, SM/J and SWR/J (see Table 9.3). If one is trying to map a newly cloned gene, it is possible to start a polymorphism search based on the detection of RFLPs among DNA samples digested with one of several different enzymes (Section 8.2). However, as discussed at length in Section 8.3, one is much more likely to uncover polymorphisms with a PCR-based protocol like SSCP.[79]

Once alternative alleles at a locus have been distinguished for the progenitors of any RI set, one can proceed to type all of the strains in that set. The information obtained from such a single locus analysis will represent the *strain distribution pattern*, or *SDP*, for that particular locus. An isolated SDP in and of itself is usually not informative. One would expect approximately half of the strains typed to carry each of the progenitor alleles at random.[80] However, with a new SDP in hand, it becomes possible to search for linkage with each of the other loci previously typed in the same RI set or group of sets. This is most easily accomplished with a computer program, such as Map Manager that compares the new SDP to each previously determined SDP in the database one by one, and applies a statistical test for evidence against or in favor of linkage (Manly, 1993; see Appendix B for further information).

#### 9.2.2.2 Concordance and discordance

The result of each pairwise comparison of SDPs is expressed in terms of the degree, or level, of *concordance* and *discordance*. When a particular RI strain has alleles from the same progenitor at two defined loci, the loci are considered to be "concordant" within that strain. When the alleles at the two loci come from the two different progenitors, they are considered to be "discordant." The probability of discordance is a function of the linkage distance that separates the two loci under analysis. This is easy to understand in terms of the likelihood that two loci will be retained, by chance, within the same genomic patch as illustrated in Fig. 9.4. At one extreme, unlinked loci—which can not possibly lie in the same genomic patch—will be just as likely to have alleles from the same progenitor, by chance, as from the two different progenitors. Thus, one would predict that ~50% of the strains within an RI set will show concordance for a pair of unlinked loci. At the other extreme, loci that are very closely linked will always be in the same genomic patch, which is equivalent to saying that 100% of RI strains will show concordance (or 0% will show discordance). Between these two extremes will be loci that are linked but less closely so. As the distance between two loci increases, the probability of discordance will increase in a calculable manner from 0% up to 50%.

Whenever one accumulates data on multiple RI strains, it is useful to express this information in terms of concordance and discordance rather than in terms of actual genotypes. The terms $N$ and $i$ are used, respectively, to denote the total number of strains typed and the number of discordant strains observed. The fraction $\hat{R} = i/N$ is used to denote the observed level of discordance. With the use of these terms, one can combine the data obtained from all sets of RI strains that show variation between progenitors for both loci under analysis. This can be accomplished even if different allele pairs are present in different RI sets. For example, at the *H-2K* locus, B6 has a *b* allele, DBA has a *d* allele, and AKR/J and A/J both have a *k* allele; even so, one can still combine *H-2K* data obtained from the BXD, AKXD, and AXB/BXA RI sets. If these three RI sets also show progenitor variation at another locus thought to be linked to *H-2K*, then one can count up the number of strains discordant between these two loci ($i$) and divide by the total number of strains ($N = 82$) to obtain the observed discordant fraction $\hat{R}$. When strains from multiple RI sets are combined for analysis in this manner, they will be referred to as an "RI group." Increasing the size of the RI group can have dramatic effects on the sensitivity and resolution with which it is possible to determine linkage and map distances, as described in the following subsections and depicted in Fig. 9.5, and Tables D1 and D2 in appendix D.

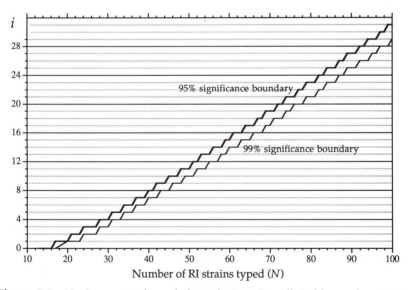

**Figure 9.5** Maximum number of discordant strains allowable to demonstrate linkage. Data are extracted from Tables 1 and 2 of Neumann (1990) and are based on a Bayesian statistical analysis. A minimum of 13 strains is required, with zero discordance, to demonstrate linkage with a probability of 95%; a minimum of 16 strains is required, with zero discordance, to demonstrate linkage with a probability of 99%. The two lines shown represent boundaries for the demonstration of linkage at each level of significance.

### 9.2.2.3  Demonstrating linkage: statistical approaches

What degree of concordance between two SDPs is required to demonstrate linkage? The problem of distinguishing a chance fluctuation above the 50% concordance expected with unlinked loci from a significant departure indicative of linkage was discussed in more general terms earlier in this chapter. Suffice it to say here that prior to 1986, researchers did not fully appreciate the more stringent requirements imposed by the Bayesian statistical approach and were misled by concordance values that passed traditional $\chi^2$ tests for significance. Mathematical formulations aimed at rectifying this situation were begun by J. Silver (1986) and were supplemented by Neumann (1990, 1991). In his 1990 paper, Neumann published tables with a complete set of maximum discordance values ($i$) that are allowed for a demonstration of linkage at various levels of significance with data obtained from RI groups up to $N = 100$ in size. Data from these tables have been extracted and shown graphically in Fig. 9.5.

With the maximum allowable discordant values that have been determined, it is possible to estimate the maximum distance over which linkage between two loci is likely to be demonstrated at a sufficient level of significance with an RI group of a particular size. This is a measure of the "swept radius," which is a concept first developed by Carter and Falconer (1951). The swept radius has been defined as the length of a chromosome interval on either side of a marker locus within which linkage can be detected with a certain level of significance. Although the swept radius was originally defined in terms of map distance, it can be readily converted into a measure of recombination fraction (with the use of an appropriate mapping function as described in Section 7.2.2.3), which is more useful for direct analysis of raw data. In this way, the swept radius can be viewed as a boundary value for the recombination fraction. If the observed rate of recombination between two loci is less than the swept radius, linkage is demonstrated at a level of significance equal to or greater than the cutoff value chosen. If the observed rate of recombination is greater than the swept radius, linkage cannot be demonstrated with the available data.[81]

The maximum discordance values allowed for each value of $N$ can be translated into linkage distances (through the use of the Haldane–Waddington equation described in the next Section) that describe swept radii at which linkage can be detected with a significance level of 95% or 99%.[82] With just 20 RI strains, one will only be likely to detect linkage with marker loci that are within 2 cM on either side of the test locus. The swept radius increases steadily as the size of the RI group climbs to 40 strains, where it becomes possible to detect linkage to markers that are within 7–8 cM of the test locus. However, even with an RI group of 100 strains, the swept radius is only 13–15 cM. In general, the distance swept by each marker locus in an RI group is only 40%–45% of the distance swept by each locus in a linkage analysis performed with an equivalent number of backcross offspring (see Fig. 9.13). This disadvantage is offset by the easy availability of the major RI sets and the ever-accumulating number of marker loci for which SDPs have been determined as discussed further below.

### 9.2.2.4 Demonstrating linkage: a practical strategy

From the preceding discussion, it should be clear that the chances of success in using RI data to demonstrate linkage for a test locus increase dramatically with both the number of strains analyzed and the number of evenly distributed SDPs that are already present in the database. As of 1993, several RI sets had been typed at over 200 loci (Table 9.3), which, if randomly distributed, would fall on the linkage map at average distances of 7 cM or less from each other. The BXD set alone has been typed for over 800 loci. Even though these marker loci are not randomly distributed,[83] their overlapping "swept diameters" of coverage are sufficient to map most new loci that are typed among all members of this set. Furthermore, RI mapping panels become ever-more efficient at detecting linkage as each new SDP is added to the database. At some point in the near future, it is likely that SDPs will be determined for all 26 BXD strains at marker loci distributed across the genome at a *maximum* interlocus distance of 5 cM. At this point, every new test locus of interest will have to lie within 2.5 cM of a previously typed marker locus and thus, by simply typing the 26 BXD strains, one will be able to determine a map position with essentially 100% probability.

Until the scenario just described is reached, it is best to maximize one's chances of demonstrating linkage by generating an SDP for the test locus over the maximal number of RI strains possible. Once data have been obtained and entered into a computer database, the first attempt to demonstrate linkage should be pursued at the highest stringency possible [equivalent to a Bayesian probability level of 99% in the Manly (1993) Map Manager program]. This will minimize the chance of picking up false linkages; if a positive result is obtained from such an analysis, one can *confidently* move on to the next task of determining a map position for the locus relative to the linked marker loci, as described in the next subsection. However, if this analysis fails to detect linkage, the stringency can be reduced step by step in subsequent runs. A positive result obtained at a lower level of significance should be considered tentative and must be confirmed (or rejected) by incorporating more RI strains into the pairwise comparison between the test locus and the putatively linked marker locus, or through an independent approach such as somatic cell hybrid analysis (Section 10.2.3), *in situ* hybridization (Section 10.2.2), or a backcross or intercross mapping panel (Sections 9.3 and 9.4). Independent confirmation of just the chromosomal assignment alone, for example, can serve to increase dramatically the significance level for an RI-based determination of linkage to a particular locus on that chromosome (Neumann, 1990). Furthermore, the failure to detect linkage to many markers distributed over large regions of the genome can provide evidence for the exclusion of the test locus from these other regions. This information, in turn, can be used to increase significance levels in cases where the direct evidence in favor of linkage is somewhat weak (Neumann, 1990).

In some cases, further evidence for, or against, linkage can also be obtained by comparing the SDP of the test locus to the SDPs associated with the two marker loci that putatively flank it on either side (Neumann, 1991). If the positioning of

| BXD- | 1 | 2 | 5 | 6 | 8 | 9 | 11 | 12 | 13 | 14 | 15 | 16 | 18 | 19 | 20 | 21 | 22 | 23 | 24 | 25 | 27 | 28 | 29 | 30 | 31 | 32 |
|---|---|---|---|---|---|---|---|---|---|---|---|---|---|---|---|---|---|---|---|---|---|---|---|---|---|---|
| *D17Leh119* | D | D | D | D | B | B | B | D | B | B | B | B | B | B | B | B | D | B | D | B | B | B | D | B | D | D |
| *D17Byu1* | D | B | D | D | B | B | B | D | B | B | B | B | B | B | B | B | D | B | D | B | B | B | D | B | D | D |
| *Plg* | D | B | D | D | B | B | B | D | B | B | B | B | D | B | B | B | D | B | D | B | D | B | B | D | D | D |
| *D17Leh66* | D | B | D | D | B | B | B | D | B | B | B | B | D | B | B | B | D | B | D | B | D | B | B | D | D | D |
| *D17Pri1* | D | B | D | D | U | B | B | D | U | B | B | B | D | B | B | B | B | D | B | D | D | B | D | B | U | D |
| *Zfp40* | D | B | D | D | B | B | B | D | B | B | B | B | D | B | B | B | D | B | D | B | D | D | B | D | U | U |
| *Hba-ps4* | D | B | D | D | B | B | B | D | B | B | B | B | D | B | B | D | D | B | D | D | B | D | B | D | D | D |
| *D17Leh12* | D | B | D | D | B | B | B | D | B | B | B | B | D | B | B | B | D | D | B | D | B | D | B | D | D | D |
| *Glo-1* | D | B | D | D | B | D | D | B | B | B | B | D | D | B | B | D | D | B | D | D | B | D | B | D | D | D |
| *Pim-1* | D | B | D | D | B | D | D | B | B | B | B | D | D | B | B | D | D | B | D | D | B | D | B | D | D | D |
| *Crya-1* | D | B | D | D | B | D | D | D | B | B | B | D | D | B | B | D | D | B | D | D | B | D | B | D | D | D |
| *H-2Eb* | U | B | D | D | B | D | D | D | B | B | B | D | D | B | B | D | D | B | D | D | D | D | B | U | U | U |
| *H-2M2* | D | B | D | D | B | D | D | D | B | B | B | D | D | B | B | D | D | B | D | D | D | D | B | D | D | D |
| *D17Leh173* | D | B | D | D | B | D | D | B | B | B | B | D | D | B | B | D | D | B | D | D | D | D | B | D | D | D |
| *D17Mit11* | U | B | D | D | B | U | D | B | B | B | B | D | D | B | B | D | D | B | D | D | D | D | B | U | U | U |
| *D17Tu16* | D | B | D | D | B | D | U | B | B | B | B | D | B | B | U | D | D | B | D | D | D | D | B | D | U | U |
| *Tpx-1* | D | B | D | D | B | D | U | B | B | B | B | D | B | B | U | D | D | B | D | D | D | D | B | D | U | U |
| *Iapls1-3* | D | B | D | D | B | B | D | B | D | B | B | D | D | B | B | B | D | B | D | D | D | D | B | D | D | D |
| *Tcte-1* | D | B | D | D | B | B | D | B | D | B | B | D | D | B | B | B | D | B | D | D | D | D | B | D | D | D |
| *D17Leh154* | D | D | D | D | B | B | D | B | D | B | B | D | D | B | B | B | D | B | D | D | D | D | B | D | D | D |
| *D17Leh23* | D | D | D | D | B | B | D | B | D | B | B | D | D | B | B | B | D | B | D | D | D | D | B | D | D | D |
| *Upg-1* | D | D | D | D | B | B | D | B | D | B | B | D | D | B | B | B | B | D | D | D | D | D | B | D | D | D |
| *D17Mit10* | U | D | D | D | B | B | D | B | D | B | B | D | D | B | B | B | D | B | D | D | D | D | B | U | U | U |
| *D17Mit6* | U | D | D | D | B | B | D | B | D | B | B | D | D | B | B | B | D | B | D | D | D | D | B | U | U | U |
| *D17Mit7* | U | D | D | D | B | B | D | B | D | B | B | D | D | B | B | B | D | D | D | B | D | D | B | U | U | U |
| *Ckb-rs2* | D | D | D | D | B | B | D | B | D | B | B | D | D | B | B | B | D | D | D | B | D | D | B | D | D | D |
| *Hprt-rs1* | D | B | B | B | B | B | D | B | D | B | B | D | D | B | B | B | D | D | D | D | D | D | B | D | D | D |

**Figure 9.6**   BXD RI data matrix for chromosome 17 strain distribution patterns. Each column shows the results obtained for a particular RI strain of the BXD set. Each row shows the results obtained for the particular locus named at the left; the data set obtained with each locus is called a strain distribution pattern or SDP. Loci are listed in order from the most centromeric at the top. A B6 allele is indicated with a "B," a DBA allele is indicated with a "D," and an unknown allele is indicated with a "U." Immortalized crossover sites are indicated with a horizontal line. Additional loci in this region of Chr 17 have been typed on the BXD RI set with SDPs that are identical to ones shown here (Himmelbauer et al., 1993).

the test locus between the flanking loci is correct, one would expect to see mostly single recombination events among the three loci within individual RI strains. If the association of the test locus with this genomic region is false, one would expect to see a substantial number of double recombination events that separate the test locus from both markers. This type of analysis is most easily visualized in the form of a data matrix, discussed just below and illustrated in Fig. 9.6. It is important to point out, however, that interference does not operate in the formation of RI strain genotypes, since the crossover events that produced each patchwork genome occurred over multiple generations. Thus, true double recombination events over short distances are not strictly forbidden; they are just less likely.

### 9.2.3 Using RI strains to determine map order

Once linkage has been demonstrated among three or more loci, one can move on to the next step of determining their order along the chromosome relative to each other. This can be accomplished computationally with a program like Map Manger (Manly, 1993), but it is also possible to carry out this analysis without a dedicated computer program. The first task is to set up an adjustable $2 \times 2$ data matrix of the kind illustrated in Fig. 9.6. This sample data matrix contains a subset of the actual data obtained with Chr 17 loci typed on the 26 RI strains present in the BXD set. Each row represents an independently determined SDP for the locus indicated at the left. Each column represents the complete genotype determined for an individual RI strain over the ~ 32 cM region that encompasses the loci shown. It is customary to order loci in the data matrix with the centromeric end of the chromosome at the top and the telomeric end at the bottom, when this information is known from other results.

A new data matrix can be initiated for any set of SDPs that have been shown to form a linkage group. It is also possible to expand an established data matrix of marker loci with the inclusion of new test locus SDPs. At the outset, SDPs can be placed into the data matrix according to a first best guess of their genetic order relative to each other. If data are entered into a computer spreadsheet, one should manually shift the order of SDP-containing rows until an arrangement is found that minimizes the total number of recombination events within the whole data set. In addition, instances of triple recombination events over short distances in individual strains should be eliminated if possible. In this manner, it is often possible to arrive at an undisputed order for an extended series of loci as is the case, for example, with the eight loci from *D17Leh119* to *D17Leh12* and the eight loci from *D17Leh173* to *D17Leh23* shown in Fig. 9.6. The computer program Map Manager will carry out this process automatically and will also allow manual adjustment of order when this is desired.

There will sometimes be cases where two or more different orders appear equally likely. For example, in Fig. 9.6, one could remove the *Glo1* and *Pim1* loci from the position shown, and place them instead between *H2M2* and *D17Leh173*. Both placements require an unsightly triple crossover: in the position shown in the figure, this occurs in strain BXD-12; in the new genetic position, this would occur in BXD-27. Both placements also require an equal number of recombination events so it is not possible to choose one order over the other based on the BXD data alone. In this case, other mapping data have been used to confirm the map order shown in the figure.

In general, determining map order with accuracy is increasingly more difficult with linked SDPs that are increasingly more discordant relative to each other. Increased discordance is indicative of increased interlocus distances, which make it more likely that multiple recombination events will have occurred along individual chromosomes; these will complicate the analysis. However, as is the case with all aspects of RI analysis, the more strains that are typed, the more accurate order determinations can become. For example, if SDP data obtained from the AKXL RI set had been considered in conjunction with BXD data, they would have provided unambiguous evidence in favor of the correct placement of *Glo1* and *Pim1* proximal to *Crya1*, as shown in Fig. 9.6.

With the accumulation of large numbers of RI SDPs and a comprehensive two by two analysis or linkage, it becomes possible to build linkage maps that span many loci distributed over large chromosomal regions as illustrated in Fig. 9.6. By looking down any column, one can clearly see the genomic patches that derive from each of the two progenitor strains. At one extreme, four strains— BXD-2, BXD-12, BXD-18, and BXD-25—appear to have fixed three separate crossover sites in their genomes leading to the presence of four alternating B6 and DBA genomic patches. At the other extreme, 12 strains—including BXD-1, BXD-8, and ten others—appear to have inherited this chromosomal region intact, without recombination, from either the B6 or DBA progenitor.[84] The remaining ten strains have fixed either one or two crossover sites leading to two or three genomic patches, respectively.

Through the visualization of RI data in this way, one can pick up suspicious results that may be due to experimental error. For example, the presence of a B6 allele at the *Upg-1* locus in BXD-22 requires a double crossover that encompasses only this locus and no others. One would be advised to go back and retype this locus on a new sample from this RI strain.

By looking horizontally across the data matrix and comparing pairs of SDPs, one can visualize the degree of concordance that exists between nearby loci. At three separate junctures—*D17Leh12/Glo-1*, *Tpx-1/Iapls1-3*, and *Ckb-rs2/Hprt-rs1*, the BXD RI data alone are not sufficient to demonstrate linkage according to the limits shown in Fig. 9.5; in all these cases, data from other mapping experiments provided evidence of linkage between two or more loci present on opposite sides of these junctures. There are also numerous examples of loci that have the same SDP; as described in the next section, loci with shared SDPs across only 26 typed strains can actually be quite distant from each other in terms of map distance (see Fig. 9.8).

The data matrix for this portion of Chr 17 also illustrates how the power of the RI approach to linkage determination increases with the number of loci that are typed. For example, the SDPs for *D17Leh119* and *Hba-ps4* are distinguished by four discordant strains (BXD-2, BXD-18, BXD-21, and BXD-25). Thus, according to Fig. 9.5, these data alone do not provide sufficient evidence for linkage between these two loci at the 95% significance level. However, when the *Plg* SDP was added to the database, it provided the required evidence for linkage of *D17Leh119* to *Hba-ps4* through their common linkage—with 99% significance—to the newly typed locus.

### 9.2.4  Using RI strains to determine map distances

#### 9.2.4.1  From discordance to linkage distance

As alluded to earlier in this Section, when two loci are known to be linked, the level of discordance that is observed between their SDPs can be equated with a mean estimate of the distance that separates them. In fact, this distance estimate can be made even in those cases where the RI data alone are not sufficient to provide evidence for linkage. Thus, RI data are useful for estimating recombination distances between loci that have been linked by non-breeding methods, such as physical mapping or cytogenetic analyses. However, it is not sufficient to simply determine the fraction of strains that show discordance and use this directly as an estimate of the recombination fraction. The problem is that, during the generation of each RI strain, an average chromosomal region will have multiple opportunities to recombine as it passes through several generations in a heterozygous state (see Fig. 9.4).

Interestingly, long before the conceptualization of RI strains, Haldane and Waddington (1931) derived a mathematical solution to this problem in the context of determining the probability with which a recombinant genotype would become fixed after successive generations of inbreeding. This solution was formulated in the following equation, where $r$ is the probability of recombination in any one gamete and R is the fraction of RI strains that are predicted to be discordant:

$$R = \frac{4r}{1 + 6r} \tag{9.7}$$

Equation 9.7 illustrates two points. First, the expected fraction of discordant strains is dependent on only a single variable—the probability of recombination between the two loci under analysis. In turn, since interference in any one generation is nearly 100% over the distances analyzed by RI analysis (see Section 7.2), the probability of recombination can be converted directly into a centimorgan linkage distance ($d$) with an $r$ value of 0.01 defined as equivalent to one centimorgan. Thus:

$$d \cong 100r \tag{9.8}$$

Second, for values of $r$ that are smaller than 0.01, equation 9.7 can be approximated by the simpler $R \approx 4r$. Thus, a 1 cM centimorgan distance becomes amplified into a predicted discordance frequency of ~4%. As Taylor (1978) pointed out, this four-fold amplification can be interpreted to mean that during RI strain development, a locus will be transmitted, on average, through four heterozygous animals (with four chances for a recombination event in its vicinity) before it is fixed to homozygosity.

The amplification of the linkage map serves to enhance the usefulness of the RI approach in the analysis of closely linked loci. For example, in a group of 100 RI strains, recombination sites will be distributed at average distances of 0.25 cM, which is four times more highly resolving than that possible with an equivalent number of backcross animals. However, this same amplification has the negative consequence of limiting the usefulness of the RI approach in studying loci that are more distantly linked to each other. For example, at a distance of 25 cM ($r \cong 0.25$), the predicted discordance level for RI strains would be 40% ($R = 0.4$), a value which is perilously close to the 50% expected with unlinked loci. As a consequence, the per locus swept radius obtained with RI strains will be much less than that obtainable with an equal number of backcross offspring.

In most cases of RI analysis, an investigator wants to go from a discordance fraction to an estimate of linkage distance. The experimentally determined discordance fraction $\hat{R} = i/N$ provides an estimate of the *true* value $R$, which is based on the *actual* probability of recombination $r$. By substituting $\hat{R}$ for $R$ in equation 9.7, one can obtain a corresponding recombination fraction estimate $\hat{r}$. This can be accomplished more easily if equation 9.7 is inverted to yield $r$ as a function of $R$ or, through direct substitution, $\hat{r}$ as a function of $\hat{R}$:

$$\hat{r} = \frac{\hat{R}}{4 - 6\hat{R}} \tag{9.9}$$

Another useful formulation of this same equation allows one to obtain the estimated recombination fraction as a function of the sample size, $N$, and the number of discordant strains, $i$:

$$\hat{r} = \frac{i}{4N - 6i} \tag{9.10}$$

Finally, a corresponding linkage distance estimate in centimorgans $(\hat{d})$ can be derived by multiplying the $\hat{r}$ value obtained in equation 9.10 by 100.

The graph in Fig. 9.7 provides a rapid means for determining a linkage distance estimate from values for $i$ and $N$ that are commonly obtained in RI analyses. Just place a ruler over the graph so that it crosses the experimental value for $N$ along the top and bottom axes, then observe the point at which the ruler crosses the curve associated with the experimentally determined value for

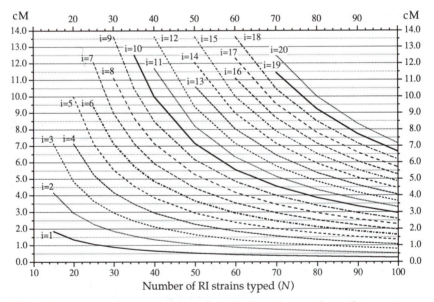

**Figure 9.7**   Linkage distance estimates from RI discordance values. The experimentally determined number of discordant strains ($i$) found in a comparison of SDPs for two loci within a total of $N$ RI strains can be used with this graph to obtain a maximum likelihood value for linkage distance as described in the text. Distances are indicated in centimorgans, with the simplifying assumption of complete interference over the interval shown.

*i.* From this point, look across to the Y axis to read off the linkage distance in centimorgans.

### 9.2.4.2 How accurate is the linkage distance estimate?

Once a value for linkage distance has been obtained from RI data, the next question an investigator will ask is: how accurate is this value? The answer to this question will be critical to investigators who want to use RI data to evaluate possible relationships between the cloned gene they have just mapped and other previously mapped loci that are defined strictly in terms of a mutant phenotype. (A detailed discussion of the general strategy used to evaluate such relationships has been left to Section 9.3.4 and is illustrated in Fig. 9.10.)

The accuracy of an experimentally determined value such as linkage distance can be quantified in terms of a "confidence interval," which is defined by lower and upper boundaries—called confidence limits. To calculate a confidence interval from a body of data, one must first choose the "confidence coefficient," or level of confidence, that one wishes to attain. The confidence coefficient represents the probability with which the associated interval is likely to contain the value of the true recombination fraction or linkage distance. Each confidence coefficient will produce a different confidence interval.

In general, two particular confidence coefficients are used most often for the evaluation of data obtained from sampling experiments aimed at estimating an absolute *real* value (like linkage distance). The first is based on the standard error of the mean ($s_{\bar{x}}$) for a normal (bell-shaped) probability distribution around a mean value $\bar{x}$. When the lower limit of the interval around a normal distribution is set at $\bar{x} - s_{\bar{x}}$, and the upper limit is set at $\bar{x} + s_{\bar{x}}$, the corresponding confidence coefficient is always equal to 68%. It is standard practice to display this confidence interval within a single term: $\bar{x} \pm s_{\bar{x}}$. Twice as often as not, the 68% confidence interval will contain the true value being estimated; this interval provides an investigator with an overall sense of the accuracy of the mean estimate predicted from experimental data.

As discussed at length in Appendix D and by other authors (Silver, 1985), data obtained from linkage studies based on small sample sizes and low levels of discordance are not well approximated by "normal" probability distributions (see Figs D1 and D2). Unfortunately, the standard deviation does not provide an accurate measure of lower and upper confidence limits in the context of non-normal probability distributions (Moore and McCabe, 1989, p. 41). In its place, I have used more appropriate estimates for lower and upper confidence limits associated with an actual 68% confidence interval.

The second confidence interval of critical importance for interpreting experimental data is one that encompasses the range of values likely to contain the *actual* recombination fraction with a probability of 95%. The more stringent 95% confidence interval is used often to define the generally accepted limits *beyond which* the actual value of *r* is unlikely to lie. A discussion of the statistical approach used to determine confidence limits for both RI strain and backcross data is presented in Appendix D along with tables of minimum and maximum values for 68% and 95% intervals. By interpolating between the numbers presented in either Table D1 or Table D2, one can derive confidence limits for pairs of $i$ and $N$ values generated in an analysis of 20–100 RI strains.

### 9.2.4.3 What is the meaning of 100% concordance?

One special case of RI strain results deserves particular attention—when complete concordance is observed between the SDP patterns obtained for two different loci. In this special case, the Haldane–Waddington formulation (equations 9.8 and 9.9) leads to a linkage distance estimate of 0 cM. However, if the two loci under analysis were independently derived and known not to be identical, an estimate of zero distance clearly makes no sense. Based on intuitive grounds alone, one would expect there to be a strong likelihood that the two loci are actually separated from each other by a significant distance, especially when the total number of RI strains typed is small.

An accurate estimate of the median expected recombination fraction $\bar{r}$ (leading directly to an estimate of the median expected linkage distance $\bar{d}$) that separates two completely concordant loci can be obtained by determining the midline of the area under the associated probability density function as discussed in appendix D and illustrated in Fig. D.1. The results of this calculation for the range of 15 to 100 RI strains are presented graphically in Figure 9.8 along with upper limits for 68% and 95% confidence intervals.

As an example, consider the implications of these statistical formulations in the case of complete concordance between two SDP patterns for the set of 26 BXD RI strains. The median estimate of linkage distance between two concordant loci is 0.66 cM (or 1.3 megabases based on a conversion of 1 cM to 2.0 mb). The 68% confidence interval extends from 0.2 cM to 1.76 cM (or 400 kb

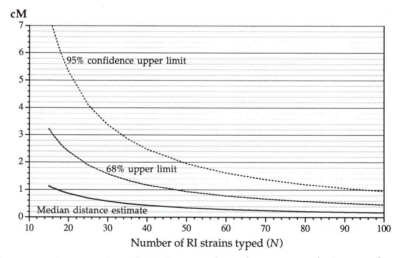

**Figure 9.8**   Interpretation of 100% concordance between two loci mapped on a panel of RI strains. The median linkage distance estimate and the upper limits of the 68% and 95% confidence intervals are indicated for complete concordance between two loci ($i = 0$) observed among a group of $N$ RI strains. Distances are shown in centimorgans.

to 3.5 mb), and the 95% confidence interval extends from 0.02 cM to 3.95 cM (or 40 kb to 7.9 mb from the computer program presented in Appendix D). These numbers confirm the intuitive suspicion that two concordant loci typed in only a small sample set probably do *not* map on top of each other.

Although a small group of RI strains is not sufficient to demonstrate close linkage, this conclusion does become more appropriate as the number of concordant RI strains increases. With 100 completely concordant strains, the median estimate of linkage distance is reduced to 0.17 cM or 340 kb, the maximum limit on the 66% interval is reduced to 0.45 cM or 900 kb, and the maximum confidence limit for the 95% interval becomes 0.95 cM or 1.9 mb.

### 9.2.4.4 A comparison of RI and backcross predictions of linkage

One can gain a better perspective on the accuracy of RI-determined linkage distance estimates by comparing a map obtained with the BXD set of 26 strains to a more accurate map obtained with a set of 374 backcross animals for the same 11 loci distributed over a span of 19–31 cM as shown in Fig. 9.9. There are five instances here in which a single RI discordance has occurred to yield an interlocus linkage distance estimate of 1.0 cM (*Plg—D17Pri1*, *D17Pri1—Hba-ps4*, *Piml—Cryal*, *Cryal—H2M2*, and *Tcte1—D17Mit10*) with a 68% confidence interval extending from 0.7 to 3.4 cM, and a 95% interval extending from 0.2 to 6.6 cM. In three of these cases, the backcross estimate lies within the 68% confidence interval and in the other two cases, it lies within the 95%

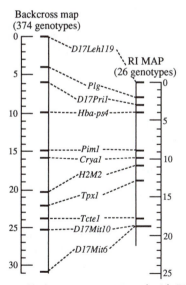

**Figure 9.9** Comparison of linkage maps generated with RI data and backcross data. Two separately generated linkage maps over the same set of loci in the proximal portion of Chr 17 are shown. The RI map is based on SDP data presented in Fig. 9.6. The backcross map is based on data obtained by Himmelbauer and Silver (1993). The scales on either side of the Figure are in centimorgans with the most centromeric locus—*D17Leh119*—arbitrarily placed at both zero points.

interval. There are two instances in which two RI discordances have occurred to yield an interlocus linkage distance estimate of 2.2 cM (*D17Leh119—Plg* and *H2M2—Tpx1*) with a 68% confidence interval extending from 1.4 to 5.3 cM and a 95% interval extending from 0.6 to 9.6 cM. In both cases, the backcross estimate lies within the 68% interval. There are another two instances in which four RI discordances yield an interlocus linkage distance of 5.0 cM (*Hba-ps4—Pim1* and *Tpx1—Tcte1*) with a 68% confidence interval extending from 3.3 to 9.7 cM, and a 95% interval extending from 1.7 to 17 cM. In one case, the backcross estimate lies within the 68% interval, and in the other case, it lies within the 95% confidence interval. Finally, there is a single instance of complete RI concordance between *D17Mit10* and *D17Mit6*. The backcross estimate of linkage distance between these two loci lies outside the 95% confidence interval (but within the 99% confidence interval according to results obtained from the computer program listed in Appendix D).

In summary, backcross estimates lie within the 68% confidence interval of the RI estimates in six (60%) of the ten pairwise comparisons, and in all but one of the remaining comparisons (90%), the backcross estimates lie within the 95% confidence interval. This level of accuracy is about as close as one can get to that predicted from statistical formulations. Furthermore, in the single case where the two mean estimates of linkage distance are significantly different from each other (0.66 cM versus 5.9 cM for *D17Mit10—D17Mit6*), their associated 95% confidence intervals (0.02—3.95 cM and 3.9—8.8 cM) do overlap (barely), and the data taken together would suggest that the actual recombination frequency is somewhere between the two extreme mean values.[85]

### 9.2.5  Using RI strains to dissect complex genetic traits

#### 9.2.5.1  Susceptibility, predisposition, penetrance, and expressivity

When Bailey first conceived of RI strains, it was with the notion that they would be useful for the analysis of the many forms of complex phenotypic variation that distinguish different inbred strains from each other. In the past, the use of RI strains for this purpose had been rather limited because of the absence of an overall framework of genetic markers on which phenotypic differences could be mapped. However, at present, the availability of highly polymorphic, rapidly typed DNA markers is allowing the construction of framework maps of marker loci that span essentially whole genomes in each of the major RI sets. These framework maps will finally open up RI strains to the use originally conceived of by Bailey.

There is an enormous reservoir of susceptibility differences to a variety of disease conditions, both chronic and infectious, among the classical inbred strains. For example, the A/J strain is relatively susceptible to various carcinogen-induced cancers (lung adenomas, sarcomas, and colorectal tumors), parasites (*Giardia, Trypanosoma,* and *Plasmodium*), bacteria (*Listeria* and *Pseudomonas*), viruses (*Ectromelia* and *Herpes*), and fungi (*Candida albicans*), as well as gallstones and teratogen-induced cleft palate; the B6 strain is relatively resistant to all of these conditions (Mu et al., 1993). On the other hand, B6 is relatively susceptible to other parasites, bacteria, viruses, and fungi as well as atherosclerosis, diabetes, and obesity; the A/J strain is relatively resistant to all of these conditions. In all, strain-specific differences in susceptibility to over 30 infectious or chronic diseases have been identified between A/J and B6, and the genetic basis for each can be approached with the use of the combined AXB/BXA set of RI strains. Differences in disease susceptibility exist among all of the traditional inbred strains, and the genes involved in many of these differences can be approached as well with the appropriate RI sets.

With most of the conditions described above, a particular genetic constitution only *predisposes* an individual to express a disease. This means that some individuals that carry the predisposing genotype will actually *not* express the disease. The fraction of genotypically identical individuals that express a particular trait defines the penetrance of that trait from that genotype. When a particular genotype guarantees the expression of a phenotype in 100% of the animals that carry it, the phenotype is considered to be *completely penetrant.* In all other cases, a phenotype is considered to be *partially penetrant* or *incompletely penetrant.* For example, a particular substrain of BALB/c mice is predisposed to a particular form of induced cancer known as a plasmacytoma. However, only 60% of these inbred animals actually produce the cancer upon induction. Thus, the penetrance of induced-plasmacytoma in these mice is 60%.

The cousin of incomplete penetrance is *variable expressivity.* Variable expressivity describes the situation in which multiple individuals all express a particular trait but in a quantitatively distinguishable manner. For example, a tumor may appear at a young age or an old age, a birth defect such as cleft palate may be more or less severe. Variable expressivity can also be measured for traits that do not show an either/or type of wild-type/mutant variation. For example, there are many strain-specific differences in physiological parameters and behavior that are strictly quantitative. Thus, blood cholesterol levels may vary among different strains as will the average number of pups that a female has in a litter (see Table 4.1).

Both incomplete penetrance and variable expressivity can be caused by genetic as well as non-genetic factors. Inbred strains allow one to distinguish genetic factors clearly, since measurements can be made of the mean level of expression or penetrance of a trait in one strain relative to another when populations of both are maintained under identical environmental conditions.

In cases where two strains differ quantitatively in penetrance levels and/or expressivity for a particular trait, it becomes difficult to design traditional breeding crosses that can uncover the loci involved. For example, if strain A shows 20% penetrance for a trait and strain B shows 80% penetrance for the same trait, then its expression in offspring from a cross between the two strains would not provide straightforward information as to which predisposing allele(s) is present. In contrast, each RI strain provides an unlimited number of animals with the same homozygous genotype. Thus, through the analysis of a sufficient number of animals, it becomes possible to quantify the levels of penetrance and expressivity, and associate distinct measurements of mean and standard deviation with each RI genotype. Furthermore, it is just as easy to map recessive traits as dominant traits, since RI strains are completely homozygous.

RI strains are also useful in those cases where multiple animals must be sacrificed in order to make a single phenotypic determination. This will be true for certain biochemical assays (although in most cases today, microtechniques allow analysis on tissues obtained from single animals) and for other assays that require a determination of multiple test points in which each point is a single animal. An example of the latter would be an $LD_{50}$ determination for a particular toxic chemical.[86]

If every RI strain in a set expresses a trait with essentially the same penetrance and expressivity as one of the two progenitor strains, and approximately half of the RI strains resemble one progenitor and half resemble the other, determining a map position for the responsible locus is no different from that described earlier in the case of DNA marker loci. Data of this type can be viewed as evidence in favor of a single major locus that is responsible for the difference in susceptibility, penetrance, or expressivity between the two progenitor strains. One can simply write out an SDP for the phenotype and then subject this SDP to concordance analysis with the SDPs obtained for all previously typed markers as described in Section 9.2.2. Once linkage is demonstrated, gene order and map distances can be determined as described in Sections 9.2.3 and 9.2.4.

### 9.2.5.2 Further genetic complexity: polygenic traits

There are two forms of RI strain data that are indicative of a more complex basis of inheritance, which may be impossible to resolve using only the RI approach. The first occurs when there is a significant departure from a balanced SDP in that the phenotype expressed by one progenitor strain is found in many more RI strains than the alternative phenotype. Data of this type would suggest that the expression of the rarer phenotype requires the simultaneous presence of two or more genes from the appropriate progenitor. One can calculate the probability of occurrence of a phenotype that requires the action of two or more unlinked loci through the law of the product as $(0.5),^n$ where $n$ is the number of loci required.[87] Thus, if two unlinked B6 loci are both required for susceptibility to a particular viral infection (relative to DBA), only $(0.5)^2 = 25\%$ of the BXD RI strains would be expected to show susceptibility. Unfortunately, for obvious reasons, unbalanced SDPs cannot be compared directly for linkage relationships with normal single-locus marker SDPs.[88]

The second form of RI data indicative of genetic complexity is the occurrence of strains that show a level of penetrance or expressivity that is significantly different from both of the progenitors. Since every RI strain can be considered homozygous for one progenitor allele or the other at every locus, data of this type will also implicate the action of multiple genes. The simplest explanation for these results is that different combinations of alleles from the two progenitors cause the different levels of phenotypic expression. For example, with the involvement of two loci, X and Y, in the expression of a trait that distinguishes the strains A/J and B6, there will be four relevant genotypes among the AXB/BXA RI strains—$X^A Y^A$, $X^A Y^B$, $X^B Y^A$, and $X^B Y^B$. Two of these genotypic combinations are different from that found in either progenitor, and one or both could be responsible for a novel phenotypic expression.[89]

Many variations upon these examples are possible. Thus, every complex trait will have to be approached independently to formulate a reasonable hypothesis for inheritance. In some cases, RI strains may still provide an appropriate tool for genetic analysis, but in most cases it will be necessary to move to a different form of analysis that may require the establishment of a new breeding cross as described in Section 9.4.

## 9.3 INTERSPECIFIC MAPPING PANELS

### 9.3.1 Overview

The "interspecific mapping" approach was conceived of by François Bonhomme (Bonhomme et al., 1979) working in Montpellier, France. Bonhomme had discovered that two clearly distinct mouse species—*M. musculus* and *M. spretus*—could be bred together in the laboratory to form fertile $F_1$ female hybrids (Bonhomme et al., 1978). The two parents involved in the generation of these $F_1$ animals are so evolutionarily divergent (Fig. 2.2) that polymorphisms in the form of RFLPs can be readily identified between them with the great majority of mouse DNA probes. Thus, by backcrossing these $F_1$ females to one parental strain, it becomes possible to follow the segregation and linkage of almost any group of cloned loci (Avner et al., 1988; Copeland and Jenkins, 1991).

For historical reasons, the *M. musculus* representative chosen for use in most interspecific crosses has been B6 (Bonhomme et al., 1979; Copeland and Jenkins, 1991; Nadeau et al., 1991), although there is no reason why other traditional inbred strains cannot be used instead (Hammer et al., 1989; Moseley and Seldin, 1989). The initial outcross is always set up between a B6 (or other traditional inbred strain) female and an inbred M. spretus male [(written as B6 × SPRET)]; the outcross is carried out in this direction because of the greater fecundity associated with traditional *M. musculus* inbred females relative to *M. spretus* females. In the subsequent generation, a backcross is performed between an $F_1$ female (since $F_1$ males are sterile) and either a *M. spretus* or B6 male; the standard written descriptions of these entire two generation protocols are: [(B6 × SPRET) × SPRET] and [(B6 × SPRET) × B6], respectively.

An "interspecific mapping panel" is typically composed of DNA samples obtained from 100–1,000 $N_2$ offspring from this backcross. Aliquots of each sample are digested typically with one restriction enzyme at a time, electrophoresed on gels, and transferred to Southern blots, which can be sequentially probed with radioactively labeled DNA clones. As more and more loci are typed, and as segregation patterns are compared, linkage groups will begin to emerge. As the number of typed markers approaches several hundred, all will begin to coalesce into a series of only 20 linkage groups that each correspond to a single mouse chromosome. (A more detailed discussion of the actual numbers of loci and animals required for linkage determination will be presented in Section 9.4.) The correct assignment of linkage groups with their associated chromosomes depends upon the incorporation into the mapping panel analysis of previously assigned anchor loci.

It should be emphasized that each member of an interspecific mapping panel typically survives only in the form of DNA. Thus, the power of these panels is limited to the analysis of cloned loci. To map loci defined solely by a variant phenotype, one would have to choose an alternative approach. In most cases, it will be necessary to set up a new cross *from scratch* as described in Section 9.4.

By the end of 1990, over 600 cloned loci had been typed on the single interspecific mapping panel maintained by Jenkins and Copeland at the Frederick Cancer Research Center (Copeland and Jenkins, 1991). By the end of 1993, the interspecific mapping panels maintained by Jenkins and Copeland (Copeland et al., 1993), as well as those maintained by several other investigators, had all been typed for at least 1,000 loci. With 1,000 or more loci in a database, one can be virtually assured of finding a correct linkage relationship for any new test locus that is put through an analysis of the same mapping panel.

### 9.3.2 A comparison: RI strains versus the interspecific cross

Many investigators will want to obtain a high-resolution map position for their newly characterized DNA clone without having to set up their own cross, and without having to invest a substantial amount of time, energy, and money. For all such investigators, typing an established mapping panel will be the method of choice. But, which mapping panel should be used? One possibility is to type one or more sets of RI strains, as discussed in the previous Section. The second possibility is to type one of the well-established interspecific backcross mapping panels. Each approach has its advantages and disadvantages.

#### 9.3.2.1 Genetic considerations

In terms of ease of polymorphism discovery, the interspecific approach provides a clear advantage over the RI approach. As discussed previously, it is often difficult to uncover RFLPs between the progenitors of RI strains. Furthermore, the identification of a RFLP between one set of RI progenitors is often not useful for the analysis of other RI sets. Thus, even when RFLPs have been uncovered, the total number of RI strains that can be analyzed is often quite limited; it can be as few as 26 and it is rarely more than 80. In contrast, the ease of RFLP identification between the progenitors of the interspecific cross was the main impetus to the initial use of this mapping approach. Furthermore, one need only identify a single type of polymorphism to type the entire interspecific panel.

With the newer PCR-based approaches to polymorphism identification discussed in Section 8.3, it is now easier to identify differences between RI progenitor strains. Of course, with these same approaches, polymorphism identification between the interspecific progenitors is even easier still.

In terms of the resolution of the genetic map that is obtained, the interspecific approach has a number of advantages over the RI approach. First, the number of samples in several of the well-characterized interspecific panels ranges from over 200 to as high as 1,000; 1,000 samples provides an average map resolution of 0.1 cM. In contrast, the total number of well-characterized RI strains is less than 140. Second, interference acts to eliminate nearby double crossover events in the interspecific backcross, and thus gene order can be determined with very high levels of confidence for any linked loci. In contrast, crossing over in multiple generations during the creation of RI strains eliminates the effect of interference and this can sometimes cause ambiguity in the determination of gene order.

There is one potential problem that could act to reduce the resolution of the interspecific cross in certain genetic regions—the existence of small inversion polymorphisms that may have arisen during the divergence of *M. spretus* and *M. musculus*. An inversion will preclude the observation of recombination across all the loci that it encompasses and, in turn, this will prevent the mapping of all of these loci relative to each other. Only one such inversion polymorphism has been identified to date (Hammer et al., 1989), however, direct tests for the existence of others have not been performed. Inversions could only be demonstrated directly by creating an intraspecific linkage map for *M. spretus* by itself and comparing the gene order on this map to the gene order on an intraspecific *M. musculus* map. Although this comparison has not been performed for any chromosome other than 17, indirect evidence for several additional inversions has come from the finding of regions of apparent recombination suppression in an interspecific linkage map in comparison to an intersubspecific (*castaneus*-B6) linkage map (Copeland et al., 1993). Cryptic inversions could have serious consequences for those who would like to use interspecific linkage distances as a gauge for estimating the physical distance between two markers as a precursor to positional cloning as described in Chapter 10.

### 9.3.2.2 Practical considerations

A unique advantage held by the established RI mapping panel sets is that individual RI samples are actually represented by strains of mice, and as such, they are immortal; RI samples from a mapping panel will never be "used up." In contrast, the amount of DNA in each sample of every interspecific panel is finite. Even under the best conditions, the amount of DNA recovered from a single whole mouse will never be more than 40 mg, and in many cases, mapping panels were previously established with samples containing only 1 or 2 mg. In the days when all typing was carried out by Southern blot analysis of genomic DNA, it was typical to use 5–10 μg aliquots for each analysis. With a total per sample size of 1 mg of DNA, one could produce 200 Southern blots, which could each be probed multiple times. Although this may sound like a large capacity, in reality, samples are spilled or transferred inefficiently and blots become ruined. For panels that are analyzed primarily by the RFLP approach, samples will be "used up" eventually and, as a consequence, the practical lifetime of such interspecific mapping panels is limited.

Today, of course, it is possible to develop a PCR protocol for typing in many cases and this allows one to use much smaller sample aliquots—of the order of nanograms. Thus, if the typing of a panel is restricted to PCR methods, one could conceivably analyze hundreds of thousands of loci on a single panel before it goes extinct.

For many investigators, a second important advantage to the RI approach is that DNA samples or animals can be purchased, without constraints, from the Jackson Laboratory. The investigator can then perform the experimental analysis in his or her own laboratory, and, by comparing the new SDP obtained with those present in a public database (provided with the Map Manager program described in Appendix B), a map position for the new locus can be established. This entire analysis can be accomplished independently, without any need to contact, consult, or collaborate with other scientists.

In contrast, each well-characterized interspecific panel is maintained in the context of an ongoing research project by a particular scientist or laboratory. Thus, an investigator with a new clone must interact, at some level, with another scientist, in order to utilize their mapping panel and private database for the purpose of determining a new map position. Some investigators may see this interaction as an advantage. For example, in a number of cases, mapping panel "owners" are willing to carry out the experimental analysis in their own laboratories, thereby alleviating the workload of the independent investigator; such extensive interactions are normally treated as collaborations. Other investigators will wish to remain independent and will view such an extensive interaction as a disadvantage.

### 9.3.3 Access to established interspecific mapping panels

At the time of writing, the laboratories listed below maintain well-characterized interspecific mapping panels. Different laboratories operate their mapping programs in very different ways—some send out DNA samples or filters from their panel, while others perform all typing in-house; the reader should make direct contact with a particular laboratory to determine the specific protocol that is followed there. The reader should be cautioned, of course, that all of these programs are maintained by funding agencies, and a change in funding or personnel may have eliminated a particular program during the period between this writing and your reading.

In the US, the best characterized interspecific mapping panels are maintained by N. Jenkins and N. Copeland at the Frederick Cancer Research Center in Frederick, Maryland (Copeland and Jenkins, 1991; Copeland et al., 1993), M. Seldin at Duke University Medical Center in Durham, North Carolina (Moseley and Seldin, 1989; Watson et al., 1992), and E. Birkenmeier at the Jackson Laboratory in Bar Harbor, Maine (Birkenmeier et al., 1994). In Europe, a very large interspecific mapping panel with 1,000 samples (the European Collaborative Interspecific Backcross or EUCIB) is maintained by the Human Genome Mapping Project (HGMP) Resource Centre (Watford Road, Harrow, Middlesex HA1 3UJ, England; fax: 0181–869 3807). EUCIB is under the joint supervision of S. Brown at St Mary's Hospital in England and Jean-Louis Guénet at the Pasteur Institute in Paris, France (Brown, 1993; Breen et al., 1994).

### 9.3.4 Is the newly mapped gene a candidate for a previously characterized mutant locus?

The main reason that many investigators will want to map a newly cloned gene is to determine whether it is equivalent to a locus that has been previously mapped but is characterized only at the level of a mutant phenotype. Cloning the genes associated with interesting phenotypes in this roundabout manner is usually a matter of luck and is referred to as the "candidate gene" approach. How does one begin to rule out or rule in possible identity to a phenotypically defined locus? Unfortunately, the mapping panel used to localize the cloned gene will usually not provide simultaneous map information for any phenotypically defined loci. Thus, one is forced to compare map positions derived from different crosses.

One should begin a search for potentially equivalent mutationally defined loci by scanning database lists of all loci that are thought to lie within 10 cM of the map position obtained for the new clone. The databases to search should include the chromosome map compiled by the appropriate mouse chromosome committee and published annually in a special issue of *Mammalian Genome* (Chromosome Committee Chairs, 1993) and the electronic databases maintained at the Jackson Laboratory (see Appendix B). Descriptions of the phenotypes associated with loci picked up in this scan can be obtained from a compendium published in the latest edition of the *Genetic Variants and Strains of the Laboratory Mouse* (Green, 1989) or electronically from the continuously updated on-line Mouse Locus Catalog in the Jackson Laboratory databases (see Appendix B). The expression pattern of the newly cloned gene and information concerning its protein product can often provide a means for evaluating the likelihood of an association with any particular mutant phenotype.

Once a particular locus has been identified for further consideration, one should begin a statistical evaluation of the likelihood of an equivalent map position with the newly cloned gene. To carry out this evaluation, it is important to look at the raw data that were used to place the locus on the map. Much of this information is compiled in the Jackson Laboratory databases. However, at times, it may be necessary to go back to the original reports that are cited. In this way, it will be possible to determine the actual marker locus, or loci, that were shown to be linked to the mutation, the nature of the cross that was used for analysis, and the number of recombinants observed. In many cases, mutant loci will have been mapped relative to "anchor loci" with well-established positions on contemporary chromosome consensus maps. If this is not the case, it may be necessary to backtrack through citations to uncover a multiple-step linkage relationship that does exist between the mutation and a well-established anchor.

Once a particular anchor locus has been identified with a direct linkage association to both the cloned gene and the mutant locus under consideration, the next task is to determine whether the confidence intervals associated with the map position of each show overlap. For backcross and RI data, this can be accomplished with the use of the confidence limit tables presented in Appendix D.

An illustration of such an analysis is presented in Fig. 9.10. In this hypothetical example, a newly cloned locus has been mapped relative to a common anchor locus with nine recombinants found in 94 backcross samples. This provides an estimated linkage distance of 9.6 cM. By consulting Table D5, one can estimate lower and upper 95% confidence limits of 5.2 and 17 cM, respectively. Next, one evaluates the linkage data associated with three mutant loci that have been identified as having the potential to be equivalent to the cloned gene. Mutation

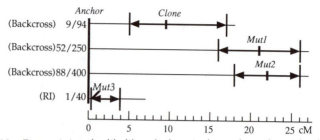

**Figure 9.10** Determining the likelihood of equivalence for independently mapped loci. In this illustration, four different loci—including a clone and three mutations—have been mapped relative to the same anchor locus in independent experiments. The type of cross used for mapping and the number of recombinants observed among the total number typed is indicated at the left of each two locus map. These numbers have been used to determine distance estimates, which are indicated by dark lines emanating from each locus symbol. The linkage data have also been used—in conjunction with Tables D2 and D6—to extrapolate minimum and maximum limits on the 95% confidence intervals associated with each map position. The confidence interval associated with the clone overlaps the interval associated with *Mut1*. This result suggests the possibility that the two loci may be equivalent although the minimal overlap observed would still make this a long shot. However, the data provide strong evidence against the possibility that the clone is equivalent to either *Mut2* or *Mut3*, which appear to map farther or closer from the anchor, respectively.

number one (*Mut1*) has been mapped relative to the same anchor locus in a backcross experiment, with 52 recombinants found among 250 samples for an estimated linkage distance of 21 cM. Extrapolation from the values given in Table D6 provides lower and upper 95% confidence limits of 16 and 26 cM, respectively. Mutation number two (*Mut2*) has also been mapped relative to the same anchor locus in a backcross, with 88 recombinants in 400 samples giving a linkage distance of 22 cM with lower and upper confidence limits of 18.2 and 26.3 cM (also from Table D6). Finally, mutation number 3 (*Mut3*) has been mapped with a group of RI strains with one discordance observed in 40 strains giving an estimated linkage distance of 0.6 cM (from Fig. 9.7) and lower and upper confidence limits of 0.2 and 4.0 cM (from Table D2).

The results of all four crosses are represented graphically in Fig. 9.10. The data make it very unlikely that the newly cloned gene is equivalent to loci defined by either mutation 2 or mutation 3 since none of these confidence intervals overlap. However, the 95% confidence intervals of the cloned gene and mutation 1 do overlap (even though absolute estimates of their map positions place them over 10 cM apart). If mutant-bearing animals are available, the potential equivalence between these two loci can be followed up with further experiments of several types. First, expression of the cloned gene can be examined in animals that carry the mutation. Second, the cloned locus itself can be examined within the mutant genome for the possible detection of easily visible alterations such as a deletion or gene-inactivating insertion. Finally, segregation of the mutant allele and the cloned gene can be followed directly in a breeding experiment (as described in the next Section). It only takes one validated recombination event[90] to rule out an equivalence between the two loci.

## 9.4 STARTING *FROM SCRATCH* WITH A NEW MAPPING PROJECT

### 9.4.1 Overview

There are two types of experimental situations in which established mapping panels may not be sufficient to the needs of an independent investigator. In the first instance, an investigator may want to pursue the mapping of a large group of cloned loci to obtain, for example, a very high resolution map for an isolated genomic region. For extended mapping projects of this and other types, it becomes both cost effective and time effective to perform an "in-house" cross for the production of a panel of samples over which the investigator has complete control.

With a second class of experimental problems, an investigator will have no choice but to perform an "in-house" cross for analysis. This will be the case in all situations where the test locus is defined only in the context of a mutant phenotype. Often, the goal of such projects will be to clone the locus of interest through knowledge of its map position. To map a mutationally defined locus, one will have to generate a special panel of samples in which segregation of the mutant and wild-type alleles can be followed phenotypically in animals prior to DNA preparation for marker locus typing. What follows in this Section is a summary of the choices that confront an investigator in the development of a mapping project *from scratch*, and the process by which an investigator should proceed through the project from start to finish.

At the outset, the investigator must make decisions concerning the form of the breeding cross itself. In particular, which parental strains will be used and what type of breeding scheme will be followed? To map a mutationally defined locus, one will obviously have to include one strain that carries the mutation. The second parental strain should be chosen based on the contrasting considerations of genetic distance (the more distant the strain, the greater the chance of uncovering polymorphisms at DNA marker loci) and the ability to generate offspring in which segregation of the mutant allele can be observed. The choice of breeding scheme is limited typically to one of two different two-generation crosses: the outcross–backcross ($F_1 \times P$, where P represents one of the original parental strains) or the outcross–intercross ($F_1 \times F_1$) illustrated in Figs 9.11 and 9.12, respectively. If the purpose of the analysis is to map loci associated with a mutant phenotype, the nature of the phenotype may limit this choice further as discussed more fully in Sections 9.4.2 and 9.4.3.

Once the strains and a breeding scheme have been chosen, one can begin to carry out the first generation cross. The number of mating pairs that should be set up need not be as large as one might think because of the expansion that will occur at the second generation. Backcrosses are usually peformed with females that are $F_1$ hybrids and intercrosses, by definition, are always based on $F_1$ hybrid females. As such, the second generation cross is likely to be highly productive

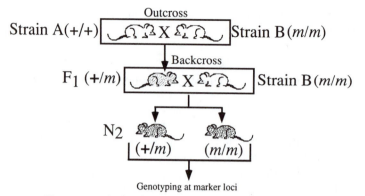

**Figure 9.11**   Illustration of a backcross mapping protocol. In this illustration, strain B is homozygous for a fully recessive mutation (m) that does not interfere with viability or fertility. All $N_2$ generation animals can be scored with either a mutant or a wild-type phenotype that translates directly into genotype at the m locus.

with larger and more frequent litters than one obtains with inbred females (see Section 4.1). Consider the goal of obtaining 1,000 offspring from several sets of an outcross–intercross or an outcross–backcross.[91] If one assumes that 90% of the second-generation mating pairs will be productive with an average of four litters with eight pups in each, one would need to set up only 35 such matings. Working backwards, to generate the 35 $F_1$ females and/or males required would entail only ten initial matings between the two parental strains with the assumption that 50% would be productive and these would each have three litters of five pups.

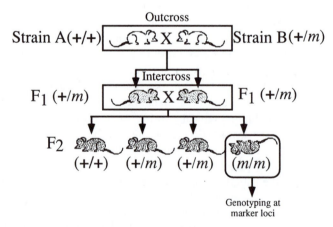

**Figure 9.12**   Illustration of an intercross mapping protocol. In this illustration, strain B is heterozygous for a fully recessive mutation (m) that causes homozygous lethality. With phenotype scoring alone, it is not possible to distinguish $F_2$ animals that carry $+/m$ or $+/+$ genotypes at the m locus. Thus, only the mutant m/m animals can be used for linkage analysis.

An alternative backcross strategy that may sometimes be even more efficient is to set up $F_1$ males with inbred females of one of the two parental strains in the second generation. This approach is only effective when the backcross parent to be used is a common inbred strain such as B6. In this situation, there is no limit to the number of females that can purchased at a modest cost from various suppliers, and individual $F_1$ males can be rotated among multiple cages of these females. Thus, what is sacrificed in terms of hybrid vigor is made up for in terms of absolute number of crosses. As few as ten males could be rotated every 5 days among cages with two females each for a total of 120 matings in a month. One should be aware, however, that an analysis of this type will be based entirely on recombination in the male germ line, which may or may not be beneficial to the investigator according to different experimental requirements as discussed at the end of Section 9.4.4.2.

When offspring from the second-generation cross are born, one will need to analyze each for expression of the mutant phenotype. In some cases, it will be possible to use both the expression and non-expression of phenotypes as direct indicators of genotype. In other cases, it will only be possible to use phenotypic expression as an indicator of genotype in a subset of animals. This will be true for all phenotypes that are only partially penetrant as well as those that are only expressed in homozygous offspring from a second-generation intercross. In both cases, the lack of phenotypic expression in any particular animal will preclude an unambiguous determination of its genotype. When it is only possible to incorporate a subset of offspring into the ultimate genetic analysis, it will obviously be necessary to generate more offspring at the outset to achieve the same level of genetic resolution. Once "phenotyping" is accomplished, animals can be converted into DNA for incorporation into the panel that will be used for analysis of marker segregation. Optimal strategies for determining map position are discussed in Sections 9.4.4 and 9.4.5.

### 9.4.2 Choosing strains

#### 9.4.2.1 For developing DNA marker maps

Upon commencing a new linkage study, an investigator will first have to decide upon the two parental mouse strains that will be used in the initial cross to generate $F_1$ animals. This choice will be informed by the goal of the linkage study. If the goal is simply to develop a new panel for mapping loci defined as DNA markers, there will be no a *priori* limitation on the strains that can be chosen. The most important considerations will be the degree of polymorphism that exists between the two parental strains and the ease with which they, and their offspring, can be bred to produce a large panel of second-generation animals for DNA typing.

As discussed earlier in this chapter and previous ones, the traditional inbred *M. musculus* strains show minimal levels of interstrain polymorphism. It was for this reason that the initial two-generation mapping panels were all based on interspecific crosses between a *M. musculus* strain and a *M. spretus* strain as described in Section 9.3. *M. spretus* is the most distant species from *M. musculus* that still allows the production of fertile $F_1$ hybrids (see Section 2.3.5). As such, the *M. musculus* × *M. spretus* cross will provide the highest level of polymorphism that is theoretically obtainable for the purpose of mapping.

It is certainly possible to replicate this interspecific cross with any one of a number of inbred *M. musculus* strains (such as B6, C3H, or DBA) and an inbred *M. spretus* strain (such as SPRET/Ei), which can be purchased from the Jackson Laboratory or another supplier. However, the interspecific cross is less than ideal for several reasons. First, breeding between *M. musculus* strains and *M. spretus* is generally poor with infrequent, small litters. As a consequence, one must begin with a larger number of initial mating pairs, and wait for a considerable length of time before obtaining a complete panel of second-generation animals. Second, only $F_1$ females are fertile. This rules out any possibility of setting up a second generation intercross. Finally, as discussed in Section 9.3.2.1, *M. musculus* and *M. spretus* differ by at least one, and perhaps many more, small inversions that will act to eliminate recombination and, as a consequence, distort the true genetic map.

These limitations have led investigators to test the practicality of using an inter*sub*specific cross as an alternative that would still show sufficient levels of polymorphism but without any of the problems inherent in the interspecific cross. In particular, several laboratories have published mapping results based on crosses between the inbred strain B6, which is derived predominantly from the subspecies *M. m. domesticus* and the inbred strain CAST/Ei (distributed by the Jackson Laboratory), which is derived entirely from the *M. m. castaneus* subspecies (Dietrich et al., 1992; Himmelbauer and Silver, 1993).

The two subspecies, *M. m. domesticus* and *M. m. castaneus*, evolved apart from a common ancestor approximately one million years before present (see Fig. 2.2 and Section 2.3.2). As a consequence, the level of polymorphism between the two is much greater than that observed among strains that are predominantly derived from just *M. m. domesticus*, but not quite as high as that observed between *M. m. domesticus* and *M. spretus*, which evolved apart three million years ago. A sense of the relative levels of polymorphism that exist in various pairwise comparisons can be achieved by looking at the frequencies with which *Taq*I RFLPs are detected at random loci. In a comparison of two predominantly *M. m. domesticus* strains—B6 and DBA—19% of the tested loci showed *Taq*I RFLPs; between B6 and the *M. m. castaneus* strain CAST/Ei, 39% of the tested loci showed RFLPs. Finally, between B6 and *M. spretus*, 63% of the tested loci showed RFLPs (LeRoy et al., 1992; Himmelbauer and Silver, 1993). Relative rates of polymorphism have also been surveyed at random microsatellite loci with the following results: among *M. m. domesticus* strains, the average rate of polymorphism is ~ 50%; between B6 and CAST/Ei, the polymorphism rate is 77%, and between B6 and *M. spretus*, the polymorphism rate is 88% (Love et al., 1990; Dietrich et al., 1992).

The bottom line from these various comparisons is that the level of polymorphism inherent in the B6 × CAST/Ei cross seems more than sufficient for generating high-resolution linkage maps, especially with the use of highly polymorphic markers like microsatellites. Furthermore, the somewhat lower rate of polymorphism is more than compensated for by various advantages that this cross has over the interspecific cross with *M. spretus*. First, the two strains, B6 and CAST/Ei, breed easily in the laboratory with the production of large numbers of offspring. Second, both male and female $F_1$ hybrids are fully fertile. Third, the single well-characterized interspecific inversion polymorphism does not exist between B6 and CAST/Ei (Himmelbauer and Silver, 1993), and it is likely that most other postulated interspecific inversion polymorphisms are also absent as well (Copeland et al., 1993). Consequently, the linkage map that one obtains with this intersubspecific cross is much more likely to represent the map that would have be derived from a cross within the *M. m. domesticus* subspecies itself.

As indicated in Fig. 2.2, and as discussed in Section 2.3.2, there are a number of other *M. musculus* subspecies that are just as divergent from *M. m. domesticus* as is *M. m. castaneus*. Inbred strains have been developed from the *M. m. musculus* subspecies, and at least two (Skive and CzechII) are available from the Jackson Laboratory. In addition, another set of inbred strains (MOLF/Ei) have been derived from the *faux* subspecies *M. m. molossinus*, which is actually a natural mixture of *M. m. musculus* and *M. m. castaneus* (see Fig. 2.2 and Section 2.3.3). It is likely that each of these inbred strains could be used in place of CAST/Ei with a similar level of polymorphism relative to *M. m. domesticus*, and with the same advantages described above. In fact, the availability of several unrelated wild-derived strains provides a means for overcoming the limitation to genetic resolution caused by recombination hotspots as described in Section 7.2.3.3 (and illustrated in Fig. 7.5). This is because $F_1$ hybrids between B6 and CAST/Ei, or MOLF/Ei or Skive are all likely to have different hotspots for recombination. Thus, by combining data from all three crosses, one will be able to "see" recombination sites that are spread out among perhaps three times as many possible hotspot locations.

### 9.4.2.2 For mapping a simple mutation

Another factor in strain choice comes into play when the goal of a breeding study is to map a locus defined solely by a mutant phenotype. In this case, it is obvious that one of the parental strains must carry the mutant allele to be mapped. Ideally, the mutation will be carried in an inbred, congenic, or coisogenic strain. In the second best situation, the mutation will be present in a genetic background that is a mixture of just two well-defined inbred strains. Finally, the most potentially difficult situation occurs when the mutation is present in a non-inbred, undefined genetic background.

In this last situation, it is advisable to use a single male as the sole representative of the mutant strain in matings to produce all $F_1$ hybrids. The advantage to this approach is that the number of alleles contributed by the mutant "strain" at any one locus in all of the $F_1$ animals will be limited to just two. If, on the other hand, one had begun with multiple males as representatives of the non-inbred mutant strain, the number of potential alleles at every locus in the panel would be twice the number of males used. The larger the number of alleles, the more complicated the analysis could become. By rotating a single male among a large set of cages containing females from the second strain, it will be possible to produce a sufficient number of $F_1$ hybrids in a reasonable period of time.

In essentially all cases, the mice that carry the mutation will be derived from the traditional inbred strains, which are themselves mostly derived from the *M.*

*m. domesticus* subspecies. For all of the same reasons discussed in the previous subSection, the best choice of a second parental strain would be one that is inbred from a different *M. musculus* subspecies such as CAST/Ei, MOLF/Ei, or *"Mus musculus"* Skive or CzechII, which are all available from the Jackson Laboratory.

### 9.4.3 Choosing a breeding scheme

The second choice that an investigator will make upon beginning a new linkage study is between the two prescribed breeding schemes. With both schemes, illustrated in Figs 9.11 and 9.12, the first mating will always be an outcross between the two parental strains chosen according to the strategies outlined above. However, once $F_1$ hybrid animals have been obtained, an investigator must decide whether to backcross them to one of the parental strains or *intercross* them with each other. There are advantages and disadvantages to each breeding scheme.

#### 9.4.3.1 The backcross

The primary advantages of the backcross approach are all based on the fact that each offspring from the backcross can be viewed as representing an isolated meiotic event. The entire set of alleles contributed by the inbred parent (strain B in Fig. 9.11) is predetermined. Thus, the only question to be resolved at each typed locus is whether the $F_1$ parent has contributed the same parental allele (from strain B) or the allele from the other parent (strain A): in the first instance, typing would demonstrate the presence of only the strain B allele, and in the second instance, typing would demonstrate the presence of both the strain A and strain B alleles.

By looking at Fig. 9.15, one can visualize the actual meiotic products contributed by the $F_1$ parent in the form of individual haplotypes. Every recombination event can be detected and the frequency of recombination between any two loci can be easily determined. The existence of strong interference over distances of 20 cM or more (Section 7.2.2.3) can be used to advantage in the determination of gene order, since any order that requires nearby double crossover events in any haplotype is likely to be incorrect.[92]

The analysis of backcross data is very straightforward, and when all loci are known to map on the same chromosome, it is possible to derive linkage relationships even in the absence of specialized computer programs. However, with the use of the Macintosh computer-based Map Manager program (described in Appendix B), data presentation and analysis become even more transparent. The major disadvantage with the backcross is that it is not universally applicable to all genetic problems. In particular, it cannot be used to map loci defined only by recessive phenotypes that interfere with viability or absolute fecundity in both males and females.

#### 9.4.3.2 The intercross

The intercross approach has two main advantages over the backcross. The first is that it can be used to map loci defined by recessive deleterious mutations, since both heterozygous $F_1$ parents will be normal, and homozygous $F_2$ offspring can be recovered at any stage (postnatal or prenatal if necessary) for use in typing further markers. The second advantage is a consequence of the fact that informative meiotic events will occur in both parents. This will lead to essentially twice as much recombination information on a per animal basis compared to the backcross approach.

The main disadvantage with the intercross is also a consequence of informative meiotic events in both parents. The problem is that the data obtained are more complex, as illustrated in Fig. 9.4 and discussed in Section 9.1.3.4, and more difficult to analyze because of the impossibility of determining which allele at each heterozygous $F_2$ locus came from which parent. Thus, while each animal will, by definition, carry two separate haplotypes for each linkage group, the assignment of alleles to each haplotype can only be accomplished retrospectively or, in some circumstances, not at all. In addition, interference is no longer as powerful a tool for ordering loci, since nearby crossover sites can be brought together into individual $F_2$ animals from the two parents. To generate *de novo* linkage maps from large-scale intercross experiments, it is essential to use computer programs such as Mapmaker that carry out multilocus maximum likelihood analysis (Lander et al., 1987; and Appendix B). However, when previously mapped codominant anchor loci are typed within an intercross, the more user-friendly Map Manager program (version 2.5 and higher) can be used for data input and analysis.

### 9.4.3.3 Making a choice

In large-scale mapping experiments with many loci spread over one or more chromosomes, the backcross is usually the breeding scheme of choice. What is sacrificed in terms of mapping resolution is made up for in terms of ease of data handling and presentation. However, when an investigator is focusing on a small genomic region (on the order of a few centimorgans or less) for very high resolution mapping as a precursor to positional cloning, the intercross may be a better choice. At this level of analysis, the data will be much less complex with only a small fraction of animals expected to show mostly single recombination events in the interval of interest; the advantage gained by doubling the frequency of such events may be critical to efforts aimed at zeroing in on the locus of interest.

Of course, as discussed above, in the case of recessive deleterious mutations, one may not have a choice but to use the intercross. Unfortunately, in situations where the mutation is strictly recessive, one will only be able to map the mutant locus with those 25% of $F_2$ animals that express the mutant phenotype because the genotype of non-expressing animals cannot be determined (see Fig. 9.12). Since two meiotic events are scored in each $F_2$ animal, the total amount of genetic information obtained will be approximately double that obtained from an equivalent number of backcross animals that can be typed. Nevertheless, this still comes out to only 50% of the information obtained from typing a complete backcross panel of the same size as the complete intercross panel. Consequently, if the trait under analysis is strictly recessive but does not seriously hinder viability or fecundity in homozygotes of at least one sex, it is more advantageous to use the backcross. In these situations, a backcross can be set up with a homozygous mutant parent, as illustrated in Fig. 9.11, and 100% of the offspring can be scored phenotypically for the contribution of either the mutant or wild-type allele from the $F_1$ parent.

### 9.4.4 The first stage: mapping to a subchromosomal interval

#### 9.4.4.1 A stratified approach to high-resolution mapping

An optimal strategy for high-resolution linkage mapping is one that proceeds in stages with nested sets of both marker loci and animals. One can see the logic of this sequential approach by considering the numbers of markers and animals required to obtain a high-resolution map in a single pass. For example, suppose one wanted to obtain a linkage map with both an average crossover resolution of ( 0.1 cM and an average marker density of one per centimorgan. In a one-pass approach, one would have to analyze 1,000 backcross animals for segregation at 1,500 marker loci (spanning 1,500 cM), which would require one and one-half million independent typings.

A much more efficient approach is to divide the protocol into two separate stages. The goal of the first stage should be to link the locus to a defined subchromosomal interval. This can be accomplished by typing a relatively small set of markers on a relatively small random subset of phenotypically typed animals from within the larger panel. Once this first stage is completed, it becomes possible to proceed to the second stage, which should focus on the construction of a high-resolution map just in the vicinity of the locus of interest with a selected set of markers and a selected set, of animal samples. The ultimate goal of this entire protocol is the identification of a handful of markers and recombinant animals that bracket a very small interval containing an interesting gene that can then be subjected to positional cloning as described in Section 10.3.

#### 9.4.4.2 How many animals and how many markers? Evaluation of the swept radius

The first step in the first stage of the protocol is to develop a framework map that is "anchored" by previously well-mapped loci spaced uniformly throughout the entire genome. To accomplish this task most efficiently, it is critical to calculate the minimum number of anchor loci required to develop this low-resolution, but comprehensive, map. This calculation is based on the length of the swept radius that extends on either side of each marker. As discussed earlier in this chapter (Section 9.2.2.3), the swept radius is a measure of the distance over which linkage can be detected between any marker and a test locus when both are typed in a set number of offspring generated with a defined breeding protocol. Although the swept radius was defined originally in terms of map distances (Carter and Falconer, 1951), it is much easier to work directly with recombination fractions, and in the following discussion, charts, and figures, this alternative metric will be used.[93]

Two measures of the backcross swept radius, determined for sample sizes that range from 20 to 100 animals, are presented in Fig. 9.13. The first measure is

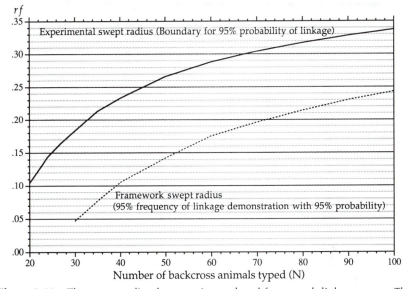

**Figure 9.13** The swept radius for experimental and framework linkage maps. The solid line indicates the *experimental swept radius* for sets of backcross animals that number between 20 and 100. The experimental swept radius should be used to evaluate the significance of experimentally derived recombination or concordance values. All experimentally determined recombination fractions or concordance values that lie below this line can be considered significant at a level of 95%. The broken line indicates the *framework swept radius*. This is a measure of the *actual* distance that a marker locus will sweep on each side with a probability of 95%. The framework swept radius should be used as a basis for choosing markers for setting up a framework map in a new linkage study. Recombination fractions ($rf$) that are less than 0.25 can be converted directly into centimorgan distances through multi-plication by 100 as discussed in Section 7.2.2.3. For larger recombination fractions, the Carter–Falconer mapping function should be used to obtain centimorgan distances (equation 7.3). The experimental swept radius was derived through a reiterative Bayesian approach based on the use of equation 9.6. The framework swept radius was derived by determining the recombination fraction having a 95% confidence interval with an upper limit equivalent to the experimental swept radius.

based on the traditional view of a swept radius as a boundary that separates significant from non-significant rates of observed recombination. This "experi-mental swept radius" is shown as the solid curve in Fig. 9.13. The graph can be used to find out quickly whether any experimentally determined recombination fraction, or concordance value, meets the strictly defined Bayesian-corrected cutoff for demonstration of linkage at a probability of 95% or greater (see Section 9.1.3.6).

Although the experimental swept radius provides a means to evaluate the significance of newly derived data, it is not useful as a means to establish the distances that should separate marker loci to be chosen for a framework map in a new cross. The problem is that marker loci that are *actually* separated by a map distance equivalent to the experimental swept radius will, by chance, recombine to a greater or lesser extent with equal probability in any particular experimental cross, and in those 50% of the crosses where a higher recombination fraction is observed, the data will not be sufficient to establish linkage at a 95% level of significance. Thus, a second, more conservative measure of swept radius is needed to determine the maximum actual recombination distance between two loci that will allow the demonstration of linkage at a probability of 95% with a frequency of 95%.[94] I will call this parameter the "framework swept radius."

The "framework swept radius" can be evaluated as a recombination fraction associated with a 95% confidence interval having an upper confidence limit equivalent to the value of the experimental swept radius for a sample set of a particular size. In the discussion that follows, this framework swept radius will be used as a means for establishing the distances that should separate markers to be used in setting up a new framework map.

With a set number of backcross samples, one can use Fig. 9.13 to find the corresponding framework swept radius associated with each anchor locus. For example, with 52 samples, the framework swept radius is 15 cM, with 72 samples, it is 20 cM, and with 94 samples, it becomes 24 cM. It is clear that once a critical number of samples has been reached (45–50), further increases in number provide only a marginal increase in the distance that is swept. Figure 9.13 can also provide a first approximation of the framework swept radius associated with a panel of intercross samples. To a first approximation, each intercross sample is equivalent to two backcross samples. Thus, a swept radius of ∼ 15 cM can be obtained with 26 intercross samples, and ∼ 20 cM can be obtained with 36 intercross samples.[95]

The framework swept radius can be used in conjunction with the lengths of each individual chromosome to determine the number of anchor loci required to provide complete coverage over the entire genome. Essentially, anchors can be chosen such that their "swept diameters" (twice the swept radius) cover directly adjacent regions that span the length of every chromosome as illustrated in Fig. 9.14. The first and last anchors on each chromosome must be placed within one swept radius of their respective ends, while the distance between adjacent anchors should be within two swept radii. The estimated lengths of all 20 mouse chromosomes are sorted into a set of ranges in Table 9.4. The number of anchors required per chromosome for a backcross analysis is calculated by dividing the chromosome length by the swept diameter defined with a sample set of a particular number (from the graph in Fig. 9.13) and rounding up to the nearest integer. As indicated in Table 9.4, with 52 backcross samples, it is possible to cover the entire mouse genome with 60 well-placed anchors. With 72 samples, the number of required anchors decreases to 46, and with 94 samples, it decreases to 43. It is clear that little is to be gained by including more than 72 samples in this initial analysis.

The minimalist approach just outlined to a comprehensive framework map has only become feasible as this chapter is being written. This feasibility is based on the availability of over 3,000 highly polymorphic microsatellite loci that span the genome with an average spacing of less than 1 cM (Copeland et al., 1993).

**Table 9.4**  Classification of mouse chromosomes based on length and number of markers required for coverage in a backcross

| Chromosomes | Range of chromosome lengths (cM)[a] | Markers required per chromosome for complete coverage at 95% confidence | | | |
|---|---|---|---|---|---|
| | | $N = 40$[b] (s.r. = 10 cM) | $N = 52$ (s.r. = 15 cM) | $N = 72$ (s.r. = 20 cM) | $N = 942$ (s.r. = 24 cM) |
| 1, 2, 5 | 103–111 | 6 | 4 | 3 | 3 |
| 3, 4, 8 | 82–88 | 5 | 3 | 3 | 2 |
| 6, 7, 9, 10, 11, 13, X | 73–80 | 4 | 3 | 2 | 2 |
| 12, 14, 15, 16 | 64–71 | 4 | 3 | 2 | 2 |
| 17, 18, 19 | 50–60 | 3 | 2 | 2 | 2 |
| | Total markers | 86 | 60 | 46 | 43 |
| | Markers × samples | 3,440 | 3,120 | 3,312 | 4,042 |

[a] The chromosome lengths indicated in this table are based on consensus linkage maps presented in the individual reports from each mouse chromosome committee published in the 1993 edition of the *Encyclopedia of the Mouse Genome* (Silver et al., 1993).
[b] $N$ is the number of backcross samples; s.r is the swept radius derived from Fig. 9.13. Coverage of each chromosome is based on the placement of markers at distances of no more than two swept radii from each other as illustrated in Fig. 9.14.

Primer pairs that define each of these loci are commercially available at a modest cost from Research Genetics, Inc. in Huntsville, Alabama. By contacting the Genome Center at the Whitehead Institute, as described in Appendix B, one can obtain chromosome-specific lists of microsatellites that are polymorphic between the particular parental strains that an investigator has used to generate his or her linkage panel. With this information, one can choose specific microsatellite loci that map to each of the general locations required to span each chromosome as illustrated in Fig. 9.14.

In the backcross linkage studies reported to date, the gender of the $F_1$ hybrid used in the second-generation cross has usually been female. In the case of the interspecific cross, there is no other choice since the $F_1$ male is sterile. However, this is not a factor in the intraspecific or intersubspecific cross. Rather, $F_1$ hybrid females are used for two other reasons. First, they have a much higher fecundity relative to inbred females, and second, they generally display higher frequencies of recombination (Section 7.2.3.2), which, in turn, will produce a higher resolution map in the second stage of linkage analysis described in the next Section. Interestingly, the lower recombination frequency associated with male mice is actually better suited to the first stage of mapping because it can act, in effect, to reduce the centimorgan length of each chromosome by 15–40%. Thus, with the use of male $F_1$ hybrids in the backcross, one would, in theory, need fewer anchor loci to span the genome. Furthermore, as discussed in Section 9.4.1, in backcrosses to a common inbred parent such as B6, the use of $F_1$ males is likely to be much more efficient and provide many more $N_2$ progeny more quickly than the reciprocal cross. Unfortunately, at the time of this writing, male-

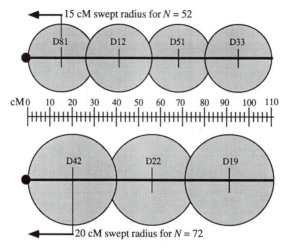

**Figure 9.14** Example of the coverage of mouse chromosome 2 with marker loci. The radius of each circle is equivalent to the swept radius around each locus as determined from Fig. 9.13. With 52 backcross animals, four marker loci would be required to span the chromosome. With 72, a completely different set of three marker loci would be required.

specific linkage maps have not been developed for the new libraries of microsatellite loci. Hence, at the current time, the spacing of microsatellites for this purpose would be a matter of guesswork.

### 9.4.4.3 Determining linkage

The first analysis of backcross data should be directed at simply determining the existence of linkage to the locus of interest. This is accomplished by comparing the pattern of allele segregation from the new locus with the patterns of allele segregation from each marker locus. Essentially, the frequency of recombination between the new locus and each marler locus is calculated, one at a time, to identify one or more markers that show a significant departure from the independent assortment frequency of 50%. This task is performed most easily by entering the accumulated allele segregation data into an electronic file that is analyzed by a special computer program developed for this type of analysis. A number of such computer programs have been described (see Appendix B). The most user-friendly of these is the Apple Macintosh-based Map Manager program developed by K. Manly (1993) and described in Appendix B.

It is also possible to determine linkage, when a backcross set is not too large, without the use of a specialized computer program. This can be accomplished by entering the allele segregation information for each locus along a separate row or line in a spreadsheet or word-processing file, where each column represents a separate animal (analogous to the RI strain data matrix illustrated in Fig. 9.6). Anchor loci should be placed in sequential rows according to their known order along each chromosome. The very first rows should be reserved for the new locus (or loci). The complete file will be a matrix of information with the number of rows equal to the number of markers and new loci typed and the number of columns equal to the number of backcross animals analyzed. For the $N = 52$ backcross typed for one new locus in addition to a minimal number of markers (from Table 9.4), this would be a $61 \times 52$ matrix of data.

Next, one would take the row representing a new locus and compare it row by row, either on the computer or on paper, for pattern similarities with each anchor locus allele distribution. Visual inspection alone will be sufficient to distinguish similar runs of alleles in two rows. The total recombination fraction between the new locus and any marker locus identified in this way can be easily calculated; if the fraction of recombinants is greater than the experimental swept radius found in Fig. 9.13 (0.27 for $N = 52$), linkage can be rejected and one can move on to the next locus. Although this process is somewhat tedious, the time that it takes is minimal compared to the time involved in actually typing DNA markers in the first place. In contrast, with whole genome data obtained from an intercross, manual determination of linkage is extremely difficult. Instead, one should use one of the limited number of programs available for this type of analysis. The most well-known of these programs is Mapmaker developed by Eric Lander (Lander et al., 1987, and Appendix B).

Ideally, linkage analysis will identify at least one, and at most two, loci that are linked at a significance level of 95% to the new locus of interest. If there are two linked loci, they should be adjacent to each other within the framework map formed on the same chromosome. With results of this type, one can move on to the next task of determining the order of the new locus relative to the framework map as discussed below.

It is possible that the data will not be sufficient to demonstrate linkage with a significance of 95% to any of the marker loci that were typed. It is critical at this point to confirm all DNA marker typings as well as phenotypic determinations for each animal. If there is still no evidence of linkage at the 95% significance level, one can attempt to uncover potential linkage relationships by reducing the required level of significance.[96] This may allow the suggestion of linkage in the middle of a particular chromosomal interval between two markers or near a chromosome end. If this approach fails, one should examine the recombination intervals that separate each marker along each chromosome (with the haplotype method described in the next section) in order to pick out intervals that are larger than anticipated. One can retype the same set of backcross animals for new markers in regions suggested by any of these approaches. If this approach fails as well, one should consider the possibility that the new locus may map very close to a centromere or telomere; to test this possibility, it would be necessary to type more centromeric and telomeric markers on each chromosome. Finally, one should consider the possibility that complex genetic interactions such as incomplete penetrance and/or polygenic effects may be acting to distort the one to one relationship between phenotype and genotype at any single locus (see Section 9.5).

### 9.4.4.4 Pooling DNA samples for the initial identification of linked markers

In essentially all mapping experiments performed today, the vast majority, if not all, of the marker loci used are typed by DNA-based techniques. At the time of this writing, the most versatile, and most commonly used, genetic marker is the microsatellite (Section 8.3.6). But other DNA markers that are useful in particular cases include those that can be assayed by the SSCP protocol (Section 8.3.3) and RFLP analysis (Section 8.2). The genotyping of all of these marker types within offspring from a mapping cross is based on the detection of "codominant" alleles recognized as different size bands after gel electrophoresis.

In the mapping approach just described in the previous section, each backcross animal is converted into a DNA sample that is typed independently for each marker locus that has been chosen to sweep the genome. The total number of PCR reactions (or restriction digests) required can be determined from Table 9.4 by multiplying the number of markers by the number of backcross animals. The smallest number is obtained with 52 animals typed for 60 markers, which comes out to 3,120 reactions (followed by an equivalent number of lanes on gels). Unless one has access to automated PCR and gel-running equipment and unlimited funds for thermostable DNA polymerase, this approach could be prohibitive in cost.

A much more efficient approach can be used when the goal of a cross is to map the locus or loci responsible for a particular mutant phenotype or polymorphic trait that is segregating in either a backcross or an intercross. The only essential prerequisite is that the parents used in the first-generation mating must be from an inbred or segregating inbred strain (see Section 3.2.4).

The basic strategy is to reduce the number of PCR reactions (or restriction digests) and subsequent gel runs through the analysis of only one or two combined DNA samples that are obtained by pooling together equivalent amounts of high-quality DNA from all second-generation animals expressing the same phenotype (Michelmore et al., 1991; Asada et al., 1994). This pooled DNA strategy works for both the backcross protocol and the intercross protocol. It works for incompletely penetrant traits and for quantitative traits controlled by segregating alleles at more than one locus (see Section 9.5.4.2). However, it requires the use of markers with segregating alleles that can be reproducibly distinguished and detected with equal levels of intensity. Thus, not all PCR-based markers will be suitable.

Let us consider the simple example of a backcross in which all $N_2$ animals can be phenotypically distinguished at a single mutant locus as illustrated in Figure 9.11. The first step of the analysis would be to classify each animal as $+/m$ or $m/m$ followed by the conversion of each individual into a high-quality DNA sample. Then, equal amounts of DNA from each $m/m$ sample would be combined into one pool, and equal amounts of DNA from each $+/m$ sample would be combined into a second pool. A third control sample would be formed by combining equal amounts of DNA from the two parents of the cross: the $F_1$ hybrid and strain B in Fig. 9.11. Finally, an aliquot from each of these three composite samples would be subjected to PCR amplification with primer pairs specific for one marker at a time (or restriction digestion), and the amplified (or digested) samples would be separated by gel electrophoresis and analyzed by ethidium bromide staining, or probing, or autoradiography.

The results expected for markers showing different linkage relationships to the mutant locus are illustrated in Table 9.5. For all markers that are not linked to the

**Table 9.5**   Relative intensities of marker alleles expected in pooled samples from a backcross or intercross[a]

| Concordance | *m/m* | | | | +/*m* | | |
|---|---|---|---|---|---|---|---|
|  | 100% | 90% | 80% | None | 80% | 90% | 100% |
| **Backcross[c]** | | | | | | | |
| Strain A allele[b] | 0.00 | 0.05 | 0.10 | 0.25 | 0.40 | 0.45 | 0.50 |
| Strain B allele[b] | 1.00 | 0.95 | 0.90 | 0.75 | 0.60 | 0.65 | 0.50 |
| **Intercross[d]** | | | | | | | |
| Strain A allele[b] | 0.00 | 0.10 | 0.20 | 0.50 | 0.80 | 0.90 | 1.00 |
| Strain B allele[b] | 1.00 | 0.90 | 0.80 | 0.50 | 0.20 | 0.10 | 0.00 |

[a] Relative intensities are indicated as fractions of the total signal present in both allelic bands of any one sample. For experimental protocols in which phenotype corresponds completely with genotype at the test locus (or loci), the percentage concordance is equivalent to the fraction of non-recombinant samples within the pool.
[b] Strains A and B are defined by the crosses illustrated in Figs 9.11 and 9.12.
[c] The ratio of alleles expected is a function of the effective recombination fraction ($r$) that separates a test locus from a marker locus. For backcross data of the type illustrated in Fig. 9.11, the equation for concordance in the *m/m* pool is: [*strain A/strain B* $= r/(2 - r)$], and in the +/*m* pool, the equation is: [*strain A/strain B* $= (1 - r)/(1 + r)$].
[d] For intercross data of the type illustrated in Fig. 9.12, the equation for concordance in the *m/m* pool is: [*strain A/strain B* $= r/(1 - r)$], and in the +/+ pool, the equation is: [*strain A/strain B* $= (1 - r)/r$].

test locus, the allele patterns obtained with the three composite DNA samples should be indistinguishable with a ratio of 1:3 in the intensities of the strain A and strain B alleles. In contrast, when a marker is very closely linked to the mutant locus, the ratio of alleles in the two pooled samples will diverge significantly in opposite directions from the control sample: in the *m/m* sample, the strain A allele will be absent or very light, while in the +/*m* sample, the intensity of the strain A allele will climb to equality with the strain B allele (whose signal will decrease proportionally). For ease of analysis, it is best to run the control sample in between the two pooled $N_2$ samples.

The power of this strategy for linkage analysis derives from the huge reduction in the number of samples that must be typed for each marker. Instead of 40, 50, 60, or more, the number is reduced to just three. However, to get a sense of the overall savings in time and cost, it is important to consider several factors: (1) the number of individual $N_2$ samples that must be included in each pool; and (2) the recombination distance over which a significant departure from the control sample can be observed.

Increasing the number of individual samples in each pool serves two purposes. First, random errors in the measurement of individual sample aliquots will tend to become evened out over a larger pool. Second, chance departures from the control ratio of alleles (i.e. false positives) will become much less frequent for unlinked markers (see Fig. 9.13). For both of these reasons, one should set a minimum pool size at 20 animals. There is no maximum to the pool size but there is nothing to be gained from pooling more than 50 samples together.

It is difficult to predict the level of concordance that must exist between the test locus and a marker before one can judge a result to be evidence of linkage.

A certain level of non-genetic variation is likely from sample to sample, and thus, a positive result must be one with a signal ratio that goes significantly beyond this normal variation. Consequently, the swept radius for markers analyzed in pooled samples will almost certainly be less than that possible with individual animal analysis as well as different from one marker to another. From the numbers shown in Table 9.5, the detection of linkage out to a distance of ~ 20 cM, but not much farther, would appear possible. Thus, up to 50% more markers may be required to sweep the entire genome.

The pooled DNA approach is maximally resolving when the nature of the phenotype under analysis allows the investigator to obtain two pools representing samples from each of the parents in the backcross (the $F_1$ and strain B in Fig. 9.11) or the two original strains used to generate the intercross (strain A and strain B in Fig. 9.12). In a situation of this type, each departure from the control ratio observed for a marker in one pool should be accompanied by a departure in the opposite direction for the other pool (see Table 9.5). This requirement for confirmation will act to reduce the frequency of false-positive results. In many experimental situations, however, it will only be possible to develop a single pool of homozygous $m/m$ samples for analysis. This will be the case for backcross studies of incompletely penetrant traits and for intercross studies of fully recessive phenotypes (Figure 9.12). In such cases, it will be necessary to generate and determine a phenotype for a larger number of animals in order to identify the smaller subset of samples that can be included within the single pool that can be made available for comparison to the control.

Once markers potentially linked to the test locus have been identified by the DNA pooling approach, it is essential to go back with each "positive" marker and individually type each sample in the pool to obtain quantitative confirmation of linkage or to rule it out. However, even with the reduction in genetic resolution and the requirement for confirmatory analysis, the DNA pooling approach can reduce the number of samples to be analyzed by at least an order of magnitude with large savings in labor and cost. If linkage to a single marker has been confirmed through individual sample analysis, the investigator can retype each of the samples with additional markers that lie within a 30 cM radius on either side to pursue the haplotype analysis described in the next section.

### 9.4.4.5 Determining gene order: generating a map

Once linkage has been demonstrated for a new locus, it is usually straightforward to determine its relative position on the chromosome framework map. For backcross data, this is accomplished by a method referred to as haplotype analysis.Haplotype analysis is performed on one linkage group at a time. For the mapping of any new locus, it is only necessary to carry out this approach for the chromosome to which the locus has been linked. The first task is to classify each backcross animal according to the alleles that it carries at the marker loci typed just on the chromosome of interest. By definition, when two or more animals carry an identical set of alleles, they have the same "haplotype" on that chromosome. By comparing the data obtained for all members of the backcross panel, one can determine the total number of different haplotypes that are present.

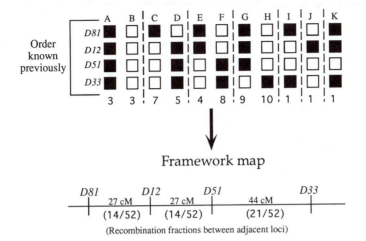

Framework map

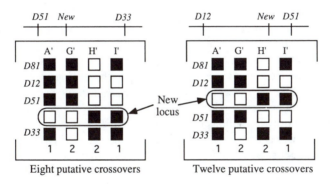

Placement of new locus in alternative positions relative to *D51*

**Figure 9.15** Haplotype analysis of backcross data. Illustration of the approach used to map a new locus relative to marker loci typed in a backcross. At the top, haplotypes are shown for the marker loci on one chromosome, which is determined to show evidence of linkage to the new locus under analysis. These haplotypes allow the generation of the framework map shown in the middle of the Figure. Recombination fractions derived directly from the haplotype data are indicated under the framework map. These fractions have been converted into centimorgan distances with the Carter–Falconer mapping function (equation 7.3). At the bottom, recombinant haplotypes are shown only for mice with crossover events between the new locus and the linked *D51* locus.

As illustrated in Fig. 9.15, each distinct haplotype is represented by a column of boxes, with one box for each locus; each box is either filled in to indicate one parental allele or left empty to indicate the other parental allele. Marker loci are placed according to their order along the chromosome from most centromeric at the top to most telomeric at the bottom. The number of animals that carry each haplotype is indicated at the bottom of each column. Haplotypes are presented in order from left to right according to the number and location of recombination events. Parental haplotypes—showing no recombination—are indicated first. Haplotypes with single recombination events are presented next, followed by those with two recombination events, and more, if they exist. Vertical lines can be used to separate haplotype pairs defined by reciprocal allele combinations. The class of single recombination haplotypes is presented in order from left to right according to the position of the breakpoint from most centromeric to most telomeric. The haplotype diagram shown in Fig. 9.15 can be generated automatically (in printable form) from recorded data in the Manly (1993) Map Manager program.

The haplotype diagram can be used to generate a linkage map by adding up the total number of animals that are recombinant between adjacent loci. For example, the G, H, I, and K haplotypes show recombination between the hypothetical *D51* and *D33* loci shown in Fig. 9.15; these haplotypes are carried by 9, 10, 1, and 1 animals respectively. Thus, in total, 21 animals are recombinant between these loci for a calculated recombination fraction (*rf*) of 0.404. When a recombination fraction is larger than 0.25, one should use the Carter–Falconer mapping function (equation 7.3) to obtain a more accurate estimate of map distance in centimorgans. The calculated $m_{FC}$ value is 44 cM. Similarly, the recombination fractions that separate *D81* from *D12*, and *D12* from *D51* are both found to be 0.269. With the Carter Falconer equation, this recombination fraction value is adjusted slightly to a map distance of 27.3 cM.

With a framework haplotype diagram and map, it becomes possible to determine the location of a new locus under analysis. Consider the hypothetical example in Fig. 9.15 where linkage has already been demonstrated between a new locus and just one marker locus—*D51*. In this case, the new locus could be in either one of two positions on the chromosome, proximal or distal to *D51*. To test these two locations, one can draw a second set of haplotype diagrams that include only those newly defined haplotypes showing recombination between the linked anchor *D51* and the new locus. In this example, a subset of animals from the previously defined haplotype classes A, G, H, and I define four new haplotypes labeled A′, G′, H′, and I′, respectively, as illustrated in Fig. 9.15. These haplotypes are drawn in two different ways with the new locus either proximal or distal to *D51*. The correct order can be determined by minimizing both the number of multiply recombinant haplotypes and the total number of implied recombination events within the sample set. In the example shown, a distal location requires a total of eight crossover events that take place within four single recombinant chromosomes and two double recombinant chromosomes. Alternatively, a proximal location requires a total of 18 crossover events with no single recombinant chromosomes, one double, six triples and one quadruple. Data of this type clearly point to a distal location for the new locus. Although any real set of data will obviously give different results, the same logical progression will almost always provide a definitive map position. With the computer program Map Manager, this analysis can be accomplished automatically.

With intercross data, whole chromosome haplotype analysis can be much less straightforward (as illustrated in Fig. 9.3). Consequently, gene order is usually determined computationally by the method of maximum likelihood analysis (Lander et al., 1987). Nevertheless, with the aid of a framework map, it is usually possible to breakdown $F_2$ genotype information into pairs of most likely haplotypes for each animal (D'Eustachio and Clarke, 1993). At this point, a new locus could be mapped according to the same logic described above.

### 9.4.5 The second stage: high-resolution mapping

The ultimate goal of the second stage of many mapping projects is to identify both DNA markers and recombination breakpoints that are closely enough linked to a new locus of interest to provide the tools necessary to begin positional cloning. This second stage can be broken down optimally into a series of steps as follows:

*Step 2.1.* The first goal of this second stage should be to narrow down the map interval as much as possible using only the small panel of samples typed in stage 1. This can normally be accomplished by selecting and typing additional microsatellite markers spaced across the 20 cM region to which the locus of interest has been mapped. With an original panel of 54 backcross samples, for example, recombination breakpoints will be distributed at average distances of about 2 cM. Thus, by typing additional markers, one should be able to reduce the size of the gene-containing interval from an original 25–40 cM down to 4–10 cM. The goal of this step is to identify the closest "limiting markers" on both sides of the locus of interest that *do* show recombination with it in order to establish an interval within which the locus must lie.

*Step 2.2.* The next step requires the breeding of a large number of animals that segregate the mutant allele. Ideally, the total number of animals bred should be at least 300 with a maximum of 1,000 spread among several crosses (see Section 7.2.3.3). However, this large set can be quickly reduced to the smaller set of samples that show recombination in the interval to which the gene has already been mapped. This can be accomplished by typing each animal for just the two "limiting markers" identified in step 2.1. If, for example, the locus-containing interval had previously been restricted to a 10 cM region bounded by these markers, this analysis would eliminate from further consideration approximately 90% of the total samples in the large cohort. If a PCR-based analysis is used to type the two markers, rapid methods for obtaining small quantities of partially purified DNA from members of the large cohort may be sufficient (Gendron–Maguire and Gridley, 1993).

*Step 2.3.* The smaller subset of animals selected in step 2.2 can now be typed with a larger set of markers previously localized to the genomic interval between the two limiting markers defined in step 2.1. At this point, it makes sense to test all segregating microsatellites that have been placed into 1 cM bins extending from one limiting marker to the other as well as any other suitably located DNA markers (Copeland et al., 1993; Appendix B). Newly tested markers that show no recombination with either one limiting marker or the other (among all animals tested) are likely to map outside the defined interval. However, all new markers that show recombination in different samples with each of the previously defined limiting markers will almost certainly map between them. Haplotype analysis can be used once again to obtain a relative order for these newly mapped markers. If the initial interval defined in step 2.1 is 10 cM or less, double recombination events will be extremely unlikely, and with this underlying assumption, it should be possible to obtain an unambiguous order for all markers that show recombination with each other and/or the phenotypically defined locus.

*Step 2.4.* As multiple new markers are mapped to the interval between the two previously defined limiting markers, it should become possible to reduce the size of the gene-containing interval even further than the one defined in step 2.1. As the size of the interval is reduced, the number of animal samples within the panel that need to be analyzed further can also be reduced to include only those that show recombination between the newly defined limiting markers. Additional markers should be typed until one reaches the ultimate goal of identifying limiting markers that each show only one (ideally) or a few recombination events on either side of the locus of interest along with one or more markers that show absolute concordance with the locus itself as illustrated in Fig. 10.1. If one exhausts the available sources of markers without coming close to this goal, it may be necessary to derive additional region-specific markers as discussed in Section 8.4.

With the identification of one or more DNA markers that show no recombination with the locus of interest, Figs 9.16 or 9.17 can be used to gain a

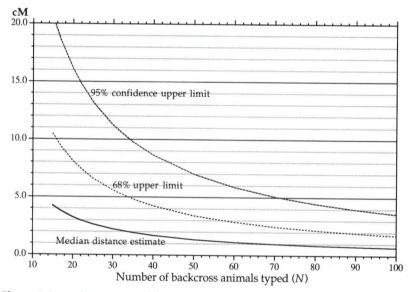

**Figure 9.16** Interpretation of the absence of recombination between two loci in a backcross analysis with 15–100 animals. Median linkage distance estimates and upper limits on the 68% and 95% confidence intervals are indicated for cases where no recombination is observed between two loci among all members of a backcross panel. Distances are shown in centimorgans.

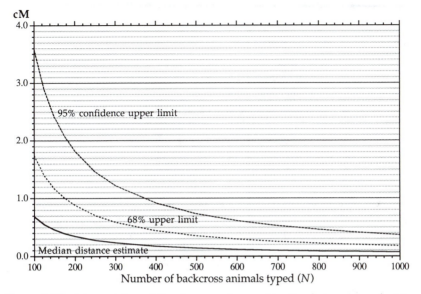

**Figure 9.17** Interpretation of the absence of recombination between two loci in a backcross analysis with 100–1,000 animals. Median linkage distance estimates and upper limits on the 68% and 95% confidence intervals are indicated for cases where no recombination is observed between two loci among all members of a backcross panel. Distances are shown in centimorgans.

sense of the distance that separates them. For example, with an initial cohort of 380 animals, the mean distance that will separate marker and locus will be 0.2 cM, and the 95% confidence interval will extend out to 1 cM. With an initial cohort of 1,000 animals, the mean distance will be less than 0.1 cM and the 95% confidence interval will extend out less than 0.4 cM. At this stage of analysis, one can move on to the task of generating a physical map that extends across the genomic region between the two closest limiting markers as described in Section 10.3.3.3 and Fig. 10.1.

## 9.5 QUANTITATIVE TRAITS AND POLYGENIC ANALYSIS

### 9.5.1 Introduction

Most of the phenotypic characteristics that distinguish different individuals within a natural population are not of the all or none variety associated with laboratory-bred mouse mutations like albino, non-agouti, brown, quaking, Kinky tail, and hundreds of others. On the contrary, easily visible human traits such as skin color, wavy hair, and height, as well as hidden traits such as blood pressure, musical talent, longevity, and many others each vary over a continuous range of phenotypes. These are "quantitative traits," so-called because their expression in any single individual can only be described numerically based on the results of an appropriate form of measurement. Quantitative traits are also called continuous traits, and they stand in contrast to qualitative, or discontinous, traits that are expressed in the form of distinct phenotypes chosen from a discrete set.

Continuous variation in the expression of a trait can be due to both genetic and non-genetic factors. Non-genetic factors can be either environmental (in the broadest definition of the term) or a matter of chance. In mice, it is relatively straightforward to separate genetic from non-genetic contributions through the analysis and comparison of animals within and between inbred strains. Variation in expression among individual members of an inbred strain must be caused by non-genetic factors. Furthermore, if one is convinced that all individuals are maintained under identical environmental conditions, then existing variation is likely to be the result of chance alone.

Geneticists are, obviously, most interested in the genetic contribution to a quantitative trait. A genetic contribution cannot be demonstrated by looking at individuals from a single inbred strain alone. Rather, a comparison of expression levels must be made on sets of animals from two different inbred strains (Fig. 9.18). The statistical approach described in Appendix D2 can be used to determine formally whether two strains differ significantly in the expression of the quantitative trait. If a significant strain-specific difference is demonstrated and all other variables have been controlled for, it becomes possible to attribute the observed difference in quantitative expression to allelic differences that distinguish the two strains.

In practice, a quantitative trait is most amenable to genetic analysis in mice and other experimental organisms with a pair of inbred strains that show non-overlapping distributions in measured levels of expression among at least 20 members of each group (Fig. 9.18). Although a significant strain-specific difference can be demonstrated under much less stringent criteria (as described in Appendix D2), it becomes more and more difficult to ferret out the quantitative trait loci (QTLs) involved as the possibility of phenotypic overlap increases.

The appearance of a quantitative trait usually signifies the involvement of multiple genetic loci, although this need not be the case. In particular, a single polymorphic locus with multiple, differentially expressed alleles can give rise to continuous variation within a natural population. There may also be some instances where the expression of a quantitative trait is controlled by a mutant allele at a single locus with a high degree of variable expressivity (Asada et al., 1994). However, if a single locus is responsible for the entire genetic contribution to a quantiative trait difference between two inbred strains, this would almost certainly become apparent in the second generation of either an outcross–backcross or outcross–intercross breeding protocol. In the first instance, half the $N_2$ animals will be identical to the $F_1$ parent, and the other half will be identical to the inbred backcross parent at the critical locus as illustrated in the top panel of Fig. 9.19. The result would be a discontinuous distribution of phenotypes that fall into two equally populated classes with separable distributions that parallel those found for each of the first-generation parents. With the intercross protocol, $F_2$ animals will be distributed among three classes (in a 1:2:1 ratio) that will parallel the phenotypic distributions found among one parental strain, the $F_1$ hybrid, and the second parental strain.

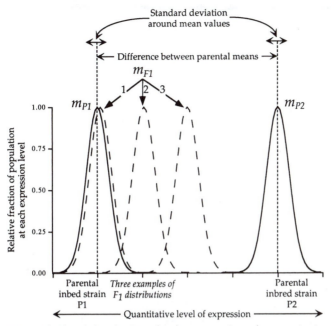

**Figure 9.18**  Idealized distributions for the expression of a quantitative trait in two inbred strains and the F1 hybrid between them. In this example, the mean levels of expression for two inbred parental populations ($m_{P1}$ and $m_{P2}$) are separable from each other by 16 standard deviations. Three examples of $F_1$ distributions are shown. The first $F_1$ distribution would be expected with a trait controlled by a series of P1 alleles that all show complete dominance to their corresponding P2 alleles. The third $F_1$ distribution would be expected with a trait controlled by P1 and P2 alleles, that all show strict semidominance. The second $F_1$ distribution would be expected with more complex scenarios including incomplete dominance skewed toward one parent or a combination of quantitative trait loci that show different dominance relationships. Since each population is genetically homogeneous, the intra-strain variation in expression must be due to environmental effects or chance. In actuality, very large numbers of animals would have to be typed to obtain distributions that so closely resemble the bell curves shown in this figure.

If a significant number of second-generation animals are found to express phenotypes intermediate to those found in the parental strains and $F_1$ hybrid,[97] it is most likely that multiple genetic differences between the progenitor strains are responsible as illustrated in the lower panels of Fig. 9.19. The term polygenic is used to describe traits that are controlled by multiple genes, each of which has a significant impact on expression. The term multifactorial is also used to describe such traits, but is more broadly defined to include those traits controlled by a combination of at least one genetic factor with one or more environmental factors.

Not all polygenic traits are quantitative traits. A second polygenic class consists of those traits associated with a discrete phenotype that requires particular alleles at multiple loci for its expression. Polygenic traits of this type

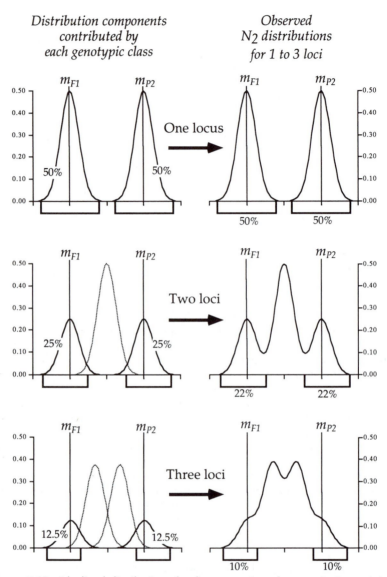

**Figure 9.19** Idealized distributions for the expression of a quantitative trait in $N_2$ backcross populations. Each pair of panels presents an example of a quantitative trait distribution that might be expected in a backcross population formed by a mating between the $F_1$ and P2 parents illustrated in Fig. 9.18. Component distributions are shown on the left, and composite distributions are shown on the right for quantitative traits controlled by one, two, or three segregating loci with strictly semidominant alleles and strictly additive effects. In the case of one locus, there are only two component distributions contributed by animals heterozygous or homozygous for alleles at this locus. In the case of two loci, there are three component distributions; one contains animals heterozygous at both loci like the $F_1$ parent, a second contains animals homozygous at both loci like the P2 parent, and a third contains animals heterozygous at one locus and homozygous at the other. Each complete parental-like

can be classified and analyzed with breeding protocols that are the same as those used for quantiative traits. For example, suppose strain DBA shows hypersensitivity to loud noises with 100% penetrance while neither strain B6 nor $F_1$ hybrid animals show any sensitivity. This result would indicate that hypersensitivity is recessive. Further analysis would proceed by backcrossing the $F_1$ animals to the homozygous recessive DBA parent. If instead, the DBA trait was expressed in a dominant manner, the backcross would be made to the homozygous recessive B6 parent. In either case, backcross offspring would be typed for hypersensitivity. If 25% or less of the backcross animals expressed the trait while all of the others were normal, this would provide evidence for the requirement of at least two DBA genes to allow phenotypic expression of a discontinuous trait.

It is important to mention that more complex scenarios are possible and likely to be the rule, rather than the exception. In particular, different members of the gene set involved in the expression of a trait may differ in their relative contribution to the trait; they may behave differently relative to their corresponding wild-type allele with some showing complete dominance or recessiveness, and others showing varying degrees of partial or semi-dominance; and they be involved in additive interactions, associative interactions or both. In some instances, a discrete trait may become quantitative upon outcrossing, or it may exhibit a threshold effect where the probability of expression in $N_2$ offspring increases with an increasing number of critical genes from the affected parental strain. The strategy described in the next section for the analysis of polygenic traits is a general one, which should be applicable to all of these situations. However, it is almost always true that the greater the genetic complexity, the larger the number of animals that will have to be bred and analzyed to obtain the same degree of genetic resolution.

### 9.5.2 A choice of breeding strategy and estimation of locus number

Whenever viability and fecundity are not a problem, it is much more efficient to analyze complex genetic traits through a backcross rather than an intercross. This is because each backcross animal will have one of only two genotypes at each locus. In contrast, offspring from an intercross can have one of three genotypes at each locus, which can combine into many more permutations with a set of multiple unlinked, but interacting, loci. Consider the situation where three loci are involved. With the backcross, all offspring

---

component contributes 25% to the total distribution. In the case of three loci, there are four components. One is identical to the $F_1$ parent and a second is identical to the P2 parent at all three loci; each of these parental components will represent 12.5% of the total distribution. The two other components of the three locus distribution contain animals heterozygous at one locus and homozygous at two, or heterozygous at two loci and homozygous at one. The bracketed portions of each distribution represent the subsets of animals that are most likely to carry the same parental genotype at each segregating locus.

will have one of $(1/2)^3 = 8$ different genotypes, whereas in the intercross, offspring can have one of $(1/3)^3 = 27$ different genotypes. Furthermore, as described below, the most efficient initial method for the analysis of polygenic traits is based on the collection and analysis of DNA from only those animals that express the most extreme forms of the phenotype since these animals are most likely to be homozgyous for all of the involved genes. If three genes are involved, $\sim 12.5\%$ of the $N_2$ animals will have a genotype equivalent to the backcross parent and an equivalent proportion will be identical to the $F_1$ parent. However, in offspring from an intercross, only 1.6% will be expected to have a genotype equivalent to that of each parental strain. Finally, as discussed in Section 9.4.3, when marker data are finally obtained, their compilation and analysis is much easier for a backcross than an intercross.

Before embarking on a detailed mapping project, it is useful to derive an estimate of the number of segregating genes involved in the expression of the trait under analysis. In complex cases of inheritance, the derivation of such an estimate will not be possible. However, an estimate can be made in two simple situations. The first is that of a discrete phenotype whose expression shows an absolute requirement for alleles at multiple unlinked loci from the affected parent. With a sufficient number of backcross animals, an estimation of gene number in this situation is trivial because the expression of the variant phentotype is absolutely correlated with the presence of a parental strain genotype at all involved loci. The probability of this occurrence is $(0.5)^n$, where $n$ is the total number of loci required for expression. Thus, if the observed proportion of affected animals is $\sim 25\%$, this would imply the action of two required genes, at $\sim 12.5\%$, the prediction would be three genes, at $\sim 6.25\%$, the prediction would be four genes, and so on. With these numbers, it is easy to see that each additional locus will require a doubling in the total number of backcross animals for which a phenotype must be determined to obtain the same number of affected animals for genotyping.

In the case of quantitative traits, it is also possible to estimate gene number if one makes the simplifying assumption that all the genes involved are unlinked and active in a strictly semidominant manner with an equivalent contribution to the phenotype. In this situation, one can use a modified form of a formula derived by Wright (1952) for an intercross analysis and known as "Wright's polygene estimate:"

$$\text{Number of loci} = \frac{(m_{P2} - m_{F1})^2}{4(|V_{N2} - V_{F1}|)} = \frac{(m_{N2} - m_{F1})^2}{(|V_{N2} - V_{F1}|)} \qquad (9.11)$$

where $m_{P2}$, $m_{N2}$ and $m_{F1}$ are the mean values of expression of the backcross parent, the $N_2$ population and the $F_1$ hybrid, respectively, and $V_{N2}$ and $V_{F1}$ are the computed variances[98] for the $N_2$ and $F_1$ populations, respectively. The two forms of the equation shown here are mathematically equivalent so long as the mean value of the $N_2$ population is halfway between the means of the $F_1$ and the backcross parent. One can see the logic behind this equation by considering the probability that a backcross animal will show an extreme phenotype associated with one of its parents. From Fig. 9.19, one can see the proportion of genotypes equivalent to either parent drop by a factor of two with each successive increase in locus number from one to two to three, and so on. As a consequence, the variance in the complete $N_2$ generation (shown in the right panels of Fig. 9.19) will also drop as values tend to cluster more around the mean. As the $N_2$ variance goes down, the denominator of equation 9.11 will decrease as well. It is important to realize that equation 9.10 will only provide a very rough, *minimum* estimate of locus number because it is unlikely that all of the assumptions that went into the use of the equation will hold true in a real biological situation.

### 9.5.3 Choices involved in setting up crosses

The first step in polygenic analysis is the same as the first step in mapping a single phenotypically defined locus—the choice of two parental strains (Section 9.4.2). Unlike the situation with single locus studies, the two parental strains to be chosen for polygenic analysis must be inbred; if not, unexpected and uninterpretable genetic complications could arise. The most important consideration in the choice of parental strains is that they should show the greatest difference possible in the expression of the trait under analysis. Other considerations are the same as those discussed in Section 9.4.2 with the caveat that an investigator may want to avoid interspecific, and perhaps intersubspecific, crosses because of the possibility that "abnormal" admixtures of alleles may not function together as they would in a normal offspring from either breeding group.

Upon choosing two inbred parental strains (called P1 and P2 in the following discussion), one should perform a cross to obtain $F_1$ hybrid offspring. However, before proceeding to a second-generation cross, it is critical to determine the expression of the trait of interest in the $F_1$ population. Figure 9.18 shows different examples of the potential results that might be obtained. If the pairs of alleles present at all "polygenetic" loci that distinguish P1 from P2 act in a strictly semidominant manner, the $F_1$ population will show a mean level of expression halfway between the means of the two parental strains (example 3 in Fig. 9.18). On the other hand, the $F_1$ population may show a distribution that is indistinguishable from one parental strain or the other if there are strong dominant effects (example 1 in Fig. 9.18). Finally, a likely result is the complex one with unequal allele strengths—but not strict dominance—that lead to a distribution differing from both parental strains, but with a mean value that is closer to one than the other (example 2 in the figure). In fact, the $F_1$ distribution can have a mean value that lies anywhere along the continuum between the two parental means. However, in all cases, the standard deviation around this mean value should be similar to that found with the parental strains, since the $F_1$ population is always genetically homogeneous.[99]

If the mean expression of the $F_1$ population lies essentially halfway between that found with the two parental strains, then the backcross can be performed with either parent. Other criteria, such as reproductive performance, should be the deciding factors (Chapter 4 and Table 4.1).[100] However, if the mean $F_1$ expression is closer to one parent (such as P1 in examples 1 and 2 shown in Fig. 9.18), one should backcross $F_1$ animals to the opposite parent (P2 in the example). As one can see from Fig. 9.18, this choice will serve to minimize the degree of phenotypic overlap between the two "parents of the backcross" and will allow a more accurate identification of $N_2$ animals with genotypes that match one parent or the other as discussed below.

It has been customary in mouse genetic studies to perform a backcross between an $F_1$ female and a male from the chosen parental strain. The main advantage to backcrossing in this direction is the higher fecundity of $F_1$ females that results from "hybrid vigor." However, as discussed in Sections 9.4.1 and 9.4.4.2, it may sometimes be more advantageous to cross the $F_1$ male with an inbred female.

### 9.5.4 An optimal strategy for mapping polygenic loci

#### 9.5.4.1 As the number of loci increases, interpreting results becomes more difficult

Once backcross ($N_2$) progeny are obtained, they can be analyzed for expression of the trait of interest with the same protocol used to measure expression in the $F_1$ and progenitor strain (P1 and P2) populations. When a sufficient number of $N_2$ animals have been tested, the distribution of expression levels can be plotted and compared to the distributions obtained with the $F_1$ and P2 populations. The right-hand side of Fig. 9.19 shows examples of the idealized distributions that one would obtain upon analysis of a trait whose expression is determined through the additive effects of semidominant alleles at one, two, or three loci that contribute equally to expression levels.

Consider the trival case in which a trait that was thought to be polygenic is actually controlled primarily by a single locus, $A$, having two semidominant alleles $A^1$ and $A^2$. There are only two potential gentoypes in the $N_2$ population obtained from backcrossing to parent P2: $A^1/A^2$ and $A^2/A^2$. Thus, the complete distribution shown in the upper right-hand panel of Fig. 9.19 can be broken down into the two separate distributions associated with each of these genotypes, as shown in the upper left-hand panel. This analysis shows that the telltale sign of involvement of only a single major locus is a biphasic distribution with peaks similar to those of the parents and a paucity of animals in between.

Next, consider the simplest case of polygenic inheritance with alleles at two major loci, $A$ and $B$, that both have additive semidominant effects on expression. In this case, the number of relevant $N_2$ genotypes doubles from two to four: $A^1/A^2$ $B^1/B^2$; $A^1/A^2$ $B^2/B^2$; $A^2/A^2$ $B^1/B^2$; and $A^2/A^2$ $B^2/B^2$. If one assumes that the pair of genotypes containing one heterozygous and homozygous locus affect expression equally, the idealized distribution pattern shown in the middle right-hand panel of Fig. 9.19 would be obtained. This idealized pattern can be broken down into the three subdistributions that correspond to the different genotypic classes as shown in the middle left-hand panel; the intermediate subdistribution is twice as high as the side distributions because of the contribution of two genotypes rather than one.

In some experimental cases, when there is a sufficient distance between the mean values of expression of the two parents, it may be possible actually to obtain a triphasic distribution pattern with a shape and peak distribution similar to that shown in the middle right-hand panel of Fig. 9.19. A result of this type would be a sign that only two major additive loci were involved in the expression of the trait.

In most experimental situations, the distribution patterns obtained for the expression of a complex trait of interest in an $N_2$ population are unlikely to show significant evidence of multiple phases and multiple peaks. Rather, the most likely distribution will be an undifferentiated continuum that extends across the range between and beyond the mean values of expression observed for the $F_1$ and P2 parental populations. There are several factors that are likely to contribute to this tendency toward a monophasic distribution. First, with each increment in the number of loci having an effect on expression, there will be a doubling in the number of different genotypes that are possible in the $N_2$ population. With just three loci, the number of genotypes will be eight. If alleles at all three loci show addititive semidominant effects, a distribution of the form shown in the bottom right-hand panel of Fig. 9.19 will be obtained. This nearly monophasic distribution results from the combination of only four subdistributions that correspond to separate genotypic classes.[101] As the number of genes involved grows beyond three, the possibility of seeing multiple distribution peaks that correspond to different genotypic classes is essentially nil.

### 9.5.4.2 Selective genotyping

For the purposes of genetic analysis, the most critical feature of polygenic, quantitative trait inheritance is the impossibility of correlating intermediate levels of phenotypic expression with particular genotypes at each of the segregating loci involved. This problem is clearly visible even in the idealized distributions of the three-locus trait shown in the bottom panels of Fig. 9.19. In this simple example, an $N_2$ phenotype halfway between the means of the $F_1$ and P2 parents could be caused by heterozygosity at any one or two of the three loci involved; thus, this halfway phenotype is almost useless in terms of providing marker linkage information. However, there will always be one or two portions of each $N_2$ distribution that will have a high level of predictability for genotype at linked markers—the tails at one or both ends.

An $N_2$ animal that shows an extreme level of phenotypic expression that is, in fact, within the normal range observed for one parental strain (either the $F_1$ or P2) is likely to have the genotype of that parent at *all* of the segregating trait loci that distinguish the $F_1$ and P2 parents. This means that a set of animals with the same extreme phenotype at one end of the $N_2$ distribution will be likely to show a significant level of concordance with the same parental genotype at all markers that are closely linked to any one of the segregating trait loci. For example, imagine that one has chosen a subset of 20 $N_2$ animals that most resemble the P2 parental strain in the expression of a trait and each animal within this set is typed for DNA markers that span the genome. A marker that is closely linked to any one of the trait loci will appear homozygous for the P2 allele in a significant majority of the animals of this subset. If possible, a second subset of animals could be collected that most resemble the $F_1$ population; markers linked to the trait loci would appear heterozygous for the P1 and P2 alleles in a significant majority of the animals of this subset.

The strategy just described, known as "selective genotyping," provides the most highly efficient means for mapping polygenic loci (Soller, 1991). Phenotypic analysis is performed on the complete set of backcross animals, which should typically number in the hundreds. This analysis allows the investigator to identify one or two smaller subsets of $N_2$ animals with the greatest amount of genotypic information content. DNA typing is performed only on these smaller subsets, each of which can be pooled together into single composite samples as described in Section 9.4.4.4.

How does one decide what proportion of $N_2$ animals to include in each extreme phenotypic subset when a continuum of expression levels is observed for the whole population? The answer is not simple. If one is too stringent, there may be too few animals to type and the power of the linkage test will suffer accordingly (Fig. 9.13). However, if one is too lax, animals without a parental genotype at each critical locus will be included at a higher frequency (Fig. 9.19). This will cause the level of discordance with truly linked markers to increase beyond the actual recombination fraction to a point that may fall beyond the level of significance shown in Fig. 9.13. There will obviously be an optimal cutoff point but it will be impossible to ascertain its position in advance without knowing how many segregating loci have a major effect on expression. As illustrated in Fig. 9.19, as the number of loci grows, so does the phenotypic overlap between each completely parental genotypic class (indicated with dark lines) and its adjacent mixed genotypic class (indicated with lighter lines).

In a first round of analysis without prior information, a reasonable fraction of backcross animals to include within each extreme subset would be 10% (Soller, 1991). Since it is important to have at least 20 individual samples within each composite sample for DNA pooling, this would entail the inital phenotypic analysis of at least 200 backcross animals. With a sample size that is this small, the swept radius is quite modest (see Fig. 9.13) and a large number of markers will be required to span the whole genome. If it is possible to pool together 30 or 40 samples, this will greatly improve the sweep of individual markers. Alternatively, if the DNA pooling method provides evidence of potential marker linkage, the results obtained upon analysis of individual samples in the two extreme classes (if there are two that can be formed) can be combined for greater statistical power.

The results obtained from the initial analysis of the 10% DNA pools will provide the investigator with a certain amount of information on the experimental direction that is best to follow. For example, if the initial analysis allows the identification of even one marker that shows 100% concordance within an extreme phenotypic class, it is likely that this class does not contain any animals with non-parental genotypes. Thus, it would be worthwhile to expand the extreme class to include a larger sample size to search more efficiently for markers linked to additional loci that affect trait expression. Furthermore, positive results with individual markers that fail to meet the most stringent requirements for significance could still be pursued through the typing of markers that are removed by 10–20 cM and may be closer to a potential trait locus. If a trait locus is, indeed, present in the vicinity of the original marker, this strategy could yield closer markers that will show higher levels of concordance and significance. Finally, once QT loci have been defined, an investigator can return to the complete set of animals and type all of those not typed already for closely linked QT marker loci. This more comprehensive data set can be subjected to advanced non-parametric statistical methods, such as the Mann–Whitney $U$ test[102] (available within most statistical software packages for desktop computers), in order to better understand the nature of the interactions among QT loci.

# 10

# Non-breeding Mapping Strategies

## 10.1 LINKAGE MAPS WITHOUT BREEDING

### 10.1.1 Single sperm cell typing

In 1988, Arnheim and his colleagues (Li et al., 1988) reported the extraordinary finding that unique DNA sequences could be amplified from isolated sperm cells. Amplification from the single DNA target locus that is present in haploid sperm cells represents the theoretical limit in sensitivity that is possible with the polymerase chain reaction (PCR). The implications of this experimental breakthrough were enormous, especially in the field of human genetics. For the first time, it became possible to consider the analysis of unlimited numbers of individual meiotic events derived from a single individual.

Linkage information could be derived by first identifying a male volunteer who was heterozygous at all loci of interest in a manner that could be distinguished with the use of PCR. The DNA within individual sperm cells donated by the volunteer would be subjected to co-amplification with primer pairs that define each of the loci. Allele determination at each locus in each cell could, in theory, be accomplished by any of the various PCR techniques described in Section 8.3, including hybridization to allele-specific oligonucleotides, restriction enzyme digestion, or single-strand confirmation polymorphism SSCP. In contrast to typical genetic studies in humans, the typing of large numbers of single sperm cells from a single volunteer provides genetic information that is simple to interpret. Linkage distances and gene order can be derived directly by counting the number of cells with each allelic combination (Goradia et al., 1991), in a manner equivalent to that described in Section 9.4 for the analysis of backcross data in the form of haplotypes. Since the number of sperm cells is essentially unlimited, the resolution of the map obtained is a function simply of the time and effort that the investigator wishes to expend in the typing of additional cells. Furthermore, with the use of a protocol for universal PCR amplification, Arnheim and his colleagues have been able to co-amplify the majority of sequences present in each sperm genome (Zhang et al., 1992). Thus, in theory, it should be possible to co-type many different loci within a single panel of sperm samples.

Although single sperm typing provides a significant new tool for mapping in humans and other large animals that also do not provide a sufficient number of offspring for typing, its importance to mouse genetics would appear to be more limited since numerous, well defined backcross mapping panels are available for analysis. Furthermore, at the time of this writing, the typing of single cells is still so technically-demanding that it has not been used in a general way even by the broader community of human geneticists.

### 10.1.2 Mitotic linkage maps

Classical linkage analysis is based on meiotic recombination events that occur in sperm and egg precursor cells. The products of these events are observed and counted in the offspring of heterozygous animals (or in sperm cells as described in the previous section). Meiotic recombination is a very frequent event—it occurs 30 times, on average, in each individual germ cell line—and it appears to have been selected by evolution to play two very different roles in the biology of higher eukaryotes. First, the physical event itself appears to be required to tether the homologs of each chromosome to each other so that they line up and segregate into opposite daughter cells during the first meiotic division. Second, the production of offspring with non-parental combinations of alleles provides a non-mutational means for the generation of diversity at the genotypic level, and this generation of diversity appears to be generally beneficial to the population as a whole.

Recombination has also been observed at the mitotic level in somatic cells (Rajan et al., 1983; Henson et al., 1991). In comparison to meiotic recombination, mitotic events are exceedingly rare, and they do not appear to have any biological function. It seems most likely that mitotic recombination events are simply accidents that happen in response to spontaneous nicks in the DNA molecule that allow migrating single strands to invade opposite homologs. Usually, mitotically recombined cells will go entirely unnoticed among the millions of nearby cells having *germ line* haplotypes. However, in individuals that are born heterozygous for null alleles at "tumor suppressor genes" such as retinoblastoma (RB), mitotic recombination can lead to the production of rare homozygous mutant cells that are released from growth control in the absence of the wild-type allele; uncontrolled division of these cells leads to tumor formation. It now appears that a large class of human tumors are caused by this *homozygosing* at a variety of tumor-suppressor genes (Marshall, 1991; Weinberg, 1991).

If rare cells that undergo mitotic recombination could be identified and recovered in clonal form from a tissue culture line, a means for generating a linkage map that was not dependent on the breeding of animals would be possible. Such a "mitotic linkage map" has been obtained for the proximal half of chromosome 17 (Henson et al., 1991). The generation of this particular map was dependent upon the ability to immuno select cells that had undergone allele loss at an H-2 complex gene. Selected cells were isolated and expanded into cultures that could be analyzed for various DNA markers that were heterozygous in the parent. All loci that map proximal to the clone-specific break point will remain heterozygous; all that map distal will have become homozygous. Through the analysis of a large number of individual clonal lines, it becomes possible to construct a linkage map with gene order and an estimate of relative distances between loci. The mitotic linkage map constructed for the proximal half of chromosome 17 corresponds well with the meiotic linkage map (Henson et al., 1991). As expected, the gene order is identical and there is some minor variation in the relative intergenic distances.

The construction of a mitotic linkage map in one chromosomal region was important for providing biological information concerning the relationship between the distribution of mitotic and meiotic crossover sites, and future experiments of this type may also provide clues to the mechanisms responsible for homologous recombination in somatic cells. However, the mitotic map did not provide any new information specific to chromosome 17, and this approach is not likely to play an important role in future gene mapping experiments for two reasons. First, with current technology, mitotic maps can only be constructed along chromosomal regions that are marked with genes encoding polymorphic cell surface antigens that are expressed codominantly in cultured cells, and are readily distinguishable from each other with specific monoclonal antibodies.[103] The second reason is that the construction of traditional meiotic linkage maps at the same resolution is likely to be much faster and easier.

## 10.2 CHROMOSOMAL MAPPING TOOLS

### 10.2.1 Conservation of synteny

As cloning and mapping of both the mouse and human genomes began in earnest during the 1980s, two important evolutionary facts became clear. First, nearly all human genes have homologs in the mouse and vice versa. Second, not only are the genes themselves conserved, but so is their order—to a certain extent—along the chromosome. In 1984, Nadeau and Taylor used linkage data obtained from 83 loci that had been mapped in both species to estimate the average length of conserved autosomal segments as 8.2 cM, in the mouse (Nadeau, 1984). In 1993, the same analysis was performed on linkage data obtained from 917 homologous loci mapped in both species to yield an average conserved chromosomal length of 8.8 cM which is not significantly different from the earlier estimate.[104] The major evolutionary implication of this result is that approximately 150 major rearrangements have occurred along the human or mouse lines as they diverged from a common ancestor that existed 65 million years ago.

The practical implication of conserved chromosomal segments is that the mapping of a gene in one species can provide a clue to the location of its homolog in the other species. One should be cautious, however, in not overinterpreting synteny information. There are many examples of smaller genomic segments that have popped out or into larger syntenic regions. Thus, even if a human gene maps between two human loci with demonstrated synteny in the mouse as well, there is still a small chance that it will have moved to another location in the mouse genome. Nevertheless, over 80% of the autosomal genomes of mice and humans have now been matched up at the subchromosomal level (Copeland et al., 1993). Thus, with map information for a gene in humans, it will often be possible to identify a corresponding mouse chromosomal segment of ~ 10 cM in length as a likely location to test first for linkage with nearby DNA markers.

## 10.2.2 *In situ* hybridization
### 10.2.2.1 *Overview*

The technique of *in situ* hybridization was conceived of by Gall and Pardue (1969) and John and his colleagues (1969). These workers demonstrated that the DNA within preparations of chromosomes attached to microscope slides could be denatured in a gentle manner so as not to disrupt the overall morphology of the chromosomes themselves. Target sequences within these chromosomes are then available for hybridization to labeled nucleic acid probes. Thus, *in situ* hybridization allows the mapping of cloned DNA sequences to specific chromosomal sites that can be visualized directly by light microscopy.

In early work, probes were labeled with radioactive isotopes and target sequences were identified by autoradiography. This method of labeling and detection limited both the sensitivity of the technique and its resolution (Lawrence, 1990). In particular, the original protocol only allowed the detection of tandemly repeated sequences such as the ribosomal genes and satellite DNA. By 1981, however, investigators had optimized the in situ protocol for use in mapping single copy mammalian sequences (Harper et al., 1981), and in 1984, an improved method was developed for better resolution of chromosome banding patterns (Cannizzarro and Emanuel, 1984). Nevertheless, the technique was still not ideal because with single-copy radioactive probes, localization could not be determined within the chromosomes of a single cell; instead, it was necessary to perform a statistical analysis of silver grain distributions in 50–100 sets of metaphase chromosomes.

Two critical changes in the protocol now allow the detection of single-copy sequences and their high-resolution mapping through the direct observation of single chromosomes. The first change was in the nature of the label; with the substitution of fluorescent tags for radioactive ones, the physical resolution of the hybridization site was dramatically improved. The modified *in situ* protocol that utilizes fluorescent tags is referred to as FISH (for fluorescent *in situ* hybridization). The second change was in the nature of the hybridization cocktail. With the inclusion of a large excess of unlabeled total genomic DNA, it is possible to block dispersed repetitive sequences—present in essentially every genomic region larger than a few kilobases in length—from hybridization to their targets throughout the genome. This allows the use of whole-phage or cosmid clones as probes leading to a substantial increase in signal strength, which will be proportional to the length of single-copy DNA in the clone. With these major changes in the protocol and other optimizations, it is now possible to use in situ hybridization to visualize the map position of any cloned locus within single chromosomes from any mammalian species (Lawrence, 1990; Trask, 1991).

Although *in situ* hybridization has played a pivotal role in the construction of the human gene map, its role in mouse gene mapping has been more limited for several reasons. First, a certain amount of specialized training and experience is required to perform this protocol, and thus, it is often not an option for independent investigators in the absence of a collaboration. Second, in humans, classical linkage analysis is not easily performed, and thus, alternative methods for human mapping are much more important. Third, whereas the human karyotype is highly amenable to direct cytogenetic analysis—chromosomes come in a variety of shapes and sizes, and staining techniques yield excellent banding resolution—the mouse karyotype is a cytologist's nightmare. All 20 chromosomes have the same shape with only a single visible arm and a centromere that appears to lie at one end (see Fig. 5.1). A continuum of chromosome lengths makes the identification of individual chromosomes more difficult, and finally, banding patterns are much less distinct and more difficult to resolve.

In the past, *in situ* hybridization had the advantage that it did not require the existence of variants between parental strains for mapping to be accomplished. However, with the advent of new methods for the detection of polymorphism discussed in Chapter 8, it has become possible to quickly identify DNA variants at essentially all cloned loci. Consequently, *in situ* hybridization is now used most often only for specific experimental problems such as those that described below.

### 10.2.2.2 Experimental usage

The power of *in situ* hybridization lies in the fact that it allows the direct localization of DNA sequences relative to all visible cytological landmarks such as centromeres, telomeres, and rearrangement break points in aberrant chromosomes. In some instances, it will be important to localize a DNA marker relative to one or more of these landmarks. For example, an investigator may have a DNA marker that maps to the beginning or end of a linkage map associated with a particular chromosome. *In situ* hybridization can be used to determine how close to the centromere or telomere the DNA marker actually is (in physical terms); this information can serve to establish the size of the chromosomal region that is not contained within the associated linkage map. In another example, an investigator may have a DNA clone, from either a wild-type or mutant animal, that is believed to extend across a cytologically visible inversion or translocation breakpoint. If the clone was derived from a wild-type genome, the *in situ* results would show hybridization to two sites in the rearranged karyotype.

As discussed in Section 7.3.2, *in situ* hybridization is also useful in the special case of mapping transgene insertion sites. The same DNA construct that was originally injected into the embryo can often be used directly as a probe. In another instance, *in situ* hybridization can be combined with classical linkage analysis using the *M. spretus* backcross system to follow the segregation of centromeres from one parental chromosome or the other as described in Section 9.1.2 (Matsuda and Chapman, 1991; Matsuda et al., 1993). Finally, *in situ* hybridization is useful in experiments aimed at questions that go beyond the simple mapping of genes. For example, the technology has revealed the unexpected finding that both LINE and SINE sequences are non-randomly distributed among bands and interband regions of all chromosomes as described in Section 5.4.4.

## 10.2.3 Somatic cell hybrid genetics
### 10.2.3.1 Overview of the classical approach

The ability to derive long-term cultures of mammalian cells was perfected during the 1950s. Cell cultures provided important experimental material for early biochemists and molecular biologists interested in molecules and processes that occur within mammalian cells, but they were of little use to geneticists since somatic cell genomes remain essentially unchanged during continual renewal through mitotic division. This situation changed dramatically during the early 1960s when investigators discovered and developed methods for the induction of cell fusion in culture (Ephrussi and Weiss, 1969).

Normal diploid cells from all species of mammals carry approximately the same amount of DNA in their nucleus (twice the haploid amount of 3,000 mb). Thus, after fusion between any two mammalian cells, the hybrid cell nucleus becomes, in effect, tetraploid, with a genome that is twice the normal size. The enlarged genomes of hybrid cells are inherently unstable. Presumably, the increased requirement for DNA replication acts to slow down the rate of cell division, and as a consequence, cells that lose chromosomes during mitotic segregation will divide more quickly and outgrow those cells that maintain a larger genome content. Eventually, after many events of this type, cells can reach a relatively stable genome size that is close to that normally found in diploid mammalian cells. For reasons that are not understood, hybrids formed between particular combinations of species will preferentially eliminate chromosomes from just one of the parental lines. In hybrids formed between mouse cells and either hamster or human cells, mouse chromosomes will be eliminated in a relatively random manner. This process has allowed the derivation and characterization of a number of somatic cell hybrid lines that stably maintain only one or a few mouse chromosomes.

The field of somatic cell genetics had its heyday in the 1970s and early 1980s when it provided the predominant methodology for mapping loci—albeit, often to the resolution of whole chromosomes. The major tools for gene detection in this era (before the recombinant DNA revolution was in full gear) were species-specific assays for various housekeeping enzymes. Somatic cell geneticists could type each member of a panel of hybrid cells for the presence of a particular enzyme and then use karyotypic analysis to demonstrate concordance with a particular chromosome. In a strictly formal sense, this type of analysis is analogous to classical two-locus linkage studies studies with one marker being the enzymatic activity and the other marker being the particular chromosome that contains the gene encoding the enzyme.

The somatic cell hybrid approach has always been more important to human geneticists than to mouse geneticists. This is because well-established somatic cell hybrid lines with one or a few mouse chromosomes are relatively rare compared to the large number of well-characterized hybrid lines with individual human chromosomes. There are several reasons for this state of affairs. First, the power of mouse linkage mapping has always been so great that somatic cell hybrid lines were never considered to be essential tools. Second, most mouse/hamster hybrid lines are chromosomally unstable and must be recharacterized each time they are grown in culture. With the difficulty of performing karyotypic analysis on mouse chromosomes, most investigators have shied away from this approach in the past. However, with alternative PCR-based methods for characterizing the chromosomal content of hybrids (Abbott, 1992), this problem may have been overcome so that the derivation of new hybrids for special situations may no longer be as formidable as it once was.

The use of somatic cell hybrid panels as a general approach to gene mapping has now been superseded by *in situ* hybridization—which resolves map positions to chromosome bands rather than whole chromosomes—and, of course, classical linkage analysis. However, there are two special cases where somatic cell hybrid lines can provide unique tools for mouse geneticists. First, their DNA can be used as a source of material for the rapid derivation of panels of DNA markers to saturate particular chromosomes or subchromosomal regions as described in Section 8.4.4 (Herman et al., 1991; Simmler et al., 1991). Second, their DNA can also be used for the rapid screening of new clones obtained from other sources to determine their presence in a particular interval of interest as described in Section 7.3.3. This can be accomplished with the use of duplicate blots containing just three lanes of restriction digested and fractionated DNA from: (1) the somatic cell hybrid line containing the chromosome of interest; (2) mouse tissue (a positive control); and (3) the host cell line without mouse chromosomes (a negative control). Each blot can be subjected to repeated probing with different potential markers. A negative result allows one to discard a particular probe immediately; a positive result can be followed-up by higher resolution linkage analysis.

### 10.2.3.2 Radiation hybrid analysis

In 1990, Cox, Meyers, and their colleagues described a novel technique for determining gene order and distance that is as highly resolving as traditional linkage analysis but does not depend upon breeding. The approach used has similarities to, as well as differences from, both recombinational mapping and physical mapping. Radiation hybrid mapping was originally developed for use with the human genome, but with appropriate starting material and a sufficient number of chromosome-specific DNA markers, it can be used in the analysis of any species (Cox et al., 1990).

The starting material is a somatic cell hybrid line that contains only the chromosome of interest within a host background derived from another species. As indicated above, a common host species used for mouse chromosomes is the hamster. A well-established, stable hamster cell hybrid line containing a single mouse chromosome can be subjected to irradiation with X-rays that shatter each chromosome into multiple fragments. The irradiated cells are then placed together with pure hamster cells under conditions that promote fusion. Approximately 100 new hybrid clones are recovered that contain fragments of the mouse chromosome present in the original hybrid line. Finally, each of these lines are analyzed for the presence of various DNA markers that had been mapped previously into the chromosomal region of interest.

The order and distance of loci from each other can be determined according to the premise that X-rays will break the chromosome at random locations. Thus, the closer two loci are together, the less likely it is that a break will occur between them. If two loci are side by side, they will either both be present or both be absent from all 100 cells with 100% concordance. If two loci are at opposite ends of the chromosome, there will still be cells that have neither or both, but there will also be a large number that have only one or the other. (A cell can carry both loci even if the frequency of breakage between the two is 100%, since it is possible for a hybrid cell to pick up more than one chromosomal fragment.) As the probability of chromosome breakage varies between 0% and 100% for various pairs of loci under analysis, the fraction of hybrid cells that carry both loci will vary from 100% down to a control value obtained for unlinked loci. Thus, by typing each of the "radiation hybrid cells" in the set of 100 for a series of DNA markers, it becomes possible to construct a linkage map that is highly analogous to traditional recombinational maps.

It is possible to obtain linkage maps at different levels of resolution through the use of different intensities of radiation to break chromosomes. For example, with high levels of radiation that break chromosomes once every 100 kb, on average, one could map loci from 10 kb to 500 kb; with lower levels of radiation, mapping could be performed over a window from 500 kb to 5 mb.

The analogy to classical recombination mapping is striking in that a determination of linkage distance in both cases is based on the probability with which chromosomes will break followed either by recombination (in the classical case) or by segregation upon cell fusion (in the radiation hybrid case). In both cases, linkage distances are determined by counting the ratio of offspring (pups or cells) that do or do not carry particular sets of DNA markers (alleles or genes). However, linkage distances obtained through radiation hybrid analysis are much more likely to be indicative of actual physical distances.

Although radiation hybrid analysis has provided a crucial tool for genetic analysis in humans, once again, it has not been as widely used by the mouse community because classical linkage analysis is so much more powerful. Nevertheless, the resolution of this protocol has been validated in a study of the region of mouse chromosome 2 surrounding the agouti locus (Ollmann et al., 1992). In this study, the radiation hybrid map that was obtained corresponded exactly with that predicted from linkage analysis, with a level of resolution that was approximately 40-fold higher. Thus, radiation hybrid mapping could serve to fill in the gap between linkage maps and physical maps, especially in "cold" regions between hotspots where distantly spaced markers cannot be separated by recombination (Section 7.2.3).

## 10.3 PHYSICAL MAPS AND POSITIONAL CLONING

There are two stages in the process of positional cloning. The first stage is the focus of a major portion of this book: to use formal linkage analysis and other genetic approaches—as tools—to find flanking DNA markers that lie very close to the locus of interest. With these markers in hand, one can move to the second stage of this pathway: obtaining clones that cover the critical region, then identifying the gene of interest apart from all other genes and non-genic sequences within this region.

This second stage will be the focus of the remaining portion of this chapter. In what follows, I will move away from the realm of the formal geneticist to that of the molecular biologist. However, for several reasons, my intention is only to provide an overview of the conceptual framework that underlies the various approaches being used at the current time. First, the topics of physical mapping and positional cloning have filled entire books and many excellent review articles. Second, these linked topics are driven by technology and new improved protocols are constantly moving old ones onto the shelves. Consequently, any detailed discussion of actual molecular techniques will quickly become outdated.

### 10.3.1 Prerequisites to positional cloning

The absolute first step in the process of positional cloning is the high-resolution mapping of the locus of interest relative to closely linked DNA markers. This process (described at length in Chapter 9) provides an investigator with two sets of complementary tools that are essential prerequisites to the actual generation of a physical map around the locus of interest. The first set of "tools" will be represented by the small number of animal samples with crossover sites in the vicinity of the locus. The second set of tools is the small group of closely linked DNA markers.

Once the phenotypically defined gene has been closely linked to one or more DNA markers, it becomes possible to consider the complete cloning of the region that must contain the gene. There are no absolute cutoffs for determining what level of linkage is necessary before one can pursue this path, but in general, linkage should be tighter than 1 cM. Ideally, it is best to start a cloning project with one, or preferably more, DNA markers that show absolute linkage to the gene of interest upon analysis of at least 300 meiotic events or 77 recombinant inbred lines. From the equations used to derive Fig. 9.8 and 9.17 (from Appendix D), one can determine that complete concordance in either of these cases provides a mean estimate for linkage distance of 0.23 cM, which translates into a mean physical distance of 460 kb between marker and locus. These data also provide a 95% confidence upper limit of 1 cM, which translates into a distance of 2,000 kb.

### 10.3.2 Pulsed field gel electrophoresis and long-range genomic restriction maps

It is possible to derive long-range restriction maps spanning genomic regions that have yet to be cloned (Barlow and Leharch, 1987). The main utility of such restriction maps is to place lower and upper limits on the physical distance that separates two or more DNA markers known to be linked from breeding studies or other methods discussed previously. With this information in hand, one can make a more informed decision as to whether it is best to proceed directly with cloning and walking between marker loci, or better to derive additional DNA markers that lie between those available.

Long-range restriction mapping requires two tools: the first is a method for separating very large DNA fragments based on size differences and the second is a set of reagents for cutting DNA at relatively rare restriction sites. The required methodology was invented by Schwartz and Cantor (1984) and is known as pulsed field gel electrophoresis (PFGE). This technique permits the physical separation of DNA molecules that vary in size up to 9 mb in practice, with no upper limit in theory. The actual "window" of separation achieved is determined by the conditions of electrophoresis: at the lower end, one can obtain separation in the range of 20–200 kb, just beyond that possible with classical electrophoresis; at the upper end, one can obtain separation in the range of 1.4–9 megabases (Barlow and Leharch, 1987).

The PFGE protocol would not be very useful for mapping mammalian chromosomes—which typically vary in size from 100 to 250 megabases—without a means for cutting these chromosomes at specific sites that are scattered from hundreds of kilobases up to a few megabases apart from each other. The means for doing just this appeared with the discovery of a special class of "rare cutting" restriction enzymes. Restriction enzymes may cut rarely within mammalian DNA for two reasons. The first is a recognition site of eight bases rather than the usual four or six. In a genome with a truly random sequence, an eight-base recognition site would appear only once in every $4^8$ bp or 64 kb. However, mammalian DNA is not truly random. In fact, one particular dinucleotide—CpG—is severely under represented by a factor of five (see Section 8.2.2). This fact provides the second reason why certain enzymes will cut genomic mouse DNA only rarely—they contain one or more CpG dinucleotides in their recognition site. One enzyme in particular—*Not*I—has an eight-base recognition site as well as two CpG dinucleotides; the average distance between *Not*I sites is estimated at over 1 mb. Other enzymes have either an eight-base recognition site (*Sfi*I) with no CpGs or a six-base recognition site with two CpG dinucleotides (*Nru*I, *Mlu*I, *Bss*HII, *Eag*I, *Sac*II, etc.), and finally, there are enzymes with a six-base recognition site and only one CpG (*Sal*I, *Cla*I, *Nar*I, *Xho*I, etc.) Taken together, experiments with these various enzymes can be used to provide a distribution of restriction fragments that vary from 20 kb to multiple megabases in length.

Long-range restriction maps are best generated by a combination of two approaches (Herrmann et al., 1987; Barlow et al., 1991). First, single or double digests can be performed on very high molecular weight genomic DNA with a panel of rare cutting enzymes. Second, the same DNA sample can be treated with individual rare cutting enzymes under conditions where partial digestion will occur (Barlow and Lehrach, 1990). All of these samples are loaded into adjacent lanes on the same gel, which is run according to the PFGE protocol, blotted, and then probed sequentially with various markers from the region of interest. The basic strategy for building up restriction maps is similar to that encountered with isolated small clones like plasmids (Sambrook et al., 1989). The physical distance between two markers can be determined by identifying and sizing those restriction fragments, or partially digested fragments, that hybridize to both markers, or only one marker or the other.

Prior to the development, and easy availability, of large insert genomic libraries, the rare cutting enzyme/PFGE approach provided the most feasible means for estimating physical distances between linked loci that are separated by hundreds of kilobases or more. However, it is now often the case that physical mapping is more readily accomplished in the context of clones. Nevertheless, there are still many situations where a region of interest is flanked by two markers that are too distant from each other to allow rapid cloning between them. Genomic restriction mapping can play a unique role in these situations.

### 10.3.3 Large insert genomic libraries

#### 10.3.3.1 YACs and other large insert cloning systems

With the availability of one or more closely linked DNA markers from a genomic region of interest, one can begin to develop a contig of overlapping clones that spans the region. A cloned contig not only provides information on physical distances but can also be used as the raw material from which positional cloning of a phenotypically defined locus can proceed. The generation of a contig is pursued most efficiently by screening and walking through a large insert genomic library. Although a number of systems for generating large insert libraries have been described, to date, the yeast artificial chromosome (YAC) cloning system remains the most important for mouse geneticists.

The YAC cloning system was first developed by David Burke and Maynard Olson at Washington University in St Louis (Burke et al., 1987). It is based on the formation of "artificial" yeast chromosomes with the ligation of random, large fragments of genomic DNA between two arms that contain, in one case, a telomere and a centromere, and in the other case, a telomere alone, with selectable drug-resistance markers on both arms. These YAC constructs are transfected back into yeast where they will move alongside host chromosomes into both daughter cells at each mitotic division.

The construction of a YAC library proceeds in a manner that is very different from that of most other types of genomic libraries. Every clone in the library must be picked individually and placed into a separate compartment (of a microtiter plate, for example). This process is extremely time consuming and labor intensive, but once a library has been formed with individual clones in individual wells, it is essentially immortal. For this reason and others, it makes good sense to screen established libraries for a gene of interest rather than to create a new library. The first mouse YAC library to be described had a 2.2-fold genomic coverage and an average insert size of ~ 265 kb, and was distributed freely to the entire scientific community (Burke et al., 1991; Rossi et al., 1992). Several other mouse YAC libraries have since been described with greater insert size and genomic coverage (Larin et al., 1991; Chartier et al., 1992; Kusumi et al., 1993). The most comprehensive, well-characterized mouse YAC library described to date contains 19,421 clones with an average insert size of 650 kb for a 4.3-fold coverage of the genome (Kusumi et al., 1993). This library is available for screening commercially through Research Genetics, Inc. (Huntsville, Alabama, USA). Screening of this library, and most others, is based on PCR analysis of a hierarchy of clone pools (Green and Olson, 1990). Detailed protocols for library preparation, screening, and analysis have been described (Larin et al., 1993; Nelson and Brownstein, 1994).

It should be mentioned that the YAC cloning system is not perfect. At the time of writing, it is still the case that a very high percentage of clones from all of the largest insert YAC libraries are chimeric; that is, their inserts are composed of two or more unrelated genomic fragments that have become co-ligated in an undefined manner. The pre-identification of chimeric clones is essential before one can begin to generate a physical map.

Two other systems for cloning large genomic inserts have been described more recently. One is based on the use of the bacteriophage P1 as a cloning vector (Pierce and Sternberg, 1992; Pierce et al., 1992). This system has been used to obtain a mouse genomic library with average inserts in the range of 75–95 kb with a maximum cloning capacity of 100 kb. The P1 cloning system has two advantages over YACs: first, it has much more efficient cloning rates, and second, like other bacterial cloning systems, it allows the efficient purification of large amounts of clone DNA away from the rest of the bacterial genome. The utility of this cloning system in the analysis of genomic organization within the *H2* region has been demonstrated (Gasser et al., 1994).

Another more recent system is derived from the well-studied *E. coli* F factor, which is essentially a naturally occurring single-copy plasmid (Shizuya et al., 1992). This plasmid has been converted into a vector that allows the cloning of inserts with more than 300 kb of DNA, and with a reported average size range of 200–300 kb. The developers have called this vector/insert system a bacterial artificial chromosome or BAC. The BAC system has the same advantages as P1 and the added advantage of a larger potential insert size. The BAC system has not been analyzed sufficiently to know whether chimerism will be a problem and whether the whole mouse genome will be fairly represented within this library.

### 10.3.3.2 *Walking and building contigs*

All positive clones from a YAC, or other large insert, library can be sized by PFGE, and fragments at both ends of each insert can be isolated rapidly by several standard protocols (Riley et al., 1990; Cox et al., 1993; Zoghbi and Chinault, 1994). End fragments from each clone should be used as probes to perform an initial test of the possibility of chimerism. This can be accomplished by probing appropriate somatic cell hybrid lines to determine whether both ends map to the same chromosome as the original DNA marker used to isolate the clone; if appropriate somatic cell hybrid lines are not available, one can also test the segregation of the end fragments on a panel of 20 interspecific (or intersubspecific) backcross samples. If the two end fragments show complete concordance in transmission, this can be taken as strong evidence for *non*-chimerism; in contrast, two or more recombination events would be highly suggestive of a chimeric clone. Chimeric clones need not be discarded; it is just necessary to be aware of their nature in any interpretation of the data that they generate.

If multiple clones have been obtained from a screen with a DNA marker, end fragments from each should be used in cross-hybridization experiments to identify the particular clones that extend furthest in each direction along the chromosome. Often this approach will reduce the number of clones worth pursuing to just two. "Walking" through the library can proceed by using the farthest end fragments for rescreening, and then analyzing the resulting clones in the same manner described above. In this manner, a "contig" will be built over the genomic region that contains the locus of interest (Zoghbi and Chinault, 1994).

The process of deriving YAC clones from a library can be brought to a halt when the clones that have already been obtained *must* include the locus being sought. It is only possible to reach this conclusion when the derived contig extends over markers that map apart from the locus on both of its sides. In other words, the contig must extend across the two closest recombination breakpoints that define the outer limits of localization. If cloning is begun with a very dense map of markers placed onto a high-resolution cross, this endpoint is likely to be reached more quickly. With real luck, it might even be reached with the first set of YACs obtained in the initial screening of the library.

### 10.3.3.3 Physical mapping: a comprehensive example

The overall strategy that one follows to move from a phenotype to a cloned contig is best explained within the context of a hypothetical example that is illustrated in Fig. 10.1. Suppose you are interested in cloning a newly identified locus that has mutated to cause a phenotype of green eyes. First you search through the literature and the various genetic databases to see if any similar mutant phenotype has been uncovered previously. When this search fails to uncover previous examples of green-eyed mice, you decide to set up an intersubspecific backcross to follow the segregation of the mutant locus relative to DNA markers spread throughout the genome as detailed in Section 9.4. An analysis of 50 backcross offspring with two to three markers taken from each chromosome demonstrates linkage to the distal region of Chromosome 3 between 2 markers that are 40 cM apart from each other. With this information in hand, you retype the same offspring with ten additional Chromosome 3 markers— spaced at approximately 2 cM intervals over the derived map position—to localize the green eyed mutation further. This step yields a map position between two limiting markers that are spaced 4 cM apart. Now you increase the number of backcross offspring in your typing set to 400 and you analyze each for the segregation of just these two limiting markers. This analysis identifies just 16 animals with recombination break points between the two limiting markers, and each member of this smaller set is analyzed for segregation of another 30 markers that were previously mapped with or between the limiting markers. This third step yields four markers that are most tightly linked to the green-eyed locus with the hypothetical haplotype data shown in panel A of Fig. 10.1. As illustrated in the Figure, the data demonstrate: (1) complete concordance between green-eyed and the marker *D3Xy55*; (2) one recombinant in 400 with the proximal marker *D3Ab34*, and two recombinants in 400 with two completely concordant distal loci *D3Xy12* and *D3Ab29*. Panel B of Fig. 10.1 shows the linkage map that is generated from these data.

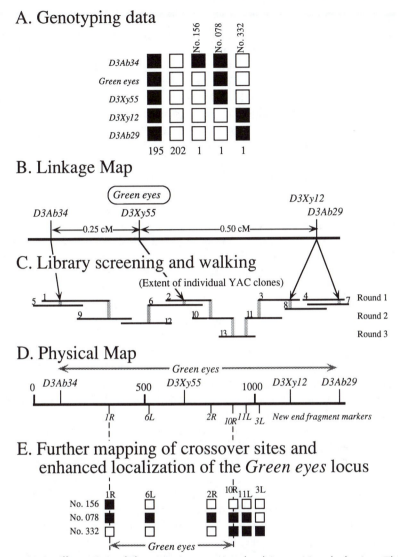

**Figure 10.1** Illustration of the genetic steps involved in positional cloning. This Figure shows the final steps involved in narrowing down the interval to which a new mutation can be mapped by a combination of genetic and molecular approaches. The initial steps required to reach this stage of analysis are illustrated in Fig. 9.15. See the text for further details.

With one concordant marker and closely flanking markers on either side of the locus of interest, one can begin to develop a physical map. All four of the nearby markers are used to screen a YAC library. As shown in panel C of Fig. 10.1, *D3Ab34* identifies two clones (1 and 5), *D3Xy55* identifies another two clones (2 and 6), *D3Xy12* identifies another two clones (3 and 8), and *D3Ab29* identifies another two clones (4 and 7).[105] End fragments are derived from all eight clones and are used first to search for overlaps by hybridization to the complete set of YACs. This search demonstrates an overlap between clone 8, and clones 4 and 7. Thus, in the first round of screening, two independent markers—*D3Xy12* and *D3Ab29*—have been physically linked into a single contig.

End fragments from clones 1, 6, 2, and 3 are used to rescreen the YAC library. This screen yields clones 9, 12, 10, and 11. Once again, end fragments are derived from these new clones and used first to search for overlaps. This search demonstrates an overlap between clones 9 and 12, which provides a physical linkup between the markers *D3Ab34* and *D3Xy55*. Thus, after two rounds of screening, two contigs have been formed with each containing two of the four markers. Finally, in an attempt to fill in the gap between YACs 10 and 11, a third round of screening is performed with an end fragment from each of these clones. Both end fragments identify the same clone (13) and thus, without further analysis, it is immediately possible to state that a single contig has been generated across the entire region of interest. Most importantly, the contig crosses recombination breakpoints both proximal and distal to the green-eyed locus. Thus, the green-eyed locus must lie within the contig.

The contig is minimally defined by 10 overlapping clones; from most proximal to most distal, they are 5, 9, 12, 2, 10, 13, 11, 3, 8, and 7. Each of these clones must be sized and restriction mapped to construct the complete physical map shown in panel D of Fig. 10.1. At this stage of the analysis, one can say only that the green-eyed locus must reside within the 1,360 kb cloned region between *D3Ab34* and *D3Ab29*. A region of this size is still quite large for undertaking gene identification studies, and thus it makes sense to try to narrow it down further. Toward this goal, one can return to the three backcross animals with recombination break points located nearest to the green-eyed gene (156, 078, and 332 from panel A). The three corresponding samples of genomic DNA can be typed at the new loci defined by the six end fragments characterized in the YAC walking protocol just completed (1R, 6L, 2R, 10R, 11L, and 3L, where R and L signify left and right ends, respectively). The results of this last genetic analysis are shown in panel E of Fig. 10.1 (where haplotypes are rotated 90° to match them up to the physical map). The data allow further localization of the proximal breakpoint between 1R and 6L, and further localization of a closest distal breakpoint (in sample no. 332) between markers 2R and 10R. These results reduce by two-fold the size of the genomic region that must contain the green-eyed locus down to 560 kb. This region is contained within just four YACs—9, 12, 2, and 10—that can now be analyzed for potential gene sequences as described in the next section.

### 10.3.4 Protocols for gene identification

Even before the entire region that encompasses the locus of interest has been cloned, it is possible to begin the search for candidate genes within the large-insert clones that first become available. It is good idea to pursue this search simultaneously with genomic walking for two reasons. First, you could get lucky and find your gene in the initial cloned region. Second, the search for candidate genes can be daunting—it took 10 years to identify the human Huntington

Disease gene (Huntington's Disease Collaborative Research Group, 1993)—so it makes good sense to start as soon as possible.

Many different protocols have been devised over the past several years to carry out this task (Parrish and Nelson, 1993). Generally-speaking, these protocols can be placed into three groups according to the underlying principle that they incorporate. First are protocols that rely upon the identification of transcribed sequences by cross-hybridization. Second are protocols that do not depend on gene activity but rather special characteristics in the DNA itself that are unique to mammalian genes. Third are computational protocols that can be used to distinguish coding regions and regulatory regions from non-functional regions within long stretches of DNA sequence.

### 10.3.4.1 Candidate gene identification based on expression

Traditional approaches to identifying expressed sequences within genomic clones rely on using these clones, or subfragments from them, as hybridization probes to screen cDNA libraries constructed from a tissue in which the locus of interest is thought to be expressed.[106] In theory, the simplest strategy would be to use YAC clones directly as probes (Marchuk and Collins, 1994).[107] In practice, this simple strategy has shown only limited success for a number of reasons including not only the high complexity of the clone—which results in a reduced signal strength for each individual transcript region within it—but also the difficulty of purifying high quality YAC DNA in sufficient quantity. These problems are circumvented by subcloning the YAC into cosmids or phage, which can each be used individually to probe the cDNA library. However, this increases the workload by at least an order of magnitude.

Another expression-based strategy that is not dependent on cDNA libraries is to use subclones from YACs as probes of Northern blots containing RNA from tissues thought to express the gene along with RNA from tissues that should not express the gene based on the mutant phenotype. Positively hybridizing subclones can be further subcloned and individual fragments can be retested to narrow down the location of the transcript-containing sequences. Although this process provides some additional expression information that can be useful for sorting out candidates, it is quite tedious, requires large amounts of tissue RNA, and is no longer the method of choice.

New approaches to detecting expressed cDNA sequences that circumvent many of the disadvantages of the methods just described are all based on the use of PCR. With one such approach, the YAC DNA—rather than the RNA or cDNA—is immobilized on filters. These filters are probed with specially engineered cDNA libraries in which all inserts are flanked by unique targets for PCR amplification. After filter hybridization and washing, those cDNAs that remain specifically attached can be eluted, amplified, and cloned (Lovett et al., 1991; Parimoo et al., 1991). Many variations upon this general theme have been described.

### 10.3.4.2 Expression-independent gene identification

There are two serious problems inherent in all attempts to locate genes based on hybridization to RNA transcripts or amplified products from these transcripts.

The first problem is that, from the phenotype alone, it may not be possible to determine the tissue specificity of gene expression. The second problem occurs even when the specific expressing-tissue can be reasonably well-identified in that the majority of transcript classes will be present at relatively low levels and will be difficult to retrieve. As a consequence, whole classes of genes will go undetected including, for example, those that are expressed only during brief periods of embryonic development, or only in a small subset of cells from complex tissues like the brain. Even genes expressed more broadly can go undetected if their corresponding transcripts are present in one or a few copies per cell.

Three general approaches to gene identification have been developed that are not dependent on gene expression. Broadly speaking, these approaches are based on three corresponding characteristics of mammalian genes: (1) the occurrence of introns in nearly all mammalian genes; (2) the presence of "CpG" islands at the 5'-ends of most mammalian genes; and (3) the evolutionary conservation of nearly all mammalian genes from mice to humans and sometimes beyond.

### 10.3.4.3 Exon trapping

The first approach is referred to as exon trapping (Buckler et al., 1991). It is based on the empirical finding that the vast majority of splice recognition sites are not cell type specific. Instead, a general splicing machinery present in all cells can act with precision upon endogenous as well as foreign transcripts. This machinery can be exploited in tissue culture to identify YAC-derived genomic fragments that contain exons. Essentially, one subjects the YAC DNA to restriction digestion with a standard six-base recognition site enzyme followed by shotgun cloning into a special eukaryotic expression vector that contains flanking splice donor and acceptor sites on either side of the insert. The clones are then transfected into a mammalian cells that allow their high-level expression into RNA containing each cloned insert. If an insert does not contain an exon, splicing will proceed directly from the splice donor site on one side of the transcribed insert to the splice acceptor site on the other side to produce a final transcript of a predefined size composed entirely of vector sequences. However, if an entire exon is contained within a particular fragment, it will be spliced into the mature transcript. The set of transcripts produced in a particular transient cell culture can be amplified by reverse transcription followed by PCR (RT-PCR) and analyzed by gel electrophoresis. All PCR products that are larger than the background splicing product should contain insert-derived exons that can be readily cloned directly from the gel. Once exons have been identified and cloned by this protocol, they can be used directly as probes to study tissue and stage specificity of expression, which is a prerequisite to the recovery of full-length cDNA clones.

In theory, a protocol of this type should allow the isolation of all of the exons present in a particular YAC clone. This collection of exons would be representative of all of the genes present on the YAC. However, in practice, sophisticated protocols of this type often fail to live up to expectations. To test the validity of exon trapping as a generalized approach to gene identification, Lehrach and his colleagues (North et al., 1993) used this strategy to search for exons on eight cosmid clones that covered a region of 185 kb from the MHC class II region. Of the eight genes that are known to be present within this contig, seven were accounted for within the exon clones that were recovered.[108] This result would imply a success rate for gene identification of ~90%.

### 10.3.4.4 CpG islands

As discussed previously, the dinucleotide CpG is severely under-represented in mammalian genomes. This under-representation results from the methylation of the cytosines on both strands of the two-basepair sequence; methylated cytosines are highly susceptible to spontaneous deamination which can cause a transitional mutation to thymidine (Barker et al., 1984). Thus, when $\frac{CG}{GC}$ sequences are present in the genome, they will mutate frequently to $\frac{TG}{AC}$ or $\frac{CA}{GT}$ and, in fact, these dinucleotide sequences are present at a frequency significantly higher than expected throughout the genome (McClelland and Ivarie, 1982). In contrast, the CpG dinucleotides present at the 5' ends of many vertebrate genes remain devoid of methylation and thus resistant to mutation. As a consequence, the distribution of CpG dinucleotides is highly non-random with high-density "islands" of multiple CpGs that mark the 5'-ends of genes in the midst of large genomic seas that contain only scattered CpGs as isolated entities (Bird, 1986, 1987; Bird et al., 1987).

Gene searchers can exploit this situation by using restriction enzymes that contain two CpG dinucleotides in their recognition sites to identify the 5' ends of genes. Lindsay and Bird (1987) have calculated that 89% of all *Not*I sites (GCGGCCGC) are located in CpG islands, as is the case for 74% of all *Eag*I (CGGCCG), *Sac*II (CCGCGG), and *Bss*HII sites (GCGCGC). Thus, the approach to identifying CpG islands within YAC clones becomes relatively straightforward. Partial digestion of the clone is performed with each of the double CpG enzymes just described, and the resulting DNA is separated by PFGE, blotted and probed sequentially with fragments from each of the YAC arms. The appearance of bands of the same size in partial digests obtained with two or more enzymes is highly suggestive of a CpG island. If *Not*I and one of these other enzymes both recognize sites within 1 or 2 kb of each other (below the resolution of PFGE), the presence of a CpG island can be assumed with a probability of 97%. If two of the six-cutter double CpG enzymes both recognize nearby sites, the likelihood of a CpG island is 93%. Once a putative CpG island is identified, various PCR-based methods can then be used to clone the DNA adjacent to the island (Parrish and Nelson, 1993), and these sequences can be examined thoroughly to characterize the associated transcription unit.

There are two main advantages to this approach to gene identification. The first is its simplicity: it is based entirely on restriction digests, gel running, and cloning. The second is that it can enable the identification of genes that may not be detectable by other approaches. The only real disadvantage is that CpG islands are only found in association with 50 to 70% of all genes.

### 10.3.4.5 Conservation from mice to humans and beyond

Less than 5% of the sequences within the mammalian genome actually contain information that is used to encode gene products. An even smaller fraction (~0.1%) of the genome accounts for all of the regulatory elements, such as promoters and enhancers that control the stage and tissue-specific expression of this genetic information. Another 5–10% of the genome consists of elements required for the construction of centromeres, telomeres and other chromosomal structures. The remaining 85–90% of the genome has no apparent sequence-specific function.

Nucleotide changes that occur within a non-functional DNA sequence are considered to be neutral. That is to say that such changes provide neither benefit nor harm to the organism within which they reside and, as such, they will not be subjected to selective forces. Instead, over a period of many generations, they will decrease or increase in allele frequency by a process of random drift, which will lead either to their extinction (in most cases) or to their fixation within the species. Since spontaneous mutations occur at a constant rate within a population and since each neutral change will have the same (very low) probability of fixation, a non-functional sequence of DNA will slowly change at a constant rate. In mammals, this rate of change has been determined empirically to be on the order of 0.5% (five changes in 1,000 nucleotides) per million years.

The constant rate of change in non-functional sequences can be used as a "molecular clock" to gauge the evolutionary distances that separate different species from each other [see, for example, Nei (1987)], and when the distance between two species is already known, the molecular clock can be used to predict the expected homology between sequences in each that are descendent from a common ancestor. Consider the consequences of genetic drift on a non-functional sequence present in the common ancestor to mice and humans some 65 million years ago. During the evolution from this common ancestor to the modern house mouse, it would have undergone changes in 65 × 0.5% or 32.5% of its nucleotides. During the separate evolution of this sequence along the line from the common ancestor to modern humans, changes would also have occurred in 32.5% of its nucleotides. With so many random changes occurring, there is a certain probability that the same nucleotide will be hit two or more times. Taking this fact into consideration yields a corrected divergence of ~27% along each evolutionary line, and a comparison between the sequences currently present in mice and humans would show a divergence of ~46%. With changes at approximately half of the nucleotides present in the derived sequences, it will often be hard to even recognize the fact that they have a common ancestor. Most importantly for physical mappers, at this level of divergence, specific cross-hybridization will not take place.

In contrast to the situation encountered with non-functional sequences, most nucleotide changes that occur within coding regions will, in fact, be subjected to selective forces. The vast majority of changes that alter protein sequence are detrimental and will not survive within a population. Thus, coding regions will evolve much more slowly than non-functional regions. Although the actual rate of evolution can vary greatly for different genes, the vast majority of mammalian genes characterized to date show specific cross-hybridization between homologous sequences in the mouse and human genomes. In addition, a subset of mammalian genes are so conserved that specific cross-species hybridization can be detected with homologs in *Drosophila* and *C. elegans*, and in a smaller subset still, cross-hybridization is detected with homologs in yeast.

With all of this information in hand, it becomes clear that cross-species Southern blot hybridization studies that use subcloned YAC fragments as probes, under low stringency conditions, can provide a tool for distinguishing between the 5% of sequences that encode proteins and the remaining 95% without coding information. Since investigators often run samples from several different species in adjacent lanes, this approach is often referred to as "zoo blotting."

There are several advantages to this approach. First, it allows detection of the vast majority of coding sequences and, thus, is more universal than the CpG island approach. Second, it is not dependent on actual gene expression, and thus, avoids all of the problems inherent in low level or restricted transcript distributions. The major problem with this approach is that the YAC clone must be subdivided into much smaller pieces that need to be tested individually for hybridization, and once a positive result is obtained, further subcloning and analysis is required. Consequently, the examination of a several hundred kilobase contig by this approach can be extremely tedious and time consuming.

### 10.3.4.6 Gene identification based on sequence

As sequencing becomes more highly automated and more accurate, the feasibility of stepping nucleotide by nucleotide across an entire YAC clone becomes increasingly realistic. The basic approach would be to begin sequencing across the insert from both sides with initial primers facing in from the two YAC arms. The maximum amount of sequence possible would be obtained moving away from these two primers and then a short segment at the end of each would be used to design a new primer for the next step in sequencing. This process would be repeated over and over again until overlap was reached in the middle of the clone. If, with improvements in technology, it becomes possible to read 1,000 bases of sequence in any single run, then each step in this procedure would provide 2 kb of information. In total, 150 steps would be required for a single pass over the complete sequence of a 300 kb clone. This is certainly not a feasible approach in 1994, but the pace of technology is such that it may well be possible within in the next 5 years.

With a long-range sequence, it becomes possible to use computational methods alone to ferret out coding regions. Sophisticated computer programs based on neural nets have been developed that can identify 90% of all exons with a 20 percent false positive rate as of 1993 (Forrest, 1993; Jan and Jan, 1993; Little, 1993; Martin et al., 1993). Once a putative exon has been identified, it can be used as a probe to search for the tissue in which its expression takes place and, with further studies, it becomes possible to identify the remaining portions of the transcription unit with which it is associated.

## 10.4 THE HUMAN GENOME PROJECT AND THE ULTIMATE MAP

The goal of determining the complete DNA sequence of the human genome has captured the imagination of many geneticists as well as other biomedical scientists (Kevles and Hood, 1992; Wills, 1992). It was this goal that originally catapulted the "Human Genome Project" into the headlines of newspapers, and the talk of politicians in the US and elsewhere during the latter part of the 1980s. In 1990, the National Institutes of Health and the Department of Energy began a coordinated effort to reach this goal within a period of 15 years by stepping through a series of intermediate goals. In 1993, Francis Collins, the new director of the National Center for Human Genome Research at NIH, outlined the newest version of a 5 year plan for this project (Collins and Galas, 1993). The 5-year goals incorporated into this plan include: (1) a high-resolution linkage map at a resolution of 2–5 cM with an established set of easy to use DNA markers; (2) a high-resolution physical map with sequence tagged site (STS) markers located every 100 kb; (3) improvements in DNA sequencing technology to allow the annual accumulation of 50 mb of sequence; (4) the development of efficient methods for gene identification within cloned sequences, and for the mapping of genes identified by other means; and (5) further development of technology for increased automation, increased use of robotics, and more sophisticated computer-based tools for information management and analysis.

Although the Human Genome Project is focused, of course, on the human, there is a uniform consensus among researchers that it is only in comparison with the genomes of model organisms such as the mouse that the human genome will reveal all of it secrets. Thus, another primary goal of the human genome project is to sequence selected segments of the mouse genome side by side with homologous regions of the human genome. By evaluating the relative level of conservation across a genomic region and combining this information with other computational assessments of sequence content, it will become possible to uncover essentially all genes, promoters, enhancers, and other regulatory elements. Unknown, and unexpected, non-coding, sequences that show conservation indicative of biological function are sure to be identified as well, and further studies will be required to understand the functions of these new elements.

Even if the Human Genome Project meets its target of a complete human sequence by the year 2005, in the minds of geneticists, the *ultimate* map of the genome will extend far beyond the sequence alone and will require many more years to attain. This ultimate map will show all genomic regions that are subject to parental imprinting, all regions that show methylation, and the locations of all DNase I hypersensitivity sites that appear in the context of the chromatin structure, which forms around the DNA molecule. Finally, the complete sequence will be used as a jumping board to map the entire network of genetic and gene product interactions that occur during the development of a human being, with a determination of the complete pattern of expression of every single gene through both time and space.

Will such an ultimate map ever be attained? It is impossible to predict in 1994 as these words are being written. However even if it is attained, will it really tell us something deeply profound about the state of being human? This will only happen if it becomes possible to visualize the manner in which the whole functional network becomes more than the sum of the parts. Will any human mind have the capacity for such a visualization? Is it our destiny to understand the molecular biology of the human soul or, rather, to search forever in vain? These are questions that may be answered in the 21st century, even later, or perhaps never.

# Appendix A

# Suppliers of Mice

**A1** Major suppliers of common inbred and outbred strains

| Supplier[a] | Services | Address | Contact numbers |
|---|---|---|---|
| Charles River Laboratories | Established strains, transgenics | 251 Ballardvale St,Wilmington, MA 01887 | Phone: 508–658–6000 Fax: 508–658–7132 |
| Harlan Sprague Dawley, Inc. | Established strains, transgenics | PO Box 29176 Indianapolis, IN 46229 | Phone: 317–894–7521 Fax: 317–894–4473 |
| Jackson Laboratory | Established strains, mutant stocks | Animal Resources, 600 Main St, Bar Harbor, ME 04609–0800 | Phone: 800–422–6423 Fax: 207–288–3398 |
| Taconic Farms, Inc. | Established strains, transgenics | 33 Hover Avenue, Germantown, NY 12526 | Phone: 518–537–6208 Fax: 518–537–7287 |

[a]Listed in alphabetical order.

**A2** Other commercial sources of mice

| Supplier[a] | Services | Address | Contact numbers |
|---|---|---|---|
| Baekon, Inc. | Transgenics | 4420 Enterprise St, Fremont, CA 94538 | Phone: 510–683–8881 Fax: 510–683–8712 |
| GenPharm International | Transgenics | 2375 Garcia Ave, Mountain View, CA 94043 | Phone: 415–964–7024 Fax: 415–964–3537 |
| Hilltop Lab, Animals, Inc. | Established strains | Hilltop Drive, Scottsdale, PA 15683 | Phone: 800–245–6921 Fax: 412–887–3582 |
| Life Sciences, Inc. | Established strains, transgenics | 2,900 72nd St, North, St. Petersburg, FL 33710 | Phone: 800–237–4323 Fax: 813–347–2957 |

[a]Listed in alphabetical order.

# Appendix B

# Computational Tools
# and Electronic Databases

## B1  ON-LINE ELECTRONIC DATABASES

### B1.1  The World Wide Web

An ever-increasing amount of genetic information can be easily accessed and selectively retrieved over the Internet through the use of newly developed software protocols that break through the arcane language that was characteristic of Internet communications up until just a few years ago. It is now possible to access information over the Internet without knowing any computer commands and by barely touching the computer keyboard. The opening of the academic information superhighway was initiated with the development of a user interface called Gopher at the University of Minnesota. Gopher provided anyone sitting at an Internet-linked computer with the ability to "surf the net" by clicking through a series of menus to move from one Internet location to another, where it was often possible to search for and recover specific information. Although Gopher opened up the Internet to the masses, it is limited in the sense that it is essentially a text-based system (although it does allow one to download images as well). Those interested in the Gopher system of software should write to the Gopher development group at the following E-mail address: Gopher@boombox.micro.umn.edu.

By the summer of 1994, a new vehicle for Internet surfing had taken hold in the form of the World Wide Web, which is also known as WWW, W3, or simply the Web. WWW is body of software and a set of protocols and conventions developed initially at CERN, a high-energy physics laboratory in Switzerland. WWW is a true multimedia system with a highly graphical interface containing imbedded hypermedia. WWW allows the flow of not only text, but also images, video animation, and sound over the Internet. Hypermedia refers to the fact that text and objects on the computer screen can be given life in the sense that pointing at them and clicking on a mouse will initiate software protocols that, among other possibilities, provide "gateways" that transport a user from one

Internet site to another which may be halfway around the world. In this way, hypermedia allows information to be organized as an interconnected web of associations rather than a linear chain.

In order to use the Web to the fullest extent, it is important to understand the meaning of three critical terms: server, client, and uniform resource locator (or URL). Any computer site on the Internet that has opened itself up to "browsing" and information retreival via WWW is called a server. A WWW server must run special server software to present its information in the form of hypermedia, but otherwise, there are few limitations to who may develop a server.

Surfing the Web is accomplished by linking into, and jumping between, WWW servers over the Internet. To begin WWW explorations, you must install special software on your computer known as client software (or a WWW client). Different types of client software have been developed for the Macintosh, IBM-compatible computers running Windows, and other computer platforms. The most popular client was developed by the National Center for Supercomputing Applications (NCSA) and is called Mosaic with separate versions for different computer types. For more information, make contact through the following E-mail address: mosaic-mac@ncsa.uiuc.edu. It is also possible to download Mosaic software directly by anonymous file transfer protocol (anonymous FTP) at the NCSA FTP server: ftp ftp.ncsa.uiuc.edu. Several other types of WWW client software have been developed independently including MacWeb for the Macintosh, and Cello and WinWeb for the Windows environment. For more information on MacWeb, write to the following E-mail address: macweb@ei-net.net. For more information on Cello, write to the following E-mail address: cellobug@fatty.law.cornell.edu. For more information on WinWeb, write to the following E-mail address: winweb@einet.net. To download either Cello or MacWeb client software directly by anonymous FTP, go to the following server: ftp.einet.net.

An understanding of the final critical term—URL—will allow you to find particular WWW servers of interest among the many thousands that exist on the Internet. The URL is a long file name that acts as an address for each individual WWW server as well as each packet of hypermedia available at that server. Once WWW client software is up and running on your computer, you can move to any WWW site by opening the File menu and choosing the "Open URL. . ." command. Then type the URL of interest into the box, hit return, and off you go. URLs of particular interest to mouse geneticists are described below. As this book is being written, WWW is still in its infancy and is expanding rapidly. It is a near-certainty that the data sources described below will be enriched and supplemented with new sources of information and modes for their retrieval. You should contact the sources of client software described above or your local computer advisor for up-to-the-date information on the most advanced system available at the time of this reading.

## B1.2 Mouse genetics

The central WWW server for mouse geneticists is located at the Jackson Laboratory and has the following URL: http://www.informatics.jax.org/. This server provides access to the Mouse Genome Database (MGD) which is a compendium of linked data sources that include locus information, genetic mapping data, polymorphisms, probes, clones, PCR primers, citations, and homology data with other mammalian genomes. Included in this compendium is the online version of Margaret Green's "Catalog of mutant genes and polymorphic loci" (Green, 1989). The online version, called the Mouse Locus Catalog (or MLC) is updated regularly by staff members of the Jackson Laboratory led by Dr M.T. Davisson and Dr D. P. Doolittle. The entire database can be searched to retrieve a list of loci (with associated descriptions and citations) that contain a particular search word or combination of words anywhere within the locus record. For further information, the MGD staff can be reached at the following E-mail address: mgi—help@Jax.org.

The *Genome Center at the Whitehead Institute, MIT* provides a highly specialized but very useful source of information through either an automated E-mail query and answer protocol or anonymous FTP. This Genome Center has characterized and mapped over 3,000 polymorphic mouse microsatellite markers (as of January, 1994) that are each defined by a unique pair of oligonucleotide primers. By filling out and transmitting a special E-mail query form, an investigator can receive an automatic response containing information about sets of microsatellite markers defined by various criteria, including chromosomal location and polymorphisms between particular strains. It is possible to retrieve primer sequences as well as a graphical representation of a chromosome map that shows the microsatellites (in Macintosh PICT format suitable for incorporation into all Mac drawing programs). To receive a blank E-mail query form and instructions for its use, send an E-mail message with the single word help to genome_database@genome.wi.mit.edu. Data can also be retrieved by anonymous ftp to genome.wi.mit.edu. For more information, contact Lincoln Stein at the Whitehead Institute [Email address: lstein@genome.wi.mit.edu].

### B1.3  Other WWW servers of interest to mouse geneticists

The central WWW server for human genetic information is maintained by the Genome Database group at the Johns Hopkins University and has the following URL: http://gdbwww.gdb.org/. A compendium of interlinked databases is available at this site where it is possible to recover information on human chromosomes, loci, maps, mutations, citations, cell lines, probes, polymorphisms, and other data items. Included in this compendium is the online version of Victor McKusick's *Mendelian Inheritance of Man* (called OMIM) which contains comprehensive textual information on all defined human loci. For further information, the GDB staff can be reached at the following E-mail address: Help@gdb.org, or the following mailing address: GDB User Support, Genome Data Base, Johns Hopkins University, 2024 E. Monument Street, Baltimore, MD 21205–2100.

Another critical WWW server is maintained by the National Center for Biotechnology Information (NCBI) which is the location of the GenBank database. The URL for this server is: http://www.ncbi.nlm.nih.gov/. GenBank searches can be conducted through this linkup. For further information, contact NCBI at the following E-mail address: www@ncbi.nlm.nih.gov.

There are numerous other databases that may be of interest to mouse geneticists including servers devoted to other experimental genetic systems. Many of the WWW servers described above provide gateways to these other servers. A very useful set of gateways to WWW servers of particular interest to biologists is called the Biologist's Control Panel and can be reached at the following URL: http://gc.bcm.tmc.edu:8088/bio/bio_home.html.

## B2  MOUSE GENETICS COMMUNITY BULLETIN BOARD

An electronic bulletin board (called MGI-LIST) has been set up at the Jackson Laboratory with E-mail addresses of mouse geneticists throughout the world. The Mouse Genome Informatics Group at the JAX will use this bulletin board to make announcements of new software related to the Mouse Genome Data Base. In addition, all subscribers will be able to broadcast to, and receive messages of interest from, all other linked-up members of the community.If you would like to subscribe to this bulletin board, send a message that reads "subscribe mgi-list < your name >" to listserver@informatics.jax.org. For further information or help, send an E-mail message to mgi-help@informatics.jax.org.

## B3  COMPUTER PROGRAMS

### B3.1  Linkage analysis

Numerous computer programs are available for the determination of linkage relationships among loci typed in large numbers of individuals. Most of these were developed for the analysis of complex human pedigrees with algorithms that incorporate maximum likelihood estimation statistics (Elston and Stewart, 1971) and provide LOD score (Morton, 1955) output. Although these pedigree-based programs can be used for the analysis of mouse linkage data, they are usually less than optimal for this task. First, it is often not possible to derive unequivocal haplotype information for all individuals studied in a human pedigree, and as such, pedigree-based programs are not oriented toward this type of analysis, which is so powerful for many mouse linkage studies (see Fig. 9.15 and Fig. 9.16). Second, for most studies, mouse geneticists will not need to avail themselves of the sophisticated maximum likelihood/LOD score estimation tools required for human pedigree analysis. Two linkage programs developed specifically for mouse linkage studies—Map Manager and GENE-LINK—are described below. In addition, the most useful multilocus pedigree-based program—Mapmaker—is also described. Other pedigree-based linkage analysis packages are listed and reviewed by Bryant (1991).

*Map Manager* is a Macintosh program that can be used for the analysis of linkage data obtained from a backcross, $F_1 \times F_1$ intercross, or from a group of recombinant inbred strains (Manly, 1993) . It was written by Dr Kenneth F.

Manly and is made available without charge from the author. Map Manager uses a standard Macintosh format, and is both extremely versatile and extremely easy to use. It allows easy storage, display, and retrieval of information from mapping experiments. The strain distribution patterns, (SDPs) obtained with newly typed loci can be evaluated rapidly to determine likely map positions relative to other loci in the database. The program is distributed with a large database of previously published RI strain distribution patterns which greatly facilitates the RI analysis of new loci. Map Manager has many other sophisticated features including a statistical evaluation package and various output options, as well as a means for importing and exporting data to and from spreadsheets or Mapmaker files (see below). It can be obtained from the author on disk or by anonymous FTP or gopher from several sites including (mcbio.med.buffalo.edu), (hobbes-.jax.org), (ftp.bio.indiana.edu), and (ftp.embl-heidelberg.de). For further information, Dr Manly can be contacted at the E-mail address Kmanly@mcbio-.med.buffalo.edu or at Roswell Park Cancer Institute, Buffalo, New York 14263.

*Gene-link* is a DOS program that can be used for the analysis of linkage data from a backcross (Montagutelli, 1990). It was written by Dr Xavier Montagutelli and has been made available to investigators without cost. It will provide best-fit map positions for entered strain distribution patterns. For further information, Dr Montagutelli can be contacted at Institut Pasteur de Paris, 25 Rue du Docteur Roux, 75724 Paris, Cedex 15, France.

*Mapmaker* and *Mapmaker/QTL* are a pair of pedigree-based programs written by Dr Eric Lander and his colleagues for constructing linkage maps from raw genotyping and phenotyping data recovered from large numbers of loci (Lander et al., 1987) . These programs use a highly efficient algorithm for "likelihood of linkage" computations. They can be run on UNIX-based workstations or VAX minicomputers running under the VMS operating system. The programs and a manual are available from the author for licensing to academic researchers. Mapmaker is most useful to mouse geneticists for the analysis of linkage data obtained from an $F_1 \times F_1$ intercross. Mapmaker/QTL can be used for the analysis of quantitative traits. For further information, contact Dr Eric Lander, Whitehead Institute, 9 Cambridge Center, Cambridge, MA 02142.

## B3.2 Mouse colony management software

*Animal House Manager (AMAN)* and MacAMAN are software packages for IBM-compatibles and Macintosh computers, that can be used for keeping track of a breeding mouse colony with records for animals, litters, cages, and samples derived from them. These programs are described more fully in Chapter 3 (Silver, 1986, 1993b). They were written by the author of this book and can be licensed for use from Princeton University. For further information, the author can be reached at the E-mail address: Lsilver@Molbiol.Princeton.Edu, or at the Department of Molecular Biology, Princeton University, Princeton, NJ 08544–1014.

# Appendix C

# Source Materials for Further Reading

## C1 SELECTED MONOGRAPHS AND BOOKS

### C1.1 General

Green, M. C. & Witham, B. A. (1991). *Handbook on Genetically Standardized JAX Mice*, 4th edition, The Jackson Laboratory, Bar Harbor. [Provided free upon request.]

Foster, H. L., Small, J. D. & Fox, J. G., eds. (1982). *The Mouse in Biomedical Research*. Academic Press, New York. [A series of four volumes with comprehensive coverage of all aspects of mouse biology and breeding.]

Altman, P. L. & Katz, D. (1979). *Inbred and Genetically Defined Strains of Laboratory Animals: Part I, Mouse and Rat.*, Biological Handbooks, 3rd edition, Federation of American Society of Experimental Biology, Bethesda. [Tables of information for different strains of mice.]

Klein, J. (1975). *Biology of the Mouse Histocompatibility-2 Complex*, Springer-Verlag, New York. [The title is misleadingly narrow; a general discussion of mouse genetics.]

Crispens, C. G. (1975). *Handbook on the Laboratory Mouse*, C.C. Thomas, Springfield, Ill. [Information on strains not commonly used today.]

Green, E. L., ed. (1966). *Biology of the Laboratory Mouse*, 2nd edition, McGraw-Hill, New York. [Reprinted by Dover (New York) in 1968 and 1975; Comprehensive for its time.]

### C1.2 Genetic information

*Mammalian Genome*: Special issue published annually by Springer-Verlag, New York, with reports from the mouse chromosome mapping committees.

Lyon, M. F. & Searle, A. G., eds. (1989). *Genetic Variants and Strains of the Laboratory Mouse.*, 2nd edition, Oxford University Press, Oxford. [The "Bible" for mouse geneticists: description and references for all known mutant loci along with other genetic information. Look for newer edition.]

Silvers, W. K. (1979). *The Coat Colors of the Mouse*, Springer-Verlag, New York. [Excellent coverage of this particular subject. Sadly, now out of print.]

### C1.3 Experimental embryology

Hogan, B., Beddington, R., Costantini, F. & Lacy, E. (1995). *Manipulating the Mouse Embryo: A laboratory Manual, 2nd edition*, Cold Spring Harbor Laboratory Press, Cold Spring Harbor. [The bible for embryological manipulations.]

Wassarman, P. M. & DePamphilis, M. L., eds. (1993). *Guide to Techniques in Mouse Development, Methods in Enzymology*, Vol. 225, Academic Press, San Diego. [Comprehensive coverage of techniques used for developmental analysis.]

Kaufman, M. H. (1992). *An Atlas of Mouse Development*, Academic Press, San Diego. [Photomicrographs of embyo sections.]

Rugh, R. (1968). *The Mouse: Its Reproduction and Development*, Oxford University Press, New York. [Source of information.]

### C1.4 Other topics

Berry, R. J. & Corti, M., eds. (1990). *Biological Journal of the Linnean Society*, volume 41, Academic Press, London. [Proceedings from a conference on the evolution, population biology, and systematics of the mouse.]

Potter, M., Nadeau, J. H. & Cancro, M. P., eds. (1986). *The Wild Mouse in Immunology*, Current Topics ib Microbiology and Immunology, Vol. 127, Springer-Verlag, New York. [Proceedings from a conference on the evolution and systematics of the mouse.]

Klein, J. (1986). *Natural History of the Major Histocompatibility Complex*, John Wiley & Sons, New York. [Comprehensive discussion of various topics in mouse genetics, evolution, and population biology.]

Green, E. L. (1981). *Genetics and Probability in Animal Breeding Experiments*, Oxford University Press, New York. [General presentation of the statistical principles of genetic analysis.]

Morse, H. C., ed. (1978). *Origins of Inbred Mice*, Academic Press, New York. [Interesting proceedings on historical aspects of mouse genetics.]

Simmons, M. L. & Brick, J. O. (1970). *The Laboratory Mouse: Selection and Management*, Prentice-Hall, Englewood Cliffs, NJ.

Cook, M. J. (1965). *The Anatomy of the Laboratory Mouse*, Academic Press, New York. [An atlas of anatomical drawings.]

## C2 SELECTED JOURNALS AND SERIALS OF PARTICULAR INTEREST

### C.2.1 Original articles

*Cell*
*Cytogenetics and Cell Genetics*

*Genes and Development*
*Genetical Research*
*Genetics*
*Genomics*
*Immunogenetics*
*Journal of Heredity*
*Mammalian Genome*
*Mouse Genome (formerly Mouse Newsletter)*
*Nature*
*Nature Genetics*
*Proceedings of the National Academy of Science USA*
*Science*

## C2.2  Reviews

*Annual Review of Genetics*
*Current Opinion in Genetics and Development*
*Trends in Genetics*

# Appendix D

# Statistics

## D1 CONFIDENCE LIMITS AND MEDIAN ESTIMATES OF LINKAGE DISTANCE

### D1.1 The special case of complete concordance

To illustrate the statistical approach used to estimate confidence limits on experimentally determined values for linkage distances, it is useful to first consider the special case where two linked loci show complete concordance or no recombination (symbolized as R = 0) in their allelic segregations among a set of $N$ samples derived either from recombinant inbred (RI) strains or from the offspring of a backcross. Let us define the *true recombination fraction*—$\Theta$— as the experimental fraction of samples expected to be discordant (or recombinant) when $N$ approaches infinity. Then the probability of recombination in any one sample is simply $\Theta$, and the probability of non-recombination, or concordance, is simply $(1 - \Theta)$. As long as multiple events are completely independent of each other, one can calculate the probability that all of them will occur by multiplying together the individual probabilities associated with each event. Thus, if the probability of concordance in one sample is $(1 - \Theta)$, then the probability of concordance in $N$ samples is: $(1 - \Theta)^N$.

In most experimental situations, the known and unknown variables are reversed in that one begins by determining the number of discordant (or recombinant) samples $i$ that occur within a total set of $N$ as a means to estimate the unknown true recombination fraction $\Theta$. When no discordant samples are observed, the probability term just derived can be used with the substitution of the random variable $\theta$ in place of $\Theta$, to provide a continuous probability density function indicative of the relative likelihoods for different values of $\Theta$ between 0.0 (complete linkage) and 0.5 (no linkage).

$$P\{\Theta \ = \ \theta\} \ = \ f(\theta) \ = \ (1 - \theta)^N \qquad \text{(D1)}$$

This equation reads "the probability that the true recombination fraction $\Theta$ is equal to a particular value $\theta$ is the function of $\theta$ given as the last term in the equation." For both RI data and backcross data, $\Theta$ can be related directly to the linkage distance in centimorgans, $d$. In the case of backcross data and for values of $\Theta$ less than 0.25 (see Section 7.2.2.3), recombination fractions are converted into centimorgan estimates through simple multiplication:

$$d = 100\Theta \tag{D2}$$

In the case of RI data, this conversion is combined with the Haldane-Waddington equation (9.8) to yield:

$$d = 100 \left( \frac{\Theta}{4 - 6\Theta} \right) \tag{D3}$$

An example of the probability density function associated with the experimental observation of complete concordance among 50 backcross samples is shown in Fig. D1. Each value of $N$ will define a different function, but in all cases, the curve will look the same with only the steepness of the fall-off increasing as $N$ increases. In all cases, the "maximum likelihood estimate" for the true recombination fraction $\hat{\Theta}$—defined as the value of $\Theta$ associated with the highest probability—will be zero. However, since this maximum likelihood value is located at one end of the probability curve, it does not provide a useful estimate for the likely linkage distance. A better estimate would be the value of $\Theta$ that defines the midpoint below which and above which the true recombination fraction value is likely to lie with equal probability; this is the definition of the median recombination fraction estimate $\overline{\Theta}$. In mathematical terms, the value of $\overline{\Theta}$ is defined at the line that equally divides the area of the complete probability density given by equation D1 (see Fig. D1).

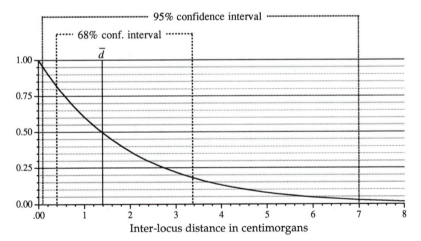

**Figure D1** No recombination between loci in 50 backcross animals. The probability density function is plotted for interlocus distances associated with the experimental finding of no recombination between two loci within a group of 50 backcross animals. This function is defined by the equation $f(\theta) = (1 - \theta)^{50}$ in which $\theta$ is the recombination fraction. Recombination fractions are converted into centimorgan distances ($d$) through the assumption of complete interference over the range shown. Minimum and maximum confidence limits are indicated for the 68% interval and the 95% interval. The median linkage distance estimate $\overline{d}$ is defined at the midline of the area under the complete density function.

Confidence limits are also defined by circumscribed portions of the entire probability density; the portion that lies outside a confidence interval is called $\alpha$. For example, in the case of a 95% confidence interval, $\alpha = (1 - 0.95) = 0.05$. It is standard practice to assign equal portions of $\alpha$ to the two "tails" of the probability density located before and after the central confidence interval. Thus, the lower confidence limit is defined as the value of $\theta$ bordering the initial $\alpha/2$ fraction of the area under the entire probability curve. The upper confidence limit is defined as the value of $\theta$ that borders the ultimate $\alpha/2$ fraction of the area under the entire probability curve; this is equivalent to saying that a "$(1 - \alpha/2)$" fraction of area lies ahead of the upper confidence limit.

In mathematical terms, the area beneath the entire probability density curve is equal to the definite integral of equation D1 over the range of legitimate values for $\theta$ between 0.0 and 0.5. To determine the *fraction* of the probability density that lies in the region between $\Theta = 0$ and any arbitrary $\Theta = x$, it is necessary to integrate over the probability density function (equation D1) between these two values, and divide the result by the total area covered by the probability density. This provides the probability that the true recombination fraction is less than or equal to x.

$$P\{\Theta \leq x\} = \frac{\int_0^x (1 - \theta)^N \, d\theta}{\int_0^{0.5} (1 - \theta)^N \, d\theta} \tag{D4}$$

By standard methods of calculus, equation D4 can be reduced analytically to the form:

$$P\{\Theta \leq x\} = 1 - (1 - x)^{N + 1} \tag{D5}$$

and this equation can be reformulated to yield x as a function of $P\{\Theta \leq x\}$.

$$x = 1 - {}^{(N+1)}\sqrt{1 - P\{\Theta \leq x\}} \tag{D6}$$

By solving equation D6 for different values of $P\{\Theta \leq x\}$, one can obtain critical values of x that define the median estimate of the recombination fraction from $P\{\Theta \leq x\} = 0.5$, lower confidence limits from $P\{\Theta \leq x\} = \alpha/2$, and upper confidence limits from $P\{\Theta \leq x\} = (1 - \alpha/2)$. Once a solution for x has been obtained, it can be converted into a linkage distance value with either equation D2 for backcross data or equation D3 for RI strain data. Solutions to equation D6 over a range of N RI strains and backcross animals are shown in Fig. 9.8, 9.16, and 9.17.

## D1.2  The general case of one or more recombinants

The statistical approach described above can be generalized to any case of $i$ discordant (or recombinant) samples observed among a total of N RI strains or backcross animals that have been typed for two loci. As in the special case above, one can arrive at a probability for the occurrence of multiple events by multiplying together the individual probabilities for each event. In the general case, there will be $i$ events of discordance, each with an individual probability equal to the true recombination fraction $\Theta$, and $(N - i)$ events of concordance, each with an individual probability of $(1 - \Theta)$. These terms are multiplied together along with a "binomial coefficient" that counts the the permutations in which the two types of events can appear to produce the "binomial formula:"

$$\frac{N!}{i!(N - i)!} \; \Theta^i \, (-\Theta)^{N-i} \tag{D7}$$

When the true recombination fraction is known, the binomial formula can be used to provide the probability that $i$ events of discordance will be observed in any set of $N$ samples. But once again, the situation encountered by geneticists is usually the reverse one in which $i$ and $N$ are discrete values determined by the experiment and the true recombination fraction $\Theta$ is unknown. In this case, one can substitute the random variable $\theta$ in place of $\Theta$ in equation D7 to generate a probability density function that provides relative likelihoods for different values of $\Theta$ between 0.0 (complete linkage) and 0.5 (no linkage). In this use of the binomial formula, the factorial fraction (known as the binomial coefficient) remains constant for all values of $\theta$ and can be eliminated since the purpose of the function is to provide relative probabilities only:

$$P\{\Theta = \theta\} = f(\theta) = \theta^i (1 - \theta)^{N-i} \tag{D8}$$

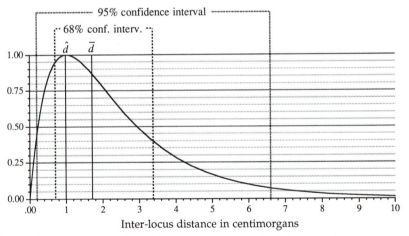

**Figure D2** One discordance between SDPs for 26 RI strains. The probability density function is shown for interlocus distances associated with the experimental finding of one discordant strain between the SDPs obtained for two loci in the typing of 26 RI strains (for example, the complete BXD set). This function is defined by the equation $f(\theta) = \theta^1(1 - \theta)^{26-1}$ in which $\theta$ is the recombination fraction; the curve shown has been normalized arbitrarily to a value of 1.0 at the maximum likelihood recombination fraction $\theta = \hat{\Theta}$. Recombination fractions are converted into centimorgan distances ($d$) with the Haldane-Waddington equations (9.7 and 9.8) and the assumption of complete interference over the range shown during each individual meiosis. The median linkage distance estimate $\overline{d}$ is defined at the midline of the area under the complete density function. Minimum and maximum confidence limits are indicated for the 68% interval and the 95% interval.

An example of the probability density function associated with the experimental observation of one discordant RI strain among a total of 26 samples is shown in Fig. D2. As one can easily see, the distribution is highly skewed toward higher recombination fractions. Each discrete pair of values $i$ and $N$ will define a different function. When both $i$ and $N$ are large, the density function will approximate a normal distribution. However, with the results typically obtained in contemporary mouse linkage studies, the density function is likely to be significantly skewed as shown in Fig. D2 and, as such, it is usually not possible to take advantage of the simplified statistical tools developed specially for use with the normal distribution.

A median estimate of linkage distance as well as lower and upper confidence limits can be obtained in the same manner described in the special case of no recombination described above. This can be accomplished by substituting equation D8 in place of the two occurrences of equation D1 within equation D4:

$$P\{\Theta \leq x\} = \frac{\int_0^x \theta^i (1 - \theta)^{N - i} \, d\theta}{\int_0^{0.5} \theta^i (1 - \theta)^{N - i} \, d\theta} \tag{D9}$$

The general form of the integral in this equation cannot be solved analytically, but a simple computer program can be used to estimate solutions and provide critical values of $x$ for defined probability values. The computer program has been written to generate minimum and maximum values in terms of centimorgan distances for discrete experimentally determined values of $i$ and $N$ from either backcross or RI data. The program was used to generate the values shown in Tables D1–D6 for 68% and 95% confidence intervals, but it is possible to generate confidence limits for any other integer percentile confidence interval as well. The program will also calculate maximum likelihood and median estimates of linkage distance.[109] It is listed below as a self-contained unit that should be ready for compiling with any standard C compiler on any computer. DOS and Macintosh versions of the executable program can be downloaded over the internet from the following anonymous FTP site: bioweb.princeton.edu. Interested investigators should look in the folder entitled pub/mouse.

**Table D1** Lower/upper limits on 68% confidence intervals for linkage distances (in cM) with $i$ discordances observed in 20–100 RI strains

| | Number of RI strains typed | | | | | | | | | | |
|---|---|---|---|---|---|---|---|---|---|---|---|
| $i$ | 20 | 25 | 30 | 35 | 40 | 45 | 50 | 60 | 70 | 80 | 100 |
| 0 | 0.2/2.4 | 0.2/1.9 | 0.1/1.6 | 0.1/1.2 | 0.1/1.3 | 0.1/1.0 | 0.1/0.9 | 0.1/0.8 | 0.1/0.7 | 0.1/0.6 | 0.0/0.5 |
| 1 | 0.9/4.8 | 0.7/3.7 | 0.6/3.0 | 0.5/2.6 | 0.4/2.2 | 0.4/1.9 | 0.4/1.7 | 0.3/1.4 | 0.3/1.2 | 0.2/1.1 | 0.2/0.8 |
| 2 | 1.8/7.5 | 1.5/5.7 | 1.2/4.6 | 1.0/3.8 | 0.9/3.3 | 0.8/2.9 | 0.7/2.6 | 0.6/2.1 | 0.5/1.8 | 0.5/1.5 | 0.3/1.2 |
| 3 | 3.0/10.8 | 2.3/7.9 | 1.9/6.2 | 1.6/5.2 | 1.4/4.4 | 1.2/3.8 | 1.1/3.4 | 0.9/2.8 | 0.8/2.3 | 0.7/2.0 | 0.5/1.6 |
| 4 | 4.4/14.8 | 3.4/10.5 | 2.7/8.2 | 2.3/6.6 | 2.0/5.6 | 1.7/4.8 | 1.5/4.3 | 1.3/3.4 | 1.1/2.9 | 0.9/2.5 | 0.7/1.9 |
| 5 | | 4.6/13.7 | 3.7/10.3 | 3.0/8.3 | 2.6/6.9 | 2.3/5.9 | 2.0/5.2 | 1.7/4.2 | 1.4/3.5 | 1.2/3.0 | 1.0/2.3 |
| 6 | | | 4.7/12.9 | 3.9/10.2 | 3.3/8.4 | 2.9/7.1 | 2.5/6.2 | 2.1/4.9 | 1.7/4.1 | 1.5/3.5 | 1.2/2.7 |
| 7 | | | | 4.8/12.3 | 4.1/10.0 | 3.5/8.4 | 3.1/7.3 | 2.5/5.7 | 2.1/4.7 | 1.8/4.0 | 1.4/3.1 |
| 8 | | | | 5.9/14.7 | 4.9/11.8 | 4.2/9.9 | 3.7/8.5 | 3.0/6.6 | 2.5/5.4 | 2.1/4.6 | 1.7/3.5 |
| 9 | | | | | 5.8/13.9 | 5.0/11.5 | 4.4/9.8 | 3.5/7.5 | 2.9/6.1 | 2.5/5.2 | 1.9/3.9 |
| 10 | | | | | | 5.8/13.2 | 5.1/11.2 | 4.0/8.5 | 3.3/6.9 | 2.8/5.8 | 2.2/4.4 |
| 11 | | | | | | 6.8/15.2 | 5.8/12.7 | 4.6/9.6 | 3.8/7.7 | 3.2/6.4 | 2.5/4.8 |
| 12 | | | | | | | 6.7/14.4 | 5.2/10.7 | 4.3/8.5 | 3.6/7.1 | 2.8/5.3 |
| 13 | | | | | | | 7.6/16.4 | 5.9/12.0 | 4.8/9.4 | 4.0/7.8 | 3.1/5.7 |
| 14 | | | | | | | | 6.6/13.3 | 5.3/10.4 | 4.5/8.5 | 3.4/6.2 |
| 15 | | | | | | | | 7.3/14.8 | 5.9/11.4 | 4.9/9.3 | 3.7/6.8 |
| 16 | | | | | | | | 8.1/16.4 | 6.5/12.5 | 5.4/10.1 | 4.0/7.3 |
| 17 | | | | | | | | 9.0/18.1 | 7.1/13.7 | 5.9/11.0 | 4.4/7.9 |
| 18 | | | | | | | | | 7.8/15.0 | 6.4/11.9 | 4.8/8.5 |
| 19 | | | | | | | | | 8.5/16.4 | 7.0/12.9 | 5.1/9.1 |
| 20 | | | | | | | | | 9.3/17.9 | 7.6/14.0 | 5.5/9.7 |
| 21 | | | | | | | | | 10.1/19.5 | 8.2/15.1 | 6.0/10.4 |
| 22 | | | | | | | | | | 8.9/16.3 | 6.4/11.1 |
| 23 | | | | | | | | | | 9.6/17.6 | 6.8/11.8 |
| 24 | | | | | | | | | | 10.3/19.0 | 7.3/12.6 |
| 25 | | | | | | | | | | 11.1/20.5 | 7.8/13.5 |
| 26 | | | | | | | | | | | 8.3/14.3 |
| 27 | | | | | | | | | | | 8.8/15.2 |
| 28 | | | | | | | | | | | 9.4/16.2 |
| 29 | | | | | | | | | | | 10.0/17.2 |
| 30 | | | | | | | | | | | 10.6/18.3 |

**Table D2**  Lower/upper limits on 95% confidence intervals for linkage distances (in cM) with $i$ discordances observed in 20–100 RI strains

| | | | | | | Number of RI strains typed | | | | | |
|---|---|---|---|---|---|---|---|---|---|---|---|
| $i$ | 20 | 25 | 30 | 35 | 40 | 45 | 50 | 60 | 70 | 80 | 100 |
| 0 | 0.03/5.3 | 0.02/4.1 | 0.02/3.4 | 0.02/2.9 | 0.01/2.5 | 0.01/2.2 | 0.01/1.9 | 0.01/1.6 | 0.01/1.4 | 0.01/1.2 | 0.01/0.9 |
| 1 | 0.3/9.3 | 0.2/7.0 | 0.2/5.6 | 0.2/4.6 | 0.2/4.0 | 0.1/3.5 | 0.1/3.1 | 0.1/2.5 | 0.1/2.1 | 0.1/1.9 | 0.1/1.5 |
| 2 | 0.8/13.9 | 0.6/10.1 | 0.5/7.9 | 0.5/6.5 | 0.4/5.5 | 0.4/4.8 | 0.3/4.2 | 0.3/3.4 | 0.2/2.9 | 0.2/2.5 | 0.2/1.9 |
| 3 | 1.5/19.8 | 1.2/13.8 | 1.0/10.5 | 0.8/8.5 | 0.7/7.1 | 0.6/6.1 | 0.6/5.4 | 0.5/4.3 | 0.4/3.6 | 0.3/3.1 | 0.3/2.4 |
| 4 | 2.3/27.1 | 1.8/18.2 | 1.5/13.5 | 1.3/10.7 | 1.1/8.9 | 1.0/7.6 | 0.9/6.6 | 0.7/5.2 | 0.6/4.4 | 0.5/3.7 | 0.4/2.9 |
| 5 | | 2.6/23.7 | 2.1/17.1 | 1.8/13.2 | 1.5/10.8 | 1.3/9.1 | 1.2/7.9 | 1.0/6.2 | 0.8/5.1 | 0.7/4.4 | 0.6/3.4 |
| 6 | | | 2.8/21.3 | 2.3/16.2 | 2.0/13.0 | 1.8/10.8 | 1.6/9.3 | 1.3/7.2 | 1.1/5.9 | 0.9/5.0 | 0.7/3.8 |
| 7 | | | | 3.0/19.6 | 2.5/15.4 | 2.2/12.7 | 2.0/10.8 | 1.6/8.3 | 1.4/6.8 | 1.2/5.7 | 0.9/4.3 |
| 8 | | | | 3.7/23.6 | 3.1/18.3 | 2.7/14.9 | 2.4/12.5 | 2.0/9.5 | 1.6/7.7 | 1.4/6.4 | 1.1/4.8 |
| 9 | | | | | 3.8/21.6 | 3.3/17.3 | 2.9/14.4 | 2.3/10.8 | 1.9/8.6 | 1.7/7.2 | 1.3/5.4 |
| 10 | | | | | | 3.9/20.0 | 3.4/16.5 | 2.7/12.1 | 2.3/9.6 | 2.0/8.0 | 1.5/5.9 |
| 11 | | | | | | 4.5/23.1 | 4.0/18.8 | 3.2/13.6 | 2.6/10.7 | 2.2/8.8 | 1.7/6.5 |
| 12 | | | | | | | 4.6/21.4 | 3.6/15.2 | 3.0/11.8 | 2.5/9.7 | 2.0/7.1 |
| 13 | | | | | | | 5.2/24.4 | 4.1/17.0 | 3.4/13.1 | 2.9/10.6 | 2.2/7.7 |
| 14 | | | | | | | | 4.6/19.0 | 3.8/14.4 | 3.2/11.6 | 2.5/8.3 |
| 15 | | | | | | | | 5.2/21.2 | 4.2/15.8 | 3.6/12.6 | 2.7/9.0 |
| 16 | | | | | | | | 5.8/23.6 | 4.7/17.4 | 3.9/13.7 | 3.0/9.7 |
| 17 | | | | | | | | 6.4/26.3 | 5.2/19.0 | 4.3/14.9 | 3.3/10.4 |
| 18 | | | | | | | | | 5.7/20.9 | 4.7/16.2 | 3.6/11.1 |
| 19 | | | | | | | | | 6.2/22.9 | 5.2/17.5 | 3.9/11.9 |
| 20 | | | | | | | | | 6.8/25.1 | 5.6/19.0 | 4.2/12.8 |
| 21 | | | | | | | | | 7.4/27.6 | 6.1/20.6 | 4.5/13.7 |
| 22 | | | | | | | | | | 6.6/22.4 | 4.9/14.6 |
| 23 | | | | | | | | | | 7.1/24.2 | 5.2/15.6 |
| 24 | | | | | | | | | | 7.7/26.3 | 5.6/16.6 |
| 25 | | | | | | | | | | 8.3/28.5 | 6.0/17.7 |
| 26 | | | | | | | | | | | 6.4/18.9 |
| 27 | | | | | | | | | | | 6.8/20.1 |
| 28 | | | | | | | | | | | 7.2/21.5 |
| 29 | | | | | | | | | | | 7.7/22.9 |
| 30 | | | | | | | | | | | 8.2/24.4 |

**Table D3** Lower/upper limits on 68% confidence intervals for linkage distances (in cM) with $i$ discordances observed in 20–100 backcross animals

| $i$ | Number of backcross animals typed | | | | | | | | | | |
|---|---|---|---|---|---|---|---|---|---|---|---|
| | 20 | 25 | 30 | 35 | 40 | 45 | 50 | 60 | 70 | 80 | 100 |
| 0 | 0.8/8.4 | 0.7/6.8 | 0.6/5.8 | 0.5/5.0 | 0.4/4.4 | 0.4/3.9 | 0.4/3.5 | 0.3/3.0 | 0.3/2.6 | 0.2/2.2 | 0.2/1.8 |
| 1 | 3.4/14.8 | 2.8/12.1 | 2.3/10.2 | 2.0/8.9 | 1.7/7.8 | 1.6/7.0 | 1.4/6.3 | 1.2/5.3 | 1.0/4.6 | 0.9/4.0 | 0.7/3.2 |
| 2 | 6.7/20.7 | 5.4/16.9 | 4.5/14.3 | 3.9/12.4 | 3.4/10.9 | 3.0/9.8 | 2.7/8.8 | 2.3/7.4 | 1.9/6.4 | 1.7/5.6 | 1.4/4.5 |
| 3 | | 8.2/21.5 | 6.9/18.2 | 5.9/15.8 | 5.2/13.9 | 4.6/12.4 | 4.2/11.3 | 3.5/9.5 | 3.0/8.2 | 2.6/7.2 | 2.1/5.8 |
| 4 | | | 9.4/21.9 | 8.1/19.0 | 7.1/16.8 | 6.2/15.0 | 5.7/13.6 | 4.7/11.4 | 4.1/9.8 | 3.5/8.7 | 2.8/7.0 |
| 5 | | | 12.0/25.5 | 10.3/22.2 | 9.0/19.6 | 8.0/17.5 | 7.2/15.9 | 6.0/13.3 | 5.2/11.5 | 4.5/10.1 | 3.6/8.1 |
| 6 | | | | 12.6/25.3 | 11.0/22.3 | 9.8/20.0 | 8.8/18.1 | 7.4/15.2 | 6.3/13.1 | 5.5/11.6 | 4.4/9.3 |
| 7 | | | | 15.0/28.3 | 13.1/25.0 | 11.6/22.4 | 10.5/20.3 | 8.7/17.1 | 7.5/14.8 | 6.6/13.0 | 5.3/10.5 |
| 8 | | | | | 15.2/27.7 | 13.5/24.8 | 12.1/22.5 | 10.1/18.9 | 8.7/16.3 | 7.6/14.4 | 6.1/11.6 |
| 9 | | | | | 17.3/30.3 | 15.4/27.2 | 13.8/24.6 | 11.5/20.7 | 9.9/17.9 | 8.6/15.8 | 6.9/12.7 |
| 10 | | | | | | 17.3/29.5 | 15.6/26.8 | 13.0/22.5 | 11.1/19.5 | 9.7/17.1 | 7.8/13.8 |
| 11 | | | | | | 19.3/31.8 | 17.3/28.9 | 14.4/24.3 | 12.3/21.0 | 10.8/18.5 | 8.6/14.9 |
| 12 | | | | | | | 19.1/30.9 | 15.9/26.1 | 13.6/22.5 | 11.9/19.9 | 9.5/16.0 |
| 13 | | | | | | | | 17.3/27.9 | 14.8/24.1 | 13.0/21.2 | 10.4/17.1 |
| 14 | | | | | | | | 18.8/29.6 | 16.1/25.6 | 14.1/22.5 | 11.2/18.2 |
| 15 | | | | | | | | 20.3/31.3 | 17.4/27.1 | 15.2/23.9 | 12.1/19.3 |
| 16 | | | | | | | | 21.8/33.0 | 18.7/28.6 | 16.3/25.2 | 13.0/20.3 |
| 17 | | | | | | | | | 20.0/30.1 | 17.4/26.5 | 13.9/21.4 |
| 18 | | | | | | | | | 21.3/31.5 | 18.6/27.8 | 14.8/22.5 |
| 19 | | | | | | | | | 22.6/33.0 | 19.7/29.1 | 15.7/23.5 |
| 20 | | | | | | | | | 23.9/34.5 | 20.8/30.4 | 16.6/24.6 |
| 21 | | | | | | | | | | 22.0/31.7 | 17.5/25.6 |
| 22 | | | | | | | | | | 23.1/33.0 | 18.5/26.7 |
| 23 | | | | | | | | | | 24.3/34.3 | 19.4/27.7 |
| 24 | | | | | | | | | | 25.5/35.5 | 20.3/28.7 |
| 25 | | | | | | | | | | 26.6/36.8 | 21.2/29.8 |
| 26 | | | | | | | | | | | 22.1/30.8 |
| 27 | | | | | | | | | | | 23.1/31.8 |
| 28 | | | | | | | | | | | 24.0/32.9 |
| 29 | | | | | | | | | | | 24.9/33.9 |
| 30 | | | | | | | | | | | 25.9/34.9 |
| 31 | | | | | | | | | | | 26.8/35.9 |
| 32 | | | | | | | | | | | 27.8/37.0 |
| 33 | | | | | | | | | | | 28.7/38.0 |

**Table D4**  Lower/upper limits on 68% confidence intervals for linkage distance determinations (in cM) based on data obtained from 100 to 1,000 backcross animals

| cM | Number of backcross animals typed | | | | | | |
|---|---|---|---|---|---|---|---|
| | 100 | 200 | 300 | 400 | 600 | 800 | 1,000 |
| 0 | 0.2/1.8 | 0.1/0.9 | 0.1/0.6 | 0.1/0.5 | 0.0/0.3 | 0.0/0.2 | 0.0/0.2 |
| 1 | 0.7/3.2 | 0.7/2.3 | 0.7/2.0 | 0.7/1.8 | 0.8/1.6 | 0.8/1.5 | 0.8/1.4 |
| 2 | 1.4/4.5 | 1.4/3.5 | 1.5/3.2 | 1.5/3.0 | 1.6/2.7 | 1.6/2.6 | 1.7/2.5 |
| 3 | 2.1/5.8 | 2.2/4.7 | 2.3/4.3 | 2.4/4.1 | 2.5/3.9 | 2.5/3.7 | 2.6/3.6 |
| 4 | 2.8/7.0 | 3.0/5.9 | 3.2/5.5 | 3.2/5.2 | 3.4/5.0 | 3.4/4.8 | 3.5/4.7 |
| 5 | 3.6/8.1 | 3.9/7.0 | 4.0/6.6 | 4.1/6.3 | 4.3/6.1 | 4.4/5.9 | 4.4/5.8 |
| 6 | 4.4/9.3 | 4.8/8.1 | 4.9/7.7 | 5.0/7.4 | 5.2/7.1 | 5.3/7.0 | 5.4/6.9 |
| 7 | 5.3/10.5 | 5.6/9.2 | 5.8/8.8 | 6.0/8.5 | 6.1/8.2 | 6.2/8.0 | 6.3/7.9 |
| 8 | 6.1/11.6 | 6.5/10.3 | 6.7/9.9 | 6.9/9.6 | 7.0/9.2 | 7.2/9.1 | 7.2/8.9 |
| 9 | 6.9/12.7 | 7.4/11.4 | 7.6/10.9 | 7.8/10.6 | 8.0/10.3 | 8.1/10.1 | 8.2/10.0 |
| 10 | 7.8/13.8 | 8.3/12.5 | 8.5/12.0 | 8.7/11.7 | 8.9/11.4 | 9.0/11.2 | 9.1/11.0 |
| 11 | 8.6/14.9 | 9.2/13.6 | 9.5/13.1 | 9.6/12.8 | 9.9/12.4 | 10.0/12.2 | 10.1/12.1 |
| 12 | 9.5/16.0 | 10.1/14.7 | 10.4/14.1 | 10.6/13.8 | 10.8/13.5 | 11.0/13.2 | 11.1/13.1 |
| 13 | 10.4/17.1 | 11.0/15.8 | 11.3/15.2 | 11.5/14.9 | 11.8/14.5 | 11.9/14.3 | 12.0/14.1 |
| 14 | 11.2/18.2 | 11.9/16.8 | 12.2/16.2 | 12.5/15.9 | 12.7/15.5 | 12.9/15.3 | 13.0/15.2 |
| 15 | 12.1/19.3 | 12.8/17.9 | 13.2/17.3 | 13.4/17.0 | 13.7/16.6 | 13.8/16.4 | 14.0/16.2 |
| 16 | 13.0/20.3 | 13.8/18.9 | 14.1/18.3 | 14.4/18.0 | 14.6/17.6 | 14.8/17.4 | 14.9/17.2 |
| 17 | 13.9/21.4 | 14.7/20.0 | 15.1/19.4 | 15.3/19.0 | 15.6/18.6 | 15.8/18.4 | 15.9/18.3 |
| 18 | 14.8/22.5 | 15.6/21.0 | 16.0/20.4 | 16.3/20.1 | 16.6/19.7 | 16.7/19.4 | 16.9/19.3 |
| 19 | 15.7/23.5 | 16.6/22.1 | 17.0/21.5 | 17.2/21.1 | 17.5/20.7 | 17.7/20.5 | 17.8/20.3 |
| 20 | 16.6/24.6 | 17.5/23.1 | 17.9/22.5 | 18.2/22.2 | 18.5/21.7 | 18.7/21.5 | 18.8/21.3 |
| 21 | 17.5/25.6 | 18.4/24.2 | 18.9/23.5 | 19.1/23.2 | 19.5/22.8 | 19.7/22.5 | 19.8/22.4 |
| 22 | 18.5/26.7 | 19.4/25.2 | 19.8/24.6 | 20.1/24.2 | 20.4/23.8 | 20.6/23.5 | 20.8/23.4 |
| 23 | 19.4/27.7 | 20.3/26.2 | 20.8/25.6 | 21.1/25.2 | 21.4/24.8 | 21.6/24.6 | 21.7/24.4 |
| 24 | 20.3/28.7 | 21.3/27.3 | 21.7/26.6 | 22.0/26.3 | 22.4/25.8 | 22.6/25.6 | 22.7/25.4 |
| 25 | 21.2/29.8 | 22.2/28.3 | 22.7/27.7 | 23.0/27.3 | 23.3/26.9 | 23.6/26.6 | 23.7/26.4 |
| 26 | 22.1/30.8 | 23.2/29.3 | 23.7/28.7 | 24.0/28.3 | 24.3/27.9 | 24.5/27.6 | 24.7/27.4 |
| 27 | 23.1/31.8 | 24.1/30.4 | 24.6/29.7 | 24.9/29.3 | 25.3/28.9 | 25.5/28.6 | 25.7/28.5 |
| 28 | 24.0/32.9 | 25.1/31.4 | 25.6/30.7 | 25.9/30.4 | 26.3/29.9 | 26.5/29.6 | 26.6/29.5 |
| 29 | 24.9/33.9 | 26.0/32.4 | 26.6/31.7 | 26.9/31.4 | 27.2/30.9 | 27.5/30.7 | 27.6/30.5 |
| 30 | 25.9/34.9 | 27.0/33.4 | 27.5/32.8 | 27.8/32.4 | 28.2/31.9 | 28.5/31.7 | 28.6/31.5 |
| 31 | 26.8/35.9 | 28.0/34.4 | 28.5/33.8 | 28.8/33.4 | 29.2/32.9 | 29.4/32.7 | 29.6/32.5 |
| 32 | 27.8/37.0 | 28.9/35.4 | 29.5/34.8 | 29.8/34.4 | 30.2/34.0 | 30.4/33.7 | 30.6/33.5 |
| 33 | 28.7/38.0 | 29.9/36.5 | 30.4/35.8 | 30.8/35.4 | 31.2/35.0 | 31.4/34.7 | 31.6/34.5 |
| 34 | | 30.9/37.5 | 31.4/36.8 | 31.7/36.4 | 32.1/36.0 | 32.4/35.7 | 32.5/35.5 |
| 35 | | 31.8/38.5 | 32.4/37.8 | 32.7/37.4 | 33.1/37.0 | 33.4/36.7 | 33.5/36.5 |

**Table D5** Lower/upper limits on 95% confidence intervals for linkage distances (in cM) with $i$ recombinants observed in 20–100 backcross animals

| | Number of backcross animals typed | | | | | | | | | | |
|---|---|---|---|---|---|---|---|---|---|---|---|
| $i$ | 20 | 25 | 30 | 35 | 40 | 45 | 50 | 60 | 70 | 80 | 100 |
| 0 | 0.1/16.1 | 0.1/13.2 | 0.1/11.2 | 0.1/9.7 | 0.1/8.6 | 0.1/7.7 | 0.1/7.0 | 0.1/5.9 | 0.1/5.1 | 0.1/4.5 | 0.1/3.6 |
| 1 | 1.2/23.8 | 1.0/19.7 | 0.8/16.7 | 0.7/14.5 | 0.6/12.9 | 0.5/11.5 | 0.5/10.5 | 0.4/8.8 | 0.3/7.6 | 0.3/6.7 | 0.2/5.4 |
| 2 | 3.1/30.4 | 2.5/25.1 | 2.0/21.4 | 1.8/18.7 | 1.5/16.5 | 1.4/14.9 | 1.2/13.5 | 1.0/11.4 | 0.9/9.8 | 0.8/8.6 | 0.6/7.0 |
| 3 | | 4.4/30.2 | 3.6/25.8 | 3.1/22.5 | 2.7/19.9 | 2.4/17.9 | 2.2/16.3 | 1.8/13.7 | 1.6/11.9 | 1.4/10.4 | 1.1/8.4 |
| 4 | | | 5.5/29.8 | 4.7/26.1 | 4.1/23.1 | 3.6/20.8 | 3.3/18.9 | 2.7/16.0 | 2.3/13.8 | 2.0/12.2 | 1.6/9.8 |
| 5 | | | 7.5/33.7 | 6.4/29.5 | 5.6/26.2 | 5.0/23.6 | 4.5/21.4 | 3.7/18.1 | 3.2/15.7 | 2.8/13.8 | 2.2/11.2 |
| 6 | | | | 8.2/32.8 | 7.2/29.2 | 6.4/26.3 | 5.7/23.9 | 4.8/20.2 | 4.1/17.5 | 3.6/15.4 | 2.8/12.5 |
| 7 | | | | 10.1/36.0 | 8.8/32.1 | 7.8/28.9 | 7.0/26.3 | 5.9/22.2 | 5.0/19.3 | 4.4/17.0 | 3.5/13.8 |
| 8 | | | | | 10.6/34.9 | 9.4/31.4 | 8.4/28.6 | 7.0/24.2 | 6.0/21.0 | 5.2/18.6 | 4.2/15.0 |
| 9 | | | | | 12.4/37.6 | 11.0/33.9 | 9.8/30.9 | 8.2/26.2 | 7.0/22.7 | 6.1/20.1 | 4.9/16.3 |
| 10 | | | | | | 12.6/36.4 | 11.3/33.1 | 9.4/28.1 | 8.0/24.4 | 7.0/21.5 | 5.6/17.5 |
| 11 | | | | | | 14.3/38.7 | 12.8/35.3 | 10.6/30.0 | 9.1/26.0 | 7.9/23.0 | 6.3/18.7 |
| 12 | | | | | | | 14.3/37.5 | 11.9/31.8 | 10.1/27.7 | 8.8/24.5 | 7.1/19.8 |
| 13 | | | | | | | | 13.2/33.7 | 11.2/29.3 | 9.8/25.9 | 7.8/21.0 |
| 14 | | | | | | | | 14.5/35.5 | 12.3/30.9 | 10.8/27.3 | 8.6/22.2 |
| 15 | | | | | | | | 15.8/37.3 | 13.5/32.4 | 11.7/28.7 | 9.3/23.3 |
| 16 | | | | | | | | 17.2/39.1 | 14.6/34.0 | 12.7/30.1 | 10.1/24.5 |
| 17 | | | | | | | | | 15.8/35.5 | 13.7/31.5 | 10.9/25.6 |
| 18 | | | | | | | | | 17.0/37.1 | 14.8/32.8 | 11.7/26.7 |
| 19 | | | | | | | | | 18.1/38.6 | 15.8/34.2 | 12.5/27.8 |
| 20 | | | | | | | | | 19.3/40.1 | 16.8/35.5 | 13.4/28.9 |
| 21 | | | | | | | | | | 17.9/36.9 | 14.2/30.0 |
| 22 | | | | | | | | | | 18.9/38.1 | 15.0/31.1 |
| 23 | | | | | | | | | | 20.0/39.5 | 15.9/32.2 |
| 24 | | | | | | | | | | 21.1/40.8 | 16.7/33.3 |
| 25 | | | | | | | | | | 22.2/42.1 | 17.6/34.3 |
| 26 | | | | | | | | | | | 18.4/35.4 |
| 27 | | | | | | | | | | | 19.3/36.5 |
| 28 | | | | | | | | | | | 20.2/37.5 |
| 29 | | | | | | | | | | | 21.0/38.6 |
| 30 | | | | | | | | | | | 21.9/39.6 |
| 31 | | | | | | | | | | | 22.8/40.7 |
| 32 | | | | | | | | | | | 23.7/41.7 |
| 33 | | | | | | | | | | | 24.6/42.7 |

**Table D6** Lower/upper limits on 95% confidence intervals for linkage distance determinations (in cM) based on data obtained from 100 to 1,000 backcross animals

| cM | Number of backcross animals typed | | | | | | |
|---|---|---|---|---|---|---|---|
| | 100 | 200 | 300 | 400 | 600 | 800 | 1,000 |
| 0 | 0.0/3.6 | 0.0/1.8 | 0.0/1.2 | 0.0/0.9 | 0.0/0.6 | 0.0/0.5 | 0.0/0.4 |
| 1 | 0.2/5.4 | 0.3/3.6 | 0.4/2.9 | 0.4/2.5 | 0.5/2.2 | 0.5/2.0 | 0.6/1.8 |
| 2 | 0.6/7.0 | 0.8/5.0 | 0.9/4.3 | 1.0/3.9 | 1.2/3.5 | 1.2/3.2 | 1.3/3.1 |
| 3 | 1.1/8.5 | 1.4/6.4 | 1.6/5.6 | 1.7/5.2 | 1.9/4.7 | 2.0/4.4 | 2.1/4.3 |
| 4 | 1.6/9.8 | 2.1/7.7 | 2.3/6.9 | 2.5/6.4 | 2.7/5.9 | 2.9/5.6 | 3.0/5.4 |
| 5 | 2.2/11.2 | 2.8/9.0 | 3.1/8.1 | 3.3/7.6 | 3.5/7.1 | 3.7/6.8 | 3.8/6.5 |
| 6 | 2.8/12.5 | 3.5/10.2 | 3.9/9.3 | 4.1/8.8 | 4.4/8.2 | 4.6/7.9 | 4.7/7.7 |
| 7 | 3.5/13.8 | 4.3/11.4 | 4.7/10.5 | 4.9/9.9 | 5.2/9.3 | 5.4/9.0 | 5.6/8.8 |
| 8 | 4.2/15.0 | 5.0/12.6 | 5.5/11.6 | 5.7/11.1 | 6.1/10.5 | 6.1/10.1 | 6.5/9.9 |
| 9 | 4.9/16.2 | 5.8/13.8 | 6.3/12.8 | 6.6/12.2 | 7.0/11.6 | 7.2/11.2 | 7.4/11.0 |
| 10 | 5.6/17.5 | 6.6/15.0 | 7.1/13.9 | 7.5/13.4 | 7.9/12.7 | 8.1/12.3 | 8.2/12.0 |
| 11 | 6.3/18.7 | 7.4/16.1 | 8.0/15.1 | 8.3/14.5 | 8.8/13.8 | 9.0/13.4 | 9.2/13.1 |
| 12 | 7.1/19.8 | 8.2/17.2 | 8.8/16.2 | 9.2/15.6 | 9.7/14.9 | 9.9/14.5 | 10.1/14.2 |
| 13 | 7.8/21.0 | 9.1/18.4 | 9.7/17.3 | 10.1/16.7 | 10.6/15.9 | 10.9/15.5 | 11.1/15.2 |
| 14 | 8.6/22.2 | 9.9/19.5 | 10.6/18.4 | 11.0/17.8 | 11.5/17.0 | 11.8/16.6 | 12.0/16.3 |
| 15 | 9.3/23.3 | 10.7/20.6 | 11.4/19.5 | 11.9/18.8 | 12.4/18.1 | 12.7/17.6 | 12.9/17.4 |
| 16 | 10.1/24.5 | 11.6/21.7 | 12.3/20.6 | 12.8/19.9 | 13.2/19.2 | 13.6/18.7 | 13.9/18.4 |
| 17 | 10.9/25.6 | 12.5/22.8 | 13.2/21.7 | 13.7/21.0 | 14.2/20.2 | 14.6/19.8 | 14.8/19.5 |
| 18 | 11.7/26.7 | 13.3/23.9 | 14.1/22.8 | 14.6/22.1 | 15.1/21.3 | 15.5/20.8 | 15.7/20.5 |
| 19 | 12.6/27.8 | 14.2/25.0 | 15.0/23.8 | 15.5/23.1 | 16.1/22.3 | 16.4/21.9 | 16.7/21.6 |
| 20 | 13.4/28.9 | 15.1/26.1 | 15.9/24.9 | 16.4/24.2 | 17.0/23.4 | 17.4/22.9 | 17.6/22.6 |
| 21 | 14.2/30.0 | 15.9/27.2 | 16.8/26.0 | 17.3/25.3 | 17.9/24.4 | 18.3/24.0 | 18.6/23.6 |
| 22 | 15.0/31.1 | 16.8/28.3 | 17.7/27.0 | 18.2/26.3 | 18.9/25.5 | 19.3/25.0 | 19.5/24.7 |
| 23 | 15.9/32.2 | 17.7/29.3 | 18.6/28.1 | 19.2/27.4 | 19.6/26.5 | 20.2/26.0 | 20.5/25.7 |
| 24 | 16.7/33.3 | 18.6/30.4 | 19.5/29.2 | 20.1/28.4 | 20.8/27.6 | 21.2/27.1 | 21.5/26.8 |
| 25 | 17.6/34.3 | 19.5/31.5 | 20.4/30.2 | 21.0/29.5 | 21.7/28.6 | 22.1/28.1 | 22.4/27.8 |
| 26 | 18.4/35.4 | 20.4/32.5 | 21.4/31.3 | 22.0/30.5 | 22.7/29.7 | 23.1/29.2 | 23.3/28.8 |
| 27 | 19.3/36.5 | 21.3/33.6 | 22.3/32.3 | 22.9/31.6 | 23.6/30.7 | 24.0/30.2 | 24.3/29.8 |
| 28 | 20.2/37.5 | 22.2/34.6 | 23.2/33.3 | 23.8/32.6 | 24.6/31.7 | 25.0/31.2 | 25.3/30.9 |
| 29 | 21.0/38.6 | 23.2/35.7 | 24.2/34.4 | 24.8/33.6 | 25.5/32.8 | 26.0/32.2 | 26.3/31.9 |
| 30 | 21.9/39.6 | 24.1/36.7 | 25.1/35.4 | 25.7/34.7 | 26.5/33.8 | 26.9/33.3 | 27.2/32.9 |
| 31 | 22.8/40.7 | 25.0/37.7 | 26.0/36.5 | 26.7/35.7 | 27.4/34.8 | 27.9/34.3 | 28.2/33.9 |
| 32 | 23.7/41.7 | 25.9/38.8 | 27.0/37.5 | 27.6/36.7 | 28.4/35.8 | 28.9/35.3 | 29.2/35.0 |
| 33 | 24.6/42.7 | 26.9/39.5 | 27.9/38.5 | 28.6/37.8 | 29.4/36.9 | 29.8/36.3 | 30.2/36.0 |
| 34 | | 27.8/40.8 | 28.9/39.5 | 29.5/38.8 | 30.3/37.9 | 30.8/37.4 | 31.1/37.0 |
| 35 | | 28.7/41.9 | 29.8/40.6 | 30.5/39.8 | 31.3/38.9 | 31.8/38.4 | 32.1/38.0 |

## D1.3  A C program for the calculation of linkage distance estimates and confidence intervals

```
/*** A C program for the calculation of linkage distance estimates and confidence intervals ***/
#include <stdio.h>
double Pin(double r,int i,int N); double pow(double x, double y); double convert(double r);
static int crosstype;
main()
{   FILE    *fopen(), *file;
    int     i = 1, istart = 1, ifin = 50, iinc = 1, N = 100, P;
    char    input;
    double  Pin(), dmin, dmax, r,rtop, dmean, smean, Nrmlize = 0.0, Sum = 0.0, convert(), min, max;
    while(1){
        printf("Enter the type of cross:1 for backcross,2 for RI analysis,or 3 to quit:");
        scanf("%d",&crosstype);
        if(crosstype != 2 && crosstype != 1) exit(0);
        printf("Enter the confidence level as an integer number(e.g. 95 for 95%%):");
        scanf("%d", &P);
        min = (1-((double)P/100.0))/2; max = 1- min;
        printf("Enter with comma delimiters->i-start,i-end,i-increment,and N,then return\n:>");
        scanf ("%d,%d,%d,%d", &istart,&ifin,&iinc,&N);
        printf(" i , dist / medn , min. / max. (values in cM assuming complete interference)\n");
        for ( i = istart; i <= ifin ; i += iinc){
            for ( r = .0001, Nrmlize = 0 ; r <.5 ; r += .0001)
                Nrmlize += Pin(r,i,N);
            for ( r = .0001, Sum = 0; Sum < min && r<.5; r += .0001)
                Sum += Pin(r,i,N)/Nrmlize;
            dmin = convert(r);
            for (; Sum <.5 && r<.5; r += .0001)
                Sum += Pin(r,i,N)/Nrmlize;
            dmean = convert(r);
            for (; Sum < max && r<.5 ; r += .0001)
                Sum + = Pin(r,i,N)/Nrmlize;
            dmax = convert(r);
            smean = convert((double)i/N);
            printf("%3d, %4.1f / %4.1f , %4.1f / %4.1f\n",i,smean,dmean,dmin,dmax);}
    }}
double convert(double r)
{   double rmean; int x = 0;
    if(crosstype == 1) return(100*r);
    if(crosstype == 2) return( r*100/(4—6*r) );}
double Pin(double r ,int i ,int N)
{   double pow();
    return ((pow(r,i))*(pow(1-r,N-i)));}
/********************* END OF PROGRAM *********************/
```

## D2  QUANTITATIVE DIFFERENCES IN EXPRESSION BETWEEN TWO STRAINS

How does one determine whether two populations of animals defined by different inbred strains are showing a significant difference in the expression of a trait? The answer is with a test statistic known as the "$t$-test" or "Student's $t$-test." To apply this test, one needs to use a pair of only three values derived from an analysis of the expression of the trait in sets of animals from each inbred strain. First is the number of animals examined in each inbred set ($N_1$ and $N_2$). Second is the mean level of expression for each set ($m_1$ and $m_2$) calculated as:

$$m = \frac{1}{N} \sum_{i=1}^{N} x_i \tag{D10}$$

where $x_i$ *refers to the expression value obtained for the* $i$th sample in the set. Third is the variance of each set of animals ($s_1^2$ and $s_2^2$) calculated as:

$$s^2 = \left( \frac{1}{N} \sum_{i=1}^{N} x_i^2 \right) - m^2 \tag{D11}$$

With values for the variance of each sample set and the size of each set, one can calculate a combined parameter refered to as the "pooled variance:"

$$s_p^2 = \frac{N_1 s_1^2 + N_2 s_2^2}{N_1 + N_2 - 2} \tag{D12}$$

Finally, one can use the value obtained for the pooled variance together with the samples sizes and sample means to obtain a "$t$ value:"

$$t = \frac{|m_1 - m_2|}{\sqrt{\dfrac{s_p^2}{N_1} + \dfrac{s_p^2}{N_2}}} \tag{D13}$$

One final combined parameter is required to convert the $t$ value into a level of significance—the number of degrees of freedom $df$

$$df = N_1 + N_2 - 2 \tag{D14}$$

With values for $t$ and $df$, one can obtain a $P$ value from a table of critical values for the $t$ distribution found in Table D7.

**Table D7**  Critical values of the $t$ distribution associated with selected $P$ values for 10–1,000 degrees of freedom ($df$)

| df | \multicolumn{5}{c}{P values} |
|---|---|---|---|---|---|
| | 0.05 | 0.01 | 0.005 | 0.001 | 0.0005 |
| 10 | 1.81 | 2.76 | 3.17 | 4.14 | 4.59 |
| 15 | 1.75 | 2.60 | 2.95 | 3.73 | 4.07 |
| 20 | 1.73 | 2.53 | 2.85 | 3.55 | 3.85 |
| 30 | 1.70 | 2.46 | 2.75 | 3.39 | 3.65 |
| 40 | 1.68 | 2.42 | 2.70 | 3.30 | 3.55 |
| 60 | 1.67 | 2.40 | 2.66 | 3.23 | 3.46 |
| 100 | 1.66 | 2.36 | 2.63 | 3.17 | 3.39 |
| 1,000 | 1.65 | 2.33 | 2.58 | 3.10 | 3.30 |

# Appendix E

# Glossary of Terms

**Allele:** An alternate form of a gene or locus. A locus can have many different alleles, which may differ from each other by as little as a single base or by the complete absence of a sequence.

**Allele-specific oligonucleotide (ASO):** An oligonucleotide designed to hybridize only to one of two or more alternative alleles at a locus. An ASO is usually designed around a variant nucleotide located at or near its center (see Chapter 8).

**Anchor locus:** A well-mapped locus that is chosen as a marker to "anchor" a particular genomic region to a framework map that is being constructed in a linkage study with a new cross (see Chapter 9).

**Anonymous locus:** An isolated DNA region with no known function but with at least two allelic states that can be followed through some form of DNA analysis in mapping studies.

**ASO:** See Allele-specific oligonucleotide.

**B1 repeat:** The most prominent SINE class of highly dispersed repetitive elements in the genome with a copy number of $\sim 150,000$ (see Chapter 5).

**B2 repeat:** The second most prominent SINE class of highly dispersed repetitive elements in the genome with a copy number of $\sim 90,000$ (see Chapter 5).

**B6:** An abbreviation for the name of the most commonly used strain of mice— C57BL/6.

**Bayesian analysis:** A statistical approach that takes prior information into account in the determination of probabilites. The Bayesian approach yields an equation that must be used to convert P values obtained from $x^2$ analysis of recombination data into actual probabilities of linkage between two loci (Chapter 9).

**Backcross:** A cross between one animal type that is heterozygous for alleles obtained from two parental strains and a second animal type from one of those parental strains. The term is often used by itself to describe the two-generation breeding protocol of an outcross followed by a backcross used frequently in linkage analysis (see Chapters 3 and 9).

**CA repeat:** The most prominent class of microsatellites found in mammalian genomes (see Chapter 8).

*castaneus:* The shortened form of *M. m. castaneus*, a subspecies within the *M. musculus* group that can be combined with a traditional inbred strain for linkage analysis (see Chapters 2 and 9).

**Centimorgan (cM):** The metric used to describe linkage distances. A centimorgan is the distance between two genes that will recombine with a frequency of exactly 1%. This term is named after Thomas Hunt Morgan, who first conceptualized linkage while working with *Drosophila*.

**Chimera:** An individual mouse, or other mammal, that is derived from the fusion of two or more preimplantation embryos, or an embryo and ES cells (see Chapter 6).

**Chi-squared:** A statistical test used most often by geneticists to ascertain whether experimental data provide significant evidence for linkage between two loci (see Chapter 9).

**Chromosome bands:** Alternative light- and dark-staining regions within chromosomes that are visualized by light microscopy (see Chapter 5).

**cM:** See Centimorgan.

**Codominance:** Defined for pairs of alleles. The situation in which an animal heterozygous for two alleles ($A^1$ and $A^2$ at the *A* locus) expresses *both* of the phenotypes observed in the two corresponding homozygotes. Thus, the heterozygote ($A^1/A^2$) and both homozygotes ($A^1/A^1$ and $A^2/A^2$) are all distinguishable from each other and $A^1$ and $A2$ would be considered to be "codominant." This term has also been co-opted to describe DNA markers defined by alternative allelic forms such as different sized restriction fragments or PCR products.

**Coisogenic:** A variant strain of mice that differs from an established inbred strain by mutation at only a single locus (see Chapter 3).

**Commensal:** Pertaining to populations of house mice that depend on human-built habitats and/or food production for survival (see Chapter 2).

**Concordance:** For two or more loci or traits typed in offspring from a backcross or RI strain, the presence of alleles (or expression of a trait) derived from the same parental chromosome (see Chapter 9).

**Congenic:** A variant strain of mice that is formed by backcrossing to an inbred parental strain for ten or more generations while maintaining heterozygosity at a selected locus (see Chapter 3).

**Conplastic:** A variation on the congenic approach in which the mitochondrial genome from one strain is transferred onto a different genetic background (see Chapter 3).

**Consomic:** A variation on the congenic approach in which an entire chromosome from one strain, usually the Y, is transferred onto a different genetic background (see Chapter 3).

**Contig:** A set of *contiguous* overlapping genomic clones that together span a larger region of the genome than that covered by any one clone (see Chapters 7 and 10).

**CpG island:** A genomic region of one or a few kilobases in length that contains a high density of CpG dinucleotides. CpG islands are associated with the

5'-ends of genes (see Chapter 10; previously called an HTF island).

**Cross:** One or more mating units set up with males and females that each have a designated genotype chosen to carry out a particular genetic analysis (see Chapter 3).

**Crossover product:** A chromosome homolog that was formed through the recombination of alleles at different loci present on opposite homologs in one of the parents of the animal in which it is observed (see Chapter 7).

**Deme:** A breeding group unit. In natural populations of mice, a deme usually consists of one breeding male with a harem of up to eight females (see Chapter 2).

**Differential segment:** In the genome of a congenic mouse, the region of chromosome surrounding the selected locus that is derived together with it from the donor genome (see Chapter 3).

**Discordance:** The opposite of concordance. Inheritance of alleles at two loci from different parental chromosomes.

**Disjunction:** The normal process by which the two homologs of each chromosome in a meiotic cell separate and move to different gametes (see Chapter 5).

**Distal:** A relative term meaning closer to the telomere; the opposite of proximal.

**DNA marker:** A cloned chromosomal locus with allelic variation that can be followed directly by a DNA-based assay such as Southern blotting or PCR (see Chapter 8).

*domesticus:* Shortened form of *M. m. domesticus*, a subspecies within the *M. musculus* group that resides in Western Europe, Africa, and throughout the New World. It is the primary component of the traditional inbred strains (see Chapter 2).

**Dominant:** A relative term describing the relationship of one allele to a second at the same locus when an animal heterozygous for these alleles expresses the same phenotype as an animal homozygous for the first allele. The second allele of the pair is considered recessive.

**ES cells:** Embryonic stem cells. Cultured cells derived from the pluripotent inner cell mass of blastocyst stage embryos. Used for gene targeting by homologous recombination (see Chapter 6).

**Expressivity:** Pertaining to observed quantitative differences in the expression of a phenotype among individuals that have the same mutant genotype. When quantitative differences are observed, a phenotype is said to show "variable expressivity," which can be caused by environmental factors, modifier genes or chance.

**$F_1$:** The first filial generation; the offspring of an outcross between different strains (see Chapter 3).

**Feral:** Pertaining to wild populations of animals derived from commensal ancestors; house mice that live apart from, and independent of, humans (see Chapter 2).

**Filial generation:** Pertaining to a particular generation in a sequence of brother–sister matings that can be carried out to form an inbred strain. The first filial generation, symbolized as $F_1$, refers to the offspring of a cross between mice having non-identical genomes. When $F_1$ siblings are crossed to each other, their offspring are considered to be members of the second filial generation or $F_2$, with subsequent generations of brother–sister matings numbered with integer increments (see Chapter 3).

**FISH:** Fluorescent *in situ* Hybridization. An enhanced from of *in situ* hybridization with high resolution and sensitivity (see Chapter 10).

**Genetic drift:** The constant tendency of genes to evolve even in the absence of selective forces. Genetic drift is fueled by spontaneous neutral mutations that disappear or become fixed in a population at random.

**Genome:** The total genetic information present within a single cell nucleus of an animal. The haploid genome content of the mouse is $3 \times 10^9$ bp.

**Genotype:** For any one animal, the set of alleles present at one or more loci under investigation. At any one autosomal locus, a genotype will be either homozygous (with two identical alleles) or heterozygous (with two different alleles).

**Giemsa:** A stain and associated protocol used to accentuate visually the difference between bands and interbands on metaphase chromosomes (see Chapter 5).

**Haplotype:** Pertaining to a particular set of alleles at linked loci (or nucleotide changes within a gene) that are found together on a single homolog. In linkage studies with backcross offspring and RI strains, the haplotypes associated with each sample provide a means for determining the order of loci (see Chapter 9).

**Heterozygote:** An animal with two distinguishable alleles at a particular locus under analysis. In this case, the locus is considered to be heterozygous.

**Histocompatible:** Pertaining to a genetic state in which cells from two animals can be cross-transplanted without immunological rejection. The opposite of histoincompatible. Histocompatibility is controlled predominantly by genes in the major histocompatiblity complex or MHC (see Chapter 3).

**Homolog:** This term is used by geneticists in two different senses: (1) one member of a chromosome pair in diploid organisms, and (2) a gene from one species, for example the mouse, that has a common origin and functions the same as a gene from another species, for example, humans, *Drosophila*, or yeast.

**Homozygote:** An animal with two identical alleles at a particular locus under analysis. In this case, the locus is considered to be homozygous.

**Hotspot, recombinational:** A localized region of chromosome, usually less than a few kilobases in length, that participates in crossover events at a very high rate relative to neighboring "Cold" regions of chromosome in a particular cross (see Chapter 7).

**House mouse:** An animal that is a member of the species *M. musculus*.

**HTF island:** See CpG island.

**Hybrid sterility:** Pertaining to the sterility of animals produced from matings between members of two different species, such as *M. musculus* and *M. spretus*. In this case, and in general, only the male hybrids are sterile while the females are fertile (see Chapter 2).

**Hybrid zone:** A narrow geographical line that separates the natural ranges of two distinct animal populations. The best-characterized house mouse hybrid zone occurs in Central Europe, and separates *M. m. domesticus* to the west and *M. m. musculus* to the east (see Chapter 2).

**Imprinting, genomic:** The situation in which the expression of a gene varies depending on its parental origin (see Chapter 5). Only a small subset of genes in the mammalian genome are imprinted.

***In situ* hybridization:** A technique for mapping cloned DNA sequences by hybridization directly to metaphase chromosomes and analysis by microscopy (see Chapter 10).

**Inbred:** Animals that result from the process of at least 20 sequential generations of brother–sister matings. This process is called inbreeding (see Chapter 3).

**Incross:** A cross between two animals that have the same homozygous genotype at designated loci; for example, between members of the same inbred strain (see Chapter 3).

**Intercross:** A cross between two animals that have the same heterozygous genotype at designated loci; for example, between sibling $F_1$ hybrids that were derived from an outcross between two inbred strains (see Chapter 3).

**Interference:** The suppression of crossing over that occurs in the extended chromosomal vicinity of an initial crossover event. Interference is responsible for a severe reduction in the expected frequency of double crossover events in 10–20 cM lengths of the genome (see Chapter 7).

**Interspecific cross:** A cross between mice from two different species, most often *M. musculus* (represented by a traditional laboratory strain) and *M. spretus* for the purpose of linkage analysis. The interspecific cross is carried out to take advantage of the high level of polymorphism between the two parents (see Chapter 9).

**Intersubspecific cross:** A cross between two subspecies (see Chapter 9). In the case of mouse genetics, this refers most often to a cross between a traditional inbred strain that is predominantly *M. m. domesticus* and one of the other subspecies in the *M. musculus* complex, usually *M. m. musculus* or *M. m. castaneus* or a combination of both (within the faux species *M. m. molossinus*).

**IRS-PCR:** Interspersed repetitive sequence PCR. A method for amplifying species-specific sequences from a complex hybrid genome (see Chapter 8).

**Karyotype:** The number of chromosomes present in a given genome and the form that they assume (including banding patterns) when they condense (see Chapter 5). A karyotype is defined entirely by microscopic observation.

**Library, genomic:** A sufficient number of genomic clones such that any sequence of interest is very likely to be present in at least one member of the set. If the library is random, the actual set of original clones must contain a cumulative length of DNA that is equal to multiple "genomic equivalents."

**Linkage:** Pertaining to the situation where two loci are close enough to each other on the same chromosome such that recombination between them is reduced to a level significantly less than 50%.

**Linkage group:** A set of loci in which all members are linked either directly or indirectly to all other members of the set. Essentially equivalent to the genetic information associated with any single chromosome.

**Locus:** Any genomic site, whether functional or not, that can be mapped through formal genetic analysis.

**Meiosis:** The process by which diploid germ cell precursors segregate their chromosomes into haploid nuclei within eggs and sperm.

**Meiotic product:** An individual haploid genome within an egg or sperm cell. Meiotic products are usually observed and analyzed within the context of diploid offspring.

**Microdissection:** A method for dissecting and cloning from defined sub-chromosomal regions by microscopic examination and manipulation (see Chapter 8).

**Microsatellite:** A very short unit sequence of DNA (2–4 bp) that is repeated multiple times in tandem. Microsatellites (also called simple sequence repeats or SSRs) are highly polymorphic and make ideal markers for linkage analysis (see Chapter 8). A polymorphism at a microsatellite locus is also referred to as a simple sequence length polymorphism (SSLP).

**Minisatellite:** A highly polymorphic type of locus containing tandemly repeated sequences having a unit length of 10–40 bp. Minisatellite polymorphisms can be assayed by RFLP analysis or by PCR (see Chapter 8). Also referred to as variable number of tandem repeat (VNTR) loci.

**Multifactorial:** A trait controlled by at least two factors, which may be genetic or environmental (see Chapter 9); polygenic traits represent a subset of multifactorial traits.

**Mus:** The name of the genus that contains all house mice (*M. musculus*) and other closely related species.

**musculus:** The abbreviated form of *M. musculus*, the species that is synonymous with the house mouse (see Chapter 2).

**Mutant allele:** This term is defined differently by formal geneticists and population biologists. The formal genetic definition is an allele that exerts a deleterious effect on phenotype. The population definition is an allele that is present at a frequency of less than 1% in a natural population; according to this definition, a mutant allele in one population may be considered non-mutant (wild-type) in another population.

**Mutation:** A new allele present in an animal that is not present in the genome of either its parents.

**$N_2$, $N_3$, $N_4$ etc.:** Symbols used to describe the generation of backcrossing and the offspring that derive from it. The $N_2$ generation describes offspring from the initial cross between an $F_1$ hybrid and one of the parental strains. Each following backcross generation is numbered in sequence (see Chapter 3).

**Outcross:** A cross between genetically unrelated animals.

**Parental:** An inbred strain of animals that is used in the initial cross of a multigenerational breeding protocol; the meiotic products and offspring in the subsequent generation that retain the same set of designated alleles as one of the parental strains.

**Penetrance:** Pertaining to the failure of some animals with a mutant genotype to express the associated mutant phenotype. In any case where less than 100% of genotypically mutant animals are phenotypically mutant, the phenotype is said to be "incompletely penetrant." Incomplete penetrance is usually a matter of chance or modifiers in the genetic background.

**PFGE:** Pulsed-field gel electrophoresis. A technique for separating very large DNA molecules from each other (see Chapter 10).

**Phenotype:** The physical manifestation of a genotype within an animal. A mutant phenotype is caused by a mutant genotype and is manifested as an alteration within an animal that distinguishes it from the wild-type.

**Phylogenetic tree:** A diagram showing the postulated evolutionary relationships that exist among related species in terms of their divergence from a series of common ancestors at different points in time.

**Polygenic:** Pertaining to a phenotype that results from interactions among the products of two or more genes with alternative alleles (see Chapter 9).

**Polymorphic:** A term formulated by population geneticists to describe loci at which there are two or more alleles that are each present at a frequency of at least 1% in a population of animals. The term has been co-opted for use in transmission genetics to describe any locus at which at least two alleles are available for use in breeding studies, irrespective of their actual frequencies in natural populations.

**Proximal:** A relative term meaning closer to the centromere; the opposite of distal.

**Quantitative trait:** A phenotype that can vary in a quantitative manner when measured among different individuals (see Chapter 9). The variation in expression can be due to combinations of genetic and environmental factors, as well as chance. Quantitative traits are often controlled by the cumulative action of alleles at multiple loci.

**Recessive:** A relative term describing the relationship of one allele to a second at the same locus when an animal heterozygous for these alleles expresses the same phenotype as an animal homozygous for the second allele. The second allele of the pair is considered dominant.

**Recombinant:** The result of a crossover in a doubly heterozygous parent such that alleles at two loci that were present on opposite homologs are brought together on the same homolog. The term is used to describe the chromosome as well as the animal in which it is present.

**Recombinant congenic strain:** A variation on recombinant inbred strains in which the initial outcross is followed by several generations of backcrossing prior to inbreeding (see Chapter 3).

**Recombinant inbred (RI) strain:** A special type of inbred strain formed from an initial outcross between two well-characterized inbred strains followed by at least 20 generations of inbreeding (see Chapter 9).

**Restriction fragment length polymorphism (RFLP):** A DNA variation that affects the distance between restriction sites (most often by a nucleotide change that creates or eliminates a site) within or flanking a DNA fragment that hybridizes to a cloned probe (see Chapter 8). RFLPs are detected upon Southern blot hybridization. The term RFLP is commonly used even in situations where the DNA variation may not represent a true polymorphism in the population-based definition of this term.

**Restriction fragment length variant (RFLV):** A more accurate term to use in place of restriction fragment length polymorphism in those cases where the frequency of the variant in natural populations is not known.

**Retroposon:** An inserted genomic element that originated from the reverse transcribed mRNA produced from another region of the genome (see Chapter 5).

**RFLP:** See restriction fragment length polymorphism.

**RFLV:** See restriction fragment length variant.

**RI strain:** See recombinant inbred strain

**Robertsonian translocation:** A fusion between the centromeres of two acrocentric chromosomes to produce a single metacentric chromosome (see Chapter 5).

**Satellite DNA:** This term was used originally to describe a discrete fraction of DNA visible in a $CsCl_2$ density gradient as a "satellite" to the main DNA band. The term now refers to all simple sequence DNA having a centromeric location, whether distinguishable on density gradients or not (see Chapter 5).

**SDP:** See strain distribution pattern.

**Simple sequence repeat (SSR):** See microsatellite.

**SINE:** Short interspersed element. Families of selfish DNA elements that are a few hundred basepairs in size and dispersed throughout the genome (see Chapter 5).

*spretus:* Abbreviated form of *M. spretus*, a species commonly used in interspecific matings for the generation of linkage maps (see Chapters 2 and 9).

**SSCP:** Single strand conformation polymorphism. A gel-based means for detecting single nucleotide changes within allelic PCR products that have been denatured and gel-fractionated as single strands.

**SSLP:** Simple sequence length polymorphism; see microsatellite.

**SSR:** Simple sequence repeat; see microsatellite.

**Strain:** Refers to a group of mice that are bred within a closed colony in order to maintain certain defining characteristics. Strains can be inbred or non-inbred (see Chapter 3).

**Strain distribution pattern (SDP):** The distribution of two segregating alleles at a single locus across a group of animal samples used for analysis in a linkage study (see Chapter 9). Used in the context of backcross data and data obtained from RI strains.

**Sympatric:** Closely related species that have overlapping ranges in nature but do not interbreed. In different parts of its range, *M. musculus* is sympatric with *M. macedonicus, M. spicilegus*, and *M. spretus* (see Chapter 2).

**Syngenic:** Literally "of the same genotype." Used most frequently by immunologists to describe interactions between cells from the same inbred strain.

**Syntenic:** Two loci known to be in the same linkage group. Conserved synteny refers to the situation where two linked loci in one species (such as the mouse) have homologs that are also linked in another species (such as humans).

**Targeting, gene:** A technology that allows an investigator to direct mutations to a specific locus in the mouse genome (see Chapter 6). Also called targeted mutagenesis.

**Transgene:** A fragment of foreign DNA that has been incorporated into the genome through the manipulation of pre-implantation embryos (see Chapter 6).

**Transgenic:** Pertaining to an animal or locus that contains a transgene (see Chapter 6).

**Translocation:** Pertaining to a novel chromosome formed by breakage and reunion of DNA molecules into a non-wild-type configuration (see Chapter 5).

**Unequal crossover:** A crossover event that occurs between non-allelic sites. It can lead to the duplication of sequences on one homolog and the deletion of sequences on the other (see Chapter 5).

**Variant:** Literally, an alternative form. Used in conjunction with locus, phenotype, or mouse strain. A 'DNA variant' is equivalent to an alternative DNA allele. A variant mouse usually refers to one that carries a mutant allele or expresses a mutant phenotype.

**VNTR:** "Variable number of tandem repeats" locus; see Minisatellite.

**Walking:** The sequential cloning of adjacent regions along a chromosome by using the ends of previously obtained clones to rescreen genomic libraries. Walking allows one to extend the length of contigs (see Chapter 10).

**Wild type:** Animal or allele that functions normally and represents a common type found in natural populations at a frequency of at least 1%.

**YAC:** Yeast artificial chromosome. A vector for cloning very large genomic inserts of 300 kb to 1 mb in length (see Chapter 10).

**Zygote:** The fertilized egg containing pronuclei from both the mother and the father.

# Notes

1. The domestic rat (*Rattus norvegicus*) was not far behind in earlier centuries, but its current census and range are much more limited than that of the mouse.
2. Phenotypes of this kind are called *Quantitative traits* and the loci that control these phenotypes are called *quantitative trait loci* or *QTLs*.
3. Several detailed histories of the early years of mouse genetics have been published (Keeler, 1931; Dunn, 1965; Morse, 1978, 1981, 1985; Russell, 1978, 1985; Klein, 1986).
4. A tree showing the scientific descendants of Castle down through several generations is presented by Morse (1978). The author is proud to trace his own scientific heritage in mouse genetics back through Karen Artzt to Dorothea Bennett to Salome Waelsch to L.C. Dunn to William Castle.
5. Many common forms of cancer susceptibility and resistance are multifactorial, and until *very* recently, it was not possible to actually separate out the various loci involved. In the absence of inbred lines, cancer appears sporadically within a population, and there is no basis on which to demonstrate the involvement of genetic factors. However, with inbreeding and selection for resistance or susceptibility to a particular form of cancer, the genetic nature of the disease can be proven beyond a doubt.
6. François Bonhomme has pointed out to the author that the lumping together of house mice and other "mouse-like" creatures is a peculiarity of English and other Germanic languages. In Latin-derived languages, a clear distinction is made; every French child, for example, knows the difference between a field mouse (*mulot*) and a house mouse (*souris*).
7. The complete definition of a commensal animal is one that "lives and feeds in close association with another species without either benefiting it or harming it directly" (Bronson, 1984).
8. Population geneticists use the relative level of heterogeneity within a population as a predictor of the time that has passed since the population was "founded."
9. Karyotypic variation does exist within the *M. m. domesticus* group but it is solely of the Robertsonian type with fusions between acrocentric centromeres to form hybrid metacentric chromosomes. This variation is discussed more fully in Chapter 5.
10. The observation that hybrid sterility is expressed in males but not females is as predicted by Haldane's rule, which states that the heterogametic sex (in this case XY males) will always be the first to become sterile in inter-specific crosses (Haldane, 1922). The rationale for this empirical observation remains unknown.
11. However, there are two as yet unpublished reports of successful laboratory matings (F. Bonhomme, personal communication).

12. The Jackson Laboratory restricts its sale of mice to investigators who will use them (and their descendants) for scientific research and not commercial purposes.

13. It is possible to deviate somewhat from a strict brother–sister mating protocol during the production of an inbred strain, so long as the animals derived can still trace their lineage back to a single breeding pair at the beginning of the process. For example, survival of a line might require a mating between an offspring and parent at one or more generations. If this variation in breeding protocol occurs, one must calculate its effect on the theoretical "inbreeding coefficient" (Green, 1981) which will allow a re-determination of the stage at which the *equivalent* of 20 sequential brother × sister matings will occur. At this stage, the strain would be considered inbred.

14. The probability calculations for these later generations are somewhat more complicated, and the interested reader should refer to the excellent exposition on the use of probability in animal breeding studies by Green (1981).

15. The exception to this general rule can be seen in children born from unions between closely related family members. These children are often sickly, and it is this consequence of consanguinity that is likely to be responsible for the taboos against marriage between siblings and cousins that are enforced by most human cultures.

16. In this case, one must remember that the genetic background of the mutant strain will begin to diverge slowly from that of the inbred strain that gave rise to it.

17. The alternative cross-intercross or M system was used in the original work performed by Snell (1978).

18. The commonly used 129/SvJ strain obviously does not follow these rules, however, like other cases of historical naming anomalies, it is accepted for use by the mouse community.

19. Italics indicate my emphasis. The symbols used for some of the first mouse genes to be characterized were *A* for agouti, *b* for brown, *c* for albino (absence of color), *d* for dilute, *T* for Brachyury (tail), etc. As the number of known genes increased, two-letter symbols were used. Nearly all genes characterized now are given primary symbols of at least three letters in length.

20. However, one must remember that *M. spretus* is a species, not a strain, and as such, different *M. spretus* lines may be polymorphic for the locus in question.

21. This would not be necessary for records maintained in a computer-based spreadsheet file.

22. Failure to produce litters could be a consequence either of a failure to mate and/or a failure to achieve pregnancy subsequent to mating.

23. For a much more detailed exposition of this topic, the reader is directed to books by Hogan and her colleagues (1994) and by Rugh (1968).

24. To be perfectly accurate, the maximal information content possible in a diploid organism is equivalent to the haploid genome size multiplied by $(1 + x)$, where $x$ is the fraction of nucleotides in the genome that are heterozygous. For mice and all other mammals, $x$ will always be less than 0.01, and thus, it can be safely ignored in complexity calculations.

25. One centimorgan (cM) is defined as the distance between two markers that are observed to recombine with a frequency of 1%. A centimorgan is synonymous with a "map unit." The relationships between map distances and recombination frequencies are discussed more fully in Chapter 7.

26. The one-to-one correspondence between chiasmata and crossing over is accepted as fact today, but originally this idea was known as the "chiasmatype theory."

27. This value is 10% less than the generally quoted, although unsupported, consensus value of 1,600 cM (Davisson and Roderick, 1989).

28. The lungfish *Lepidosiren* has a genome that is more than 20 times larger than the

human/mouse genome. The genome of the plant *Tradescantia* is almost ten times larger.

29. I am taking broad liberties here with the definition of "typical." This calculation is just a *back of the envelope* exercise to obtain a very approximate number.

30. Although a centromere *appears* to be at one end of all normal mouse chromosomes, this cannot be the case at the molecular level because all DNA molecules must be capped at both ends with telomeres that maintain their integrity. Since there must be at least a short stretch of DNA that extends beyond the centromere, mouse chromosomes are formally acrocentric rather than telocentric.

31. Of course, once a karyotype has passed through this intermediate stage, and homozygosity for each Robertsonian translocation is attained, normal fertility is completely restored.

32. There have been numerous modifications of the Holliday model —including those proposed by Meselson and Rading—that allow a better fit to the actual data, and there is still lack of consensus on some of the details involved. However, the central feature of the Holliday model—single-strand invasion, branch migration, and duplex resolution—is still considered to provide the molecular basis for gene conversion.

33. The equation is $\rho = 1.660 + 0.098(GC)$, where $\rho$ is the buoyant density, and GC is the molar fraction of nucleotides that are G or C.

34. The 234 bp unit actually evolved through the initial duplication of an even smaller unit (116/118 bp), which in turn was derived from duplications upon duplications and divergence of three ancestral nonanucleotides (9 bp sequences) with a combined consensus of GAAAAA(T/A/C)(C/G))(A/T).

35. The terms LINE and SINE were originally coined by Maxine Singer (1982) as abbreviations for long interspersed elements and short interspersed elements, respectively.

36. The best age for maximal production of embryos by superovulation is between 3 and 4 weeks before the female has actually reached sexual maturity (Hogan et al., 1994). Once sexual maturity has been reached naturally, the response to superovulation falls off, although it is not eliminated. Another factor affecting productivity is the general health and nutritional status of the females. If one intends to use females from a litter for superovulation, the litter should be culled of males early on to increase the milk supply to the female pups that remain.

37. Out of simple curiosity, it is interesting to consider why human beings have evolved away from an inductive link between the female orgasm and hormonal changes required to initiate a successful pregnancy. It seems likely that this occurred in response to the use of sex by people for purposes other than procreation—such as pair bonding—which can occur at times outside the estrus period of the female cycle.

38. In humans, implantation is the signal that alters the course of the estrus cycle. Thus, humans can not become pseudopregnant.

39. Whenever an interesting phenotype does arise, the transgene provides a tag for the cloning of flanking sequences from the disrupted gene and its wild-type counterpart to allow further characterization of the new mutant.

40. This calculation was performed by $\chi^2$ analysis with the Yates correction factor of 0.5. The $\chi^2$ statistical test is discussed in detail in Section 9.1.3. With a null hypothesis of a $Tg/+$ genotype and 13 offspring, one would expect each of the two classes (transgenic and non-transgenic) to be represented at equal levels of 6.5. If all of the observed offspring carry the transgene, the $\chi^2$ equation is solved as follows: $(13 - 6.5 - 0.5)^2/6.5 + (|(0 - 6.5)| - 0.5)^2/6.5 = 11.1$. With a single degree of

freedom, this $\chi^2$ value leads to a $P$ value of less than 0.001 (from Table 9.2).

41. The original congenic strain carried the $Sl$ allele in an obligatory heterozygous state since homozygotes are non-viable. The 129 agouti lines maintained by many investigators have since lost the $Sl$ mutation and differ knowingly from the original 129 line only in the 15 cM region on chromosome 7 that encompasses the $c$ and $p$ loci.

42. The designation "anonymous locus" does not imply a lack of function but rather an ignorance as to whether a function exists. Presumably, most anonymous loci will actually be devoid of function. However, in a small number of cases, a function may be determined at a later date, at which time the anonymous locus will no longer be considered anonymous.

43. A fascinating exception to Mendel's first law is displayed by a variant portion of mouse chromosome 17 that acts as a selfish chromosomal region and is known as a $t$ haplotype (Silver, 1993a). Ten to 20% of the animals in wild populations of $M.$ $musculus$ (of all known subspecies) carry a $t$ haplotype. Heterozygous males, with a genotype of $+/t$, can transmit their "$t$-allele" to 99% or more of their offspring. This unusual phenomenon is called segregation distortion or transmission ratio distortion.

44. A statistical analysis suggests that more than luck is required to explain two aspects of Mendel's data. First, the seven genes that he chose to study each happened to lie on a different one of the seven pea chromosomes. Second, the data that Mendel collected showed an incredible "goodness of fit" with the outcome that he expected (Fisher, 1936). Suggestions have been made by Dunn (1965) and Wright (1966) to explain this early example of what today might be grounds for an investigation into experimental fraud. In the case of the ratios that are too good to be true, Mendel may have padded his numbers to fit the theory, not aware that small deviations from the expected were *to be expected*. Wright believes that Mendel could have made subconscious errors in favor of expectation as he was counting his samples, whereas Fisher suggests that Mendel could have concocted his numbers without actually doing the experiments. An alternative possibility is that one of Mendel's workers, rather than the great man himself, may have actually carried out the subterfuge in a misguided attempt to please his boss. In the case of the seven genes that happen to map to seven different chromosomes, it appears most reasonable to assume that Mendel eliminated all genes that failed to show independent assortment with all other genes in his study. Although this would clearly be considered fraudulent behavior in today's world, Mendel may not have seen it as such in the simpler world of the 19th century.

45. There are some important classes of DNA markers where this is not the case exemplified by the random amplification of polymorphic DNA (RAPD) loci, which are typed based on the presence or absence of a particular PCR product. In these cases, the allele that can be amplified into a PCR product should be considered as having a dominant relationship to the allele that cannot be amplified.

46. Alternative homologs are separated by a "/;" coupled alleles are placed adjacent to each other in order of linkage when this is known. If the order of linked loci is known relative to the centromere, then the most centromere-proximal locus is displayed at the left of a grouping. Different linkage groups are separated by commas. For example, if a third locus, $C$, *not linked* to $A$ or $B$ is under observation in the same breeding experiment, the complete genotype would be written as $A$ $B/a$ $b$, $C/c$.

47. The value calculated for quadruple recombination events $(0.2)^4 = 0.0016$ has only a negligible effect on the final percentage and can be safely ignored.

48. One morgan is equivalent to 100 cM.

49. Evaluation of interference through analysis of linkage data is confounded by genotyping errors, which will suggest falsely that double crossovers have occurred. This is true for error rates even as low as 1%. It is only when such errors are controlled for that the true extent of interference becomes apparent in mammalian genomes (Weber et al., 1993).

50. If you use a hand-held calculator to find a specific solution to the Carter-Falconer equation, be sure the calculator is in "radial" mode and not in "degrees" mode to obtain a proper solution to the inverse trigonometric function $\tan^{-1}$.

51. It is important to remember that interference only operates on genetic events that occur in the same meiotic cell and it is most readily apparent in genetic analyses based on the backcross. It does not apply to the analysis of recombinant inbred lines, where crossover events accumulate over multiple generations, or even to a traditional outcross–intercross analysis where crossover events can occur in both parents.

52. Since the original publication of the Maniatis manual, a second edition has appeared, other competing manuals have been published, and most suppliers of molecular biology reagents now also provide detailed accounts of molecular techniques.

53. The term "polymorphic" was originally defined by population geneticists as a locus with at least two alleles that are both present at a frequency of greater than 1% in a particular population. The term has been co-opted by geneticists who work in the laboratory to refer to any locus with two alleles that can be utilized in an experimental cross. The opposite of polymorphic is monomorphic.

54. The term RFLP includes the word polymorphism, which, as indicated in the previous footnote, is not always accurate for the locus at hand from a strict interpretation according to population geneticists. Purists have substituted "variant" for "polymorphism" to produce the acronym RFLV; however, the term RFLP is still more commonly used.

55. In the absence of breeding data, alternative interpretations of these results are possible. For example, animals with a single restriction fragment could be heterozygous with a deletion of this locus on one chromosome and animals with two restriction fragments could be homozygous with the presence of a restriction site in the middle of the region that hybridizes to the probe. Other more complex interpretations are also possible.

56. This is not true for special hypervariable classes of RFLPs such as minisatellites discussed in Section 8.2.3.

57. It is more difficult to obtain "clean" DNA from animal tissue than from cells grown in tissue culture. Enzymes that are more sensitive to minor contaminants may work fine on tissue culture samples but not on spleen, liver, or tail DNA.

58. The percentage of CpGs in the mammalian genome that are methylated has been estimated at ~90%. Interestingly, the class of non-methylated CpGs are not randomly distributed, but rather, are clustered into islands that distinguish the 5'-ends of many genes. This fact has been exploited by molecular genetic sleuths searching for cryptic genes within large cloned regions for the purpose of positional cloning as described in Section 10.3.4.

59. This number is calculated from the known base composition of mammalian DNA. Nucleotides A and T are each present at a frequency of 0.29, and C and G are each present at a frequency of 0.21. The expected frequency of occurrence of the tetranucleotide TCGA can be calculated as $(0.29)(0.21)(0.21)(0.29) = 1/270$ bp. Similarly, the expected frequency of *Msp*I sites can be calculated as 1/514 bp.

60. It is also possible, although less likely considering the short length of the genomic

fragments that are amplified, that some polymorphisms may arise through insertions or deletions confined to the region between two primer sites. In this case, alternative RAPD alleles might both be recognized as PCR fragments of different sizes. In one large data set from the mouse, polymorphisms of this type were not observed (Woodward et al., 1992).

61. The calculations are greatly simplified by assuming that: (1) the unique copy component of the genome is equivalent to random sequence with no preference for, or avoidance of, any particular nucleotide combinations; and (2) PCR amplification will be limited to regions of 2 kb or less in length. The frequency with which any sequence of $N$ nucleotides will occur is once in every $4^N$ bases *on each strand*. The total number of occurrences of any one oligomer in the single copy component of a genome of complexity $C$ will be $2C/4^N$, where the multiplication by two accounts for the two strands of the double helix. The probability that a second oligomer of the same composition will be present on the opposite strand within 2 kb ($2 \times 10^3$ bases) in one direction from the first oligomer can be calculated as $2 \times 10^3/4^N$. The average total number of PCR amplifiable products can be calculated as the total number of sites at which the primer will hybridize multiplied by the frequency with which a second site will be present within a short enough distance (here set at 2 kb) to allow amplification to occur: $(2C/4^N)(2 \times 10^3/4^N) = 4,000 \ C/16^N$.

62. A least 90% of the fragments observed in experiments performed to date are less than 2 kb in length (Welsh and McClelland, 1990; Williams et al., 1990; Welsh et al., 1991; Nadeau et al., 1992; Serikawa et al., 1992; Woodward et al., 1992).

63. In actuality, the repeat families B1, B2, and L1 are not distributed in a random manner, but instead, are each located preferentially in different chromosomal domains as discussed in Section 5.4.4. As a consequence, it is not possible to derive a mathematical formula that can be used to predict the numbers of PCR products to be expected with different sets of primers.

64. The CA-repeat subclass of microsatellites appears to form an altered DNA structure known as Z-DNA (Hamada and Kakunaga, 1982) that was thought to have special properties at one time, but not now.

65. Microsatellite polymorphisms have also been referred to as simple sequence length polymorphisms or SSLPs. However, the acronym SSLP is so similar to the unrelated acronym SSCP (for single-strand conformational polymorphism) that I have not used the term here to avoid confusion.

66. Stable elongation across TATA ... stretches would require lower reaction temperatures, which, in turn, would result in higher levels of non-homologous pairing and non-specific PCR products.

67. Pentamer and hexamer repeat loci are also present in the genome at much lower frequencies.

68. The stated goal of the Whitehead/MIT Genome Center is to define a total of 6,000 microsatellite loci that can be placed into 1 cM "bins." When this goal is reached, each 1 cM bin will hold approximately four microsatellites, and the average interlocus distance will be 500 kb.

69. This simplifying assumption is only valid when there is no distortion of equal segregation of alleles from either of the two loci under analysis.

70. The author is indebted to Earl Green for pointing out this simplification of the $\chi^2$ equation in the special case of the backcross as shown in equation 1.40 in his book (Green, 1981).

71. Many statistics textbooks have graphs that allow a more precise determination of the $P$ values associated with a continuum of $\chi^2$ values.

72. As indicated in Fig. 9.13, this "swept radius" is attained in all backcross experiments

with 45 or more samples.

73. The probability of *linkage* is equal to one minus the probability of *no linkage*. In actual usage, this formulation is easier to work with since the probability of *no linkage* will be a linear function of the $P$ value obtained from $\chi^2$ analysis in all cases where $P$ is small ($< 0.001$). According to Bayes' theorem, the probability of *no linkage* is equal to the "joint probability" of *no linkage* divided by the sum of the joint probabilities of the two possible outcomes, *linkage* and *no linkage*. The joint probability of *no linkage* is equal to the "prior probability that linkage will not exist" (equivalent to the fraction of the genome outside the swept radius on either side of the marker) multiplied by "the conditional probability that the statistical test will indicate *linkage* when, in fact, *linkage* does not exist" (equal to the $P$ value derived from the $\chi^2$ test). Even if the swept radius is as large as 35 cM, the fraction of the genome not covered within this distance on either side of a marker locus is still close to one $[(1500—70)/1,500 = 0.95]$. Thus, the prior probability of *no linkage* can be safely estimated as one, and the joint probability can be approximated as the $P$ value. The joint probability of *linkage* is equal to the "prior probability of *linkage*" (equivalent to the fraction of the genome swept within two radii of the marker locus) multiplied by the "conditional probability that the statistical test will indicate *linkage* when, in fact, *linkage* does exist" (equivalent to the power of the $\chi^2$ test, $1-\beta$, where $\beta$ is defined as the probability of type II error). The power of the $\chi^2$ test is dependent on the actual data set and is not easily estimated. For the sake of simplicity, this value has been set equal to one. For a more detailed account of Bayesian analysis, the reader should consult an advanced statistics text.

74. The fraction of the genome swept is equal to twice the swept radius in centimorgans (see Fig. 9.13 and 9.14) divided by the total genome length of $\sim 1500$ cM.

75. In actuality, it is prudent to set up at least five cages of brother–sister pairs for mating at each generation in case some are not fertile. However, the offspring from only a single pair should be chosen to continue the construction of the RI strain. Each original pair of $F_2$ progenitor animals can only be used to produce a single new RI strain.

76. In the case of the very first set of RI strains established from a cross between BALB/cAnNBy and C57BL/6JNBy, upper case letters were used to distinguish different members of the set from each other with individuals strain names of CXBD, CXBE, CXBG and so on (Bailey, 1971). Most other RI sets conform to the use of numbers as distinguishing symbols.

77. At the time of writing, 31 strains from this double set had been verified genetically, two original strains were still undergoing tests, and eight original strains had become extinct but DNA samples were still available (Marshall et al., 1992).

78. Of course, like all inbred strains, new mutations can and will arise at a low frequency, especially at hypermutagenic loci like microsatellites (see Section 8.3.6.4). However, new mutant alleles are usually distinguishable as such since they are likely to be different from the alleles present in both progenitor strains.

79. If these direct protocols fail to detect polymorphisms, one could use the probe in hand to recover a cosmid or YAC clone containing the gene of interest, and then search this larger cloned fragment for a nearby microsatellite that can be developed as a linked DNA marker as described in Section 8.3.6. On the other hand, it may be more efficient to forget the RI approach at this point, and use a backcross mapping panel instead as described in Section 9.3.

80. As discussed in Section 9.2.5, this may not be true for SDPs associated with complex phenotypes rather than DNA markers.

81. This measure of the swept radius is based on the assumption that the RI- determined

values for linkage distance are accurate representations of actual linkage distances. A more stringent measure of swept radius is discussed in Section 9.4.4 for backcross data and presented in Fig. 9.13.

82. The discordant strain number used in these calculations is set at 0.5 units above the maximum allowable strain number shown in Fig. 9.5 in order to determine the map distance above which *actual* pairs of loci are more likely than not to produce *integer* discordance values that exceed the maximum number allowable for linkage determination.

83. Particular regions of the genome that are of special interest to mouse geneticists are likely to be more densely populated with marker loci than regions that are not under investigation in and of themselves. A good example of this bias in marker distribution is in the 2 cM long *H*-2 region of chromosome 17 with 16 loci that have been typed in the BXD set; all 16 loci have an identical SDP.

84. Statements concerning the observed number of genomic patches present along a chromosome must be qualified because of the possibility that additional patches lie hidden in regions between two adjacent typed loci for which SDPs are available. As the density of typed loci increases, the probability of hidden patches goes down but never reaches zero.

85. An alternative interpretation of these data is that both mean estimates of linkage distance are relatively true to the particular pair of parental strains used in each cross. As discussed in Section 7.2.3.3, different crosses can vary in their propensity to undergo recombination in individual chromosome intervals.

86. The $LD_{50}$ point for a chemical is the dosage that is lethal to 50% of the treated animals in a designated period of time.

87. The law of the product states that the probability of occurrence of $n$ independent events is equal to the product of the probabilities of each individual event. For RI strains, the probability of occurrence of either progenitor allele at any locus is simply 0.5. Thus, the probability of occurrence of $n$ loci from the same parent is $(0.5)^n$.

88. In theory, it is possible to make use of unbalanced SDPs to search for the multiple loci involved in expressing a trait. In the example described in the text, susceptible BXD strains would have to have B6 alleles at both susceptibility loci, but resistance would mean only that a DBA allele was present at either or both loci. One could fill in an SDP using only susceptible strains and search for linkage relationships with this subset of data points, but the numbers may be too small to be significant. However, potential intervals of linkage suggested by this first step analysis could be vetted in pairs (since there will be two loci) with the requirement that all non-susceptible strains must have at least one DBA allele at these two loci. The main problem with this type of analysis is that the small numbers of strains present in each characterized RI set will usually preclude one from obtaining statistically significant results.

89. In theory, it is also possible to use RI data of this type in a search for the responsible loci. Let us assume that only two loci are involved in the phenotypic expression of interest, and further, the two mixed genotypes cause an intermediate phenotype while the two single-progenitor genotypes cause a corresponding progenitor phenotype. Then an SDP can be written with three possible values for each strain: $A$, $B$ and $\frac{1}{2}A\frac{1}{2}B$. As in the case above, one could fill-in only the $A$ and $B$ values of an SDP (together expected in 50% of the strains) and search for a pair of potential linkage relationships that could be vetted by the requirement for the presence of alternative alleles in the strains that show the intermediate phenotype.

90. In a real-life study, geneticists usually wait to see at least two recombination events before concluding non-equivalence because of the possibility of isolated aberrant

events or experimental errors.

91. As discussed in Section 7.2.3.3, a sample size of ~500 per individual cross represents an upper boundary beyond which there are greatly diminishing returns in resolving nearby loci.

92. There are two other explanations for apparent double crossover events that cover a single locus in a single animal. The first is the trivial possibility of mistyping. The second is the infrequent but real possibility of gene conversion without recombination at an isolated genetic site (Hammer et al., 1991). Neither of these explanations is likely to be valid when the region between the two breakpoints contains more than one typed locus.

93. Values for recombination fraction and map distance can be readily transformed into each other through the use of a mapping function as described in Section 7.2.2.3.

94. Of course, most previously determined inter-locus map distances are far from accurate. If the true map distance between markers was previously underestimated, this would obviously lead to a reduction in the frequency at which linkage could be demonstrated in any future cross that used these two markers.

95. Direct calculations of swept radii for the intercross are confounded by the greater complexity inherent in the data that can be generated. As described in Section 9.1.3.4, intercross data show three degrees of freedom in contrast to the one degree of freedom associated with backcross data. The intercross-specific $\chi^2$ test described in this earlier section can be used to evaluate significance in all comparisons of allelic segregation from two potentially linked loci.

96. The computer program provided in Appendix D can be used to derive recombination limits for any level of confidence.

97. The statistical approach described in Appendix D2 can be used to ascertain whether a significant difference in expression exists between any two sets of animals.

98. The formula for computing variances is given in Appendix D2.

99. To ensure complete genetic homogeneity of the $F_1$ population, it is important that all of the initial matings be conducted in the same gender direction, with one strain fixed as the maternal progenitor and the other strain fixed as the paternal progenitor. The choice of gender can be informed by the reproductive characteristics of strains described in Chapter 4 and Table 4.1.

100. It may also be useful in this case to set up independent crosses with each parental strain. It may turn out that the data obtained from one of these backcrosses will be easier to analyze.

101. Although the idealized distribution for three polygenic loci shows a small dip in the middle (Fig. 9.19 bottom right-hand panel), with a typical-sized $N_2$ test population ($\leq 200$ animals), there is likely to be a finite standard error on top of the natural variance that will cause the two peaks to merge into one.

102. Non-parametric tests of this type are used to determine whether the actual genotype at a particular marker locus shows a significant level of correlation with the rank order of trait expression in individual animals.

103. The vast majority of cells that have failed to undergo a mitotic crossover in the chromosomal region centromeric to the marker gene can be eliminated with alloantigen-specific antibody and complement. Only cells that have lost the alloantigen will survive. Of these, only a subset will have undergone allele loss as a result of a mitotic recombination event.

104. The X chromosome is essentially conserved in its entirety across all eutherian mammals, and is excluded from estimates of conserved synteny.

105. To keep this hypothetical problem simple, I have assumed that all of the derived YAC clones are non-chimeric and that end fragments from all of these clones are

easily isolated. Unfortunately, the real-life situation is likely to be less straightforward.

106. The efficacy of this approach and all others dependent on cDNA libraries is greatly increased by using a normalized library in which all transcripts are represented equally irrespective of their different relative abundancies within the tissue itself (Patanjali et al., 1991). In theory, the normalization process should even out the representation of all transcribed sequences so that clones of actin mRNA, for example, are no more frequent than clones of rare messengers from the same tissue. In practice, normalization only succeeds part way; very rare messengers will always be under-represented.

107. Whenever genomic fragments are used to probe for expressed sequences, it is essential to pre hybridize the probe or target with unlabeled total mouse DNA in order to block highly repetitive sequences, which are present in the non-coding regions of a subset of mammalian transcripts.

108. In addition, two classes of artifactual products were also recovered. One class contained genomic fragments with random sequences having coincidental homology to splice sites. The other class contained true exons but with flanking intronic sequences on one side.

109. The program works by estimating solutions to definite integrals over segments of the appropriate probability density function.

# References

Abbott, C. (1992). Characterization of mouse–hamster somatic cell hybrids by PCR: A panel of mouse-specific primers for each chromosome. *Mammal. Genome* 2: 106–109.

Adolph, S., and Klein, J. (1981). Robertsonian variation in Mus musculus from Central Europe, Spain, and Scotland. *J. Hered.* 72: 139–142.

Agulnik, A. I., Agulnik, S. I., and Ruvinsky, A. O. (1991). Two doses of the paternal *Tme* gene do not compensate the lethality of the $T^{hp}$ deletion. *J. Hered.* 82: 351–353.

Aitman, T. J., Hearne, C. M., McAleer, M. A., and Todd, J. A. (1991). Mononucleotide repeats are an abundant source of length variants in mouse genomic DNA. *Mammal. Genome* 1: 206–210.

Altman, P. L., and Katz, D., eds. (1979). *Inbred and Genetically Defined Strains of Laboratory Animals: Part I, Mouse and Rat.*(Federation of American Societies of Experimental Biology, Bethesda).

Alvarez, W., and Asaro, F. (1990). An extraterrestrial impact. *Sci. Amer.*, p. 78–84.

Ammerman, A. J., and Cavalli-Sforza, L. L. (1984). *The Neolithic Transition and the Genetics of Populations in Europe.* (Princeton University Press, Princeton, NJ).

Armour, J. A. L., and Jeffreys, A. J. (1992). Biology and application of human minisatellite loci. *Curr. Opin. Genet. Devel.* 2: 850–856.

Arnheim, N. (1983). Concerted evolution of multigene families. In *Evolution of Genes and Proteins*, Nei, M., and Koehn, R. K., eds. (Sinauer New York), pp. 38–61.

Arnheim, N., Li, H., and Cui, X. (1991). Review: PCR analysis of DNA sequences in single cells: Single sperm gene mapping and genetic disease diagnosis. *Genomics* 8: 415–419.

Artzt, K., Calo, C., Pinheiro, E. N., DiMeo, T. A., and Tyson, F. L. (1987). Ovarian teratocarcinomas in LT/Sv mice carrying t-mutations. *Dev. Genet.* 8: 1–9.

Asada, Y., Varnum, D. S., Frankel, W. N., and Nadeau, J. H. (1994). A mutation in the *Ter* gene causing increased susceptibility to testicular teratomas maps to mouse chromosome 18. *Nature Genetics* 6: 363–368.

Atchley, W. R., and Fitch, W. M. (1991). Gene trees and the origins of inbred strains of mice. *Science* 254: 554–558.

Auffray, J.-C., Vanlerberghe, F., and Britton-Davidian, J. (1990). The house mouse progression in Eurasia: A palaeontological and archaeozoological approach. *Biol. J. Linnean Soc.* 41: 13–25.

Auffray, J.-C., Marshall, J. T., Thaler, L., and Bonhomme, F. (1991). Focus on the nomenclature of European species of mus. *Mouse Genome* 88: 7–8.

Avner, P. (1991). Genetics: Sweet mice, sugar daddies. *Nature* 351: 519–520.

Avner, P., Amar, L., Dandolo, L., and Guénet, J. L. (1988). Genetic analysis of the mouse using interspecific crosses. *Trends Genet.* 4: 18–23.

Bahary, N., Pachter, J. E., Felman, R., Leibel, R. L., Albright, K., Cram, S., and Friedman, J. M. (1992). Molecular mapping of mouse chromosomes 4 and 6: Use of a flow-sorted Robertsonian chromosome. *Genomics* 13: 761–769.

Bailey, D. W. (1971). Recombinant-inbred strains. An aid to finding identity, linkage and funcion of histocompatibility and other genes. *Transplantation* 11: 325–327.

Bailey, D. W. (1978). Sources of subline divergence and their relative importance for sublines of six major inbred strains of mice. In *Origins of Inbred Mice*, Morse, H. C., ed. (Academic Press, New York), pp. 423–438.

Bailey, D. W. (1981). Recombinant inbred strains and bilineal congenic strains. In *The Mouse in Biomedical Research*, Vol. 1, Foster, H. L., Small, J. D., and Fox, J. G., eds. (Academic Press, New York), pp. 223–239.

Barany, F. (1991). Genetic disease detection and DNA amplification using cloned thermostable ligase. *Proc. Nat. Acad. Sci.* USA 88: 189–193.

Barker, D., Schafer, M., and White, R. (1984). Restriction sites containing CpG show a higher frequency of polymorphism in human DNA. *Cell* 36: 131–138.

Barlow, D., and Lehrach, H. (1990). Partial Not I digests generated by low enzyme concentration or by the presence of ethidium bromide can be used to extend the range of physical mapping. *Technique* 2: 79–87.

Barlow, D. P., and Leharch, H. (1987). Genetics by gel electrophoresis: the impact of pulsed field gel electrophoresis on mammalian genetics. *Trend. Genet.* 3: 167–171.

Barlow, D. P., Stoger, R., Herrmann, B. G., Saito, K., and Schweifer, N. (1991). The mouse insulin-like growth factor type-2 receptor is imprinted and closely linked to the *Tme* locus. *Nature* 349: 84-87.

Bartolomei, M. S., Zemel, S., and Tilghman, S. (1991). Parental imprinting of the mouse H19 gene. *Nature* 351: 153–155.

Barton, N. H., and Hewitt, G. M. (1989). Adaptation, speciation and hybrid zones. *Nature* 341: 497–502.

Bateson, W., Saunders, E. R., and Punnett, R. C. (1905). Experimental studies in the physiology of heredity. *Rep. Evolution Comm. Roy. Soc.* 2: 1–55, 80–99.

Beier, D. R. (1993). Single-strand conformation polymorphism (SSCP) analysis as a tool for genetic mapping. *Mammal. Genome* 4: 627–631.

Beier, D. R., Dushkin, H., and Sussman, D. J. (1992). Mapping genes in the mouse using single-strand conformation polymorphism analysis of recombinant inbred strains and interspecific crosses. *Proc. Nat. Acad. Sci.* USA 89: 9102–9106.

Bellvé, A. R., Cavicchia, J. C., Millette, C. F., O'Brien, D. A., Bhatnagar, Y. M., and Dym, M. (1977). Spermatogenic cells of the prepuberal mouse. Isolation and morphological characterization. *J. Cell Biol.* 74: 68–85.

Berry, R. J. (1981). Population dynamics of the house mouse. *Symp. Zool. Soc. Lon.* 47: 395–425.

Berry, R. J., and Corti, M., eds. (1990). *Biological Journal of the Linnean Society* (Academic Press, London).

Berry, R. J., and Jakobson, M. E. (1974). Vagility and death in an island population of the house mouse. *J. Zool.* 173: 341–354.

Berry, R. J., and Peters, J. (1975). Macquarie Island house mice: a genetical isolate on a sub-Antarctic island. *J. Zool.* 176: 375–389.

Berry, R. J., Jakobson, M. E., and Peters, J. (1987). Inherited differences within an island population of the House mouse (*Mus domesticus*). *J. Zool. Lon.* 211: 605–618.

Bickmore, W. A., and Sumner, A. T. (1989). Mammalian chromosome banding—an expression of genome organization. *Trends Genet.* 5: 144–148.

Bird, A., Lavia, P., MacLeod, D., Lindsay, S., Taggart, M., and Brown, W. (1987). Mammalian genes and islands of non-methylated CpG-rich DNA. In *Human Genetics*, Vogel, F., and Sperling, K., eds. (Springer-Verlag, Berlin), pp. 182–186.

Bird, A. P. (1986). CpG-rich islands and the function of DNA methylation. *Nature* 321: 209–213.

Bird, A. P. (1987). CpG islands as gene markers in the vertebrate nucleus. *Trends. Genet.* 3: 342–347.

Birkenmeier, E., Rowe, L., and Nadeau, J. (1994). The Jackson Laboratory backcross. *Mammal. Genome* 5: 253–274.

Bishop, C. E. (1992). Mouse Y chromosome. *Mammal. Genome* 3: S289-S293.

Bishop, C. E., Boursot, P., Baron, B., Bonhomme, F., and Hatat, D. (1985). Most classical *Mus musculus domesticus* laboratory mouse strains carry a *Mus musculus musculus* Y chromosome. *Nature* 325: 70–72.

Blanchetot, A., Price, M., and Jeffreys, A. J. (1986). The mouse myoglobin gene. *Eur. J. Biochem.* 159: 469–474.

Bode, V. C. (1984). Ethylnitrosourea mutagenesis and the isolation of mutant alleles for specific genes located in the t-region of mouse chromosome 17. *Genetics* 108: 457–470.

Boer, P. H., Adra, C. N., Lau, Y.-F., and McBurney, M. W. (1987). The testis-specific phosphoglycerate kinase gene *pgk*-2 is a recruited retroposon. *Mol. Cell. Biol.* 7: 3107–3112.

Bohlander, S. K., Espinosa, R., LeBeau, M. M., Rowley, J. D., and Diaz, M. O. (1992). A method for the rapid sequence-independent amplification of microdissected chromosomal material. *Genomics* 13: 1322–1324.

Bonhomme, F. (1986). Evolutionary relationships in the genus *Mus. Curr. Topics Micro. Immunol.* 127: 19–34.

Bonhomme, F., Benmehdi, F., Britton-Davidian, J., and Martin, S. (1979). Analyse génétique de croisements interspécifiques *Mus musculus L.* × *Mus spretus* Lataste: liaison de *Adh-1* avec *Amy-1* sur le chromosome 3 et de *Es-14* avec *Mod-1* sur le chromosome 9. *C. R. Acad. Sci. Paris* 289: 545–548.

Bonhomme, F., Britton-Davidian, J., Thaler, L., and Triantaphyllidis, C. (1978a). Sur l'existence en Europe de quatre groupes de souris (genre Mus L.) du rang espèce et semi-espèce, démontrée par la génétique biochimique. *C.R. Acad. Sci.* Paris 287: 631–633.

Bonhomme, F., Catalan, J., Britton-Davidian, J., V.M., C., Moriwaki, K., Nevo, E., and Thaler, L. (1984). Biochemical diversity and evolution in the genus *Mus. Biochem. Genet.* 22: 275–303.

Bonhomme, F., Catalan, J., Gerasimov, S., Orsini, P., and Thaler, L. (1983). Le complexe d'espèces du genre Mus en Europe centrale et orientale. I. Génétique. *Z. Säugetierkunde* 48: 78–85.

Bonhomme, F., and Guénet, J.-L. (1989). The wild house mouse and its relatives. In *Genetic Variants and Strains of the Laboratory Mouse.*, Lyon, M. F., and Searle, A. G., eds. (Oxford University Press, Oxford), pp. 649–662.

Bonhomme, F., Guénet, J.-L., and Catalan, J. (1982). Présence d'un facteur de stérilité mâle, *Hst-2*, segrégant dans les croisements interspécifiques *M. musculus L.* × *M. spretus* Lastaste et lié à *Mod-1* et *Mpi-1* sur le chromosome 9. *C. R. Acad. Sci. Paris, Ser. III* 294: 691–693.

Bonhomme, F., Guénet, J.-L., Dod, B., Moriwaki, K., and Bulfield, G. (1987). The polyphyletic origin of laboratory inbred mice and their rate of evolution. *J.*

*Linnean Soc.* 30: 51–58.

Bonhomme, F., Martin, S., and Thaler, L. (1978b). Hybridation en laboratoire de *Mus musculus* L. et *Mus spretus* Lataste. *Experientia* 34: 1140–1141.

Botstein, D., White, R. L., Skolnick, M., and Davis, R. W. (1980). Construction of a genetic linkage map in man using restriction fragment length polymorphisms. *Am. J. Hum. Genet.* 32: 314–331.

Boursot, P., Bonhomme, F., Britton-Davidian, J., Catalan, J., Yonekawa, H., Orsini, P., Guerasimov, S., and Thaler, L. (1984). Introgression différentielle des génomes nucléaires et mitochondriaux chez deux semi-espèces européennes de Souris. *C. R. Acad. Sci.* Paris, Sér. III 299: 365–370.

Boursot, P., Auffray, J.-C., Britton-Davidian, J., and Bonhomme, F. (1993). The evolution of house mice. *Ann. Rev. Ecol. System.*: (in press).

Boyle, A. L., Ballard, S. G., and Ward, D. C. (1990). Differential distribution of long and short interspersed element sequences in the mouse genome: Chromosome karyotyping by fluorescence in situ hybridization. *Proc. Nat. Acad. Sci.* 87: 7757–7761.

Boyse, E. A., Miyazawa, M., Aoki, T., and Old, L. J. (1968). Two systems of lymphocyte isoantigens in the mouse. *Proc. Roy. Soc. Lond., Sec. B.* 170: 175–193.

Breen, M., Deakin, L., Macdonald, B., Miller, S., Sibson, R., Tarttelin, E., Avner, P., Bourgade, F., Guenet, J. L., Montagutelli, X., Poirier, C., Simon, D., Tailor, D., Bishop, M., Kelly, M., Rysavy, F., Rastan, S., Norris, D., Shepherd, D., Abbott, C., Pilz, A., Hodge, S., Jackson, I., Boyd, Y., Blair, H., Maslen, G., Todd, J. A., Reed, P. W., Stoye, J., Ashworth, A., Mccarthy, L., Cox, R., Schalkwyk, L., Lehrach, H., Klose, J., Gangadharan, U., and Brown, S. (1994). Towards high resolution maps of the mouse and human genomes—a facility for ordering markers to 0.1 cm resolution. *Hum. Mol. Genet.* 3: 621–627.

Britton, J., and Thaler, L. (1978). Evidence for the presence of two sympatric species of mice (genus *Mus*) in southern France based on biochemical genetics. *Biochem. Genet.* 16: 213–225.

Broccoli, D., Miller, O. J., and Miller, D. A. (1992). Isolation and characterization of a mouse subtelomeric-sequence. *Chromosomal* 101: 442–447.

Bronson, F. (1984). The adaptability of the house mouse. *Sci. Amer.* 250(3): 90–97.

Bronson, F. H., Dagg, C. P., and Snell, G. D. (1966). Reproduction. In *Biology of the Laboratory Mouse*, Green, E. L., ed. (McGraw-Hill, New York), pp. 187–204.

Brown, S. (1993). The European Collaborative Interspecific Backcross (EUCIB). *Mouse Genome* 91: v-xii.

Brown, S. D. M., Avner, P., and Herman, G. (1992). Mouse X chromosome. *Mammal. Genome* 3: S274-S288.

Bruce, H. M. (1959). An exteroreceptive block to pregnancy in the mouse. *Nature* 184: 105.

Bruce, H. M. (1968). Absence of pregnancy block in mice when stud and test males belong to an inbred strain. *J. Reprod. Fert.* 17: 407-408.

Bryant, S. P. (1991). Software for genetic linkage analysis. In *Protocols in Human Molecular Genetics*, Mathew, C. G., ed. (Humana Press, Clifton, NJ), Bryda, pp. 120–142.

Bryda, E. C., DePari, J. A., SantAngelo, D. B., Murphy, D. B., and Passmore, H. C. (1992). Multiple sites of crossing over within the Eb recombinational hotspot in the mouse. *Mammal Genome* 2: 123–129.

Buckler, A. J., Chang, D. D., Graw, S. L., Brook, J. D., Haber, D. A., Sharp, P. A., and Housman, D. E. (1991). Exon amplification: A strategy to isolate mammalian genes based on RNA splicing. *Proc. Natl. Acad. Sci. USA* 88: 4005–4009.

Burke, D. T., Carle, G. F., and Olson, M. V. (1987). Cloning of large segments of exogenous DNA into yeast by means of artificial chromosome vectors. *Science* 236: 806–812.

Burke, D. T., Rossi, J. M., Koos, D. S., and Tilghman, S. M. (1991). A mouse genomic library of yeast artificial chromosome clones. *Mammal. Genome* 1: 65.

Cannizzaro, L. A., and Emanuel, B. E. (1984). An improved method for G-banding chromosomes after in situ hybridization. *Cytogenet. Cell Genet.* 38: 308–309.

Capecchi, M. (1989). The new mouse genetics: Altering the genome by gene targeting. *Trends Genet.* 5: 70–76.

Carter, A. T., Norton, J. D., Gibson, Y., and Avery, R. J. (1986). Expression and transmission of a rodent retrovirus-like VL30 gene family. *J. Mol. Biol.* 188: 105–108.

Carter, T. C. (1954). The estimation of total genetical map lengths from linkage test data. *J. Genet.* 53: 21–28.

Carter, T. C., and Falconer, D. S. (1951). Stocks for detecting linkage in the mouse and the theory of their design. *J. Genet.* 50: 307–323.

Castle, W. E. (1903). The laws of Galton and Mendel and some laws governing race improvement by selection. *Proc. Amer. Acad. Arts Sci.* 35: 233–242.

Cattanach, B. M., and Kirk, M. (1985). Differential activity of maternally and paternally derived chromosome regions in mice. *Nature* 315: 469–498.

Ceci, J. D., Matsuda, Y., Grubber, J. M., Jenkins, N. A., Copeland, N. G., and Chapman, V. M. (1994). Interspecific backcrosses provide an important tool for centromere mapping of mouse chromosomes. *Genomics* 19: 515–524.

Chartier, F. L., Keer, J. T., Sutcliffe, M. J., Henriques, D. A., Mileham, P., and Brown, S. D. M. (1992). Construction of a mouse yeast artificial chromosome library in a recombination-deficient strain of yeast. *Nature Genetics* 1: 132–136.

Chevret, P., Denys, C., Jaeger, J.-J., Michaux, J., and Catzeflis, F. M. (1993). Molecular evidence that the spiny mouse (*Acomys*) is more closely related to gerbils (Gerbillinae) than to true mice (Murinae). *Proc. Natl. Acad. Sci. USA* 90: 3433–3436.

Chisaka, O., and Capecchi, M. R. (1991). Regionally restricted developmental defects resulting from targeted disruption of the mouse homeobox gene hox-1.5. *Nature* 350: 473–479.

Chromosome Committee Chairs (1993). Encyclopedia of the Mouse Genome III. *Mammal. Genome* 4: special issue: S1–S284.

Chumakov, I., Rigault, P., Guillou, S., Ougen, P., Billaut, A., Guasconi, G., Gervy, P., LeGall, I., Soularue, P., Grinas, L., Bougueleret, L., Bellanne-Chantelot, C., Lacroix, B., Barillot, E., Gesnouin, P., Pook, S., Vaysseix, G., Frelat, G., Schmitz, A., Sambucy, J.-L., Bosch, A., Estivill, X., Weissenbach, J., Vignal, A., Riethman, H., Cox, D., Patterson, D., Gardiner, K., Hattori, M., Sakaki, Y., Ichikawa, H., Ohki, M., Le Paslier, D., Heilig, R., Antonarakis, S., and Cohen, D. (1992). Continuum of overlapping clones spanning the entire human chromosome 21q. *Nature* 359: 380–387.

Cochran, W. G. (1954). Some methods for strenghening the common $\chi^2$ tests. *Biometrics* 10: 417–451.

Collins, F., and Galas, D. (1993). A new five-year plan for the U.S. human genome project. *Science* 262: 43–46.

Committee on Standardized Genetic Nomenclature for Mice (1989). Rules and guidelines for gene nomenclature. In *Genetic Variants and Strains of the Laboratory Mouse.*, Lyon, M. F. and Searle, A. G., eds. (Oxford University Press, Oxford), pp. 1–12.

Cook, M. J. (1983). Anatomy. In *The Mouse in Biomedical Research*, Vol. 3, Foster, H. L., Small, J. D., and Fox, J. G., eds. (Academic Press, NY), pp. 101–120.

Copeland, N. G., and Jenkins, N. A. (1991). Development and applications of a molecular genetic linkage map of the mouse genome. *Trends Genet* 7: 113–8.

Copeland, N. G., Jenkins, N. A., Gilbert, D. J., Eppig, J. T., Maltais, L. J., Miller, J. C., Dietrich, W. F., Weaver, A., Lincoln, S. E., Steen, R. G., Stein, L. D., Nadeau, J. H., and Lander, E. S. (1993). A genetic linkage map of the mouse: current applications and future prospects. *Science* 262: 57–66.

Corbet, G. B., and Hill, J. E. (1991). *A World List of Mammalian Species*, 3rd edition (Oxford University Press, New York).

Cornall, R. J., Aitman, T. J., Hearne, C. M., and Todd, J. A. (1991). The generation of a library of PCR-analyzed microsatellite variants for genetic mapping of the mouse genome. *Genomics* 10: 874–881.

Costantini, F., and Lacy, E. (1981). Introduction of a rabbit beta-globin gene into the mouse germ line. *Nature* 294: 92–94.

Courtney, M. G., Elder, P. K., Steffen, D. L., and Getz, M. J. (1982). Evidence for an early evolutionary origin and locus polymorphism of mouse VL30 DNA sequences. *J. Virol.* 43: 511–518.

Cox, D. R., Burmeister, M., Price, E. R., Kim, S., and Myers, R. M. (1990). Radiation hybrid mapping: A somatic cell genetic method for constructing high-resolution maps of mammalian chromosomes. *Science* 250: 245–250.

Cox, R. D., Copeland, N. G., Jenkins, N. A., and Lehrach, H. (1991). Interspersed repetitive element polymerase chain reaction product mapping using a mouse interspecific backcross. *Genomics* 10: 375–384.

Cox, R. D., Meier-Ewert, S., Ross, M., Larin, Z., Monaco, A. P., and Leharch, H. (1993). Genome mapping and cloning of mutations using yeast artificial chromosomes. In *Guides to Techniques in Mouse Development*, Methods in Enzymology 225, Wassarman, P. M., and DePamphilis, M. L., eds. (Academic Press, San Diego), pp. 623-637.

Craig, J. M., and Bickmore, W. A. (1993). Chromosome bands—flavours to savour. *Bioessays* 15: 349–354.

Crow, J. F. (1990). Mapping functions. *Genetics* 125: 669–671.

Cuénot, L. (1902). La loi de Mendel et l'hérédité de la pigmentation chez les souris. *Arch. Zool. exp. gén., 3e sér.* 3: 27–30.

Cuénot, L. (1903). L'hérédité de la pigmentation chez les souris, 2me note. *Arch. Zool. Exp. Gén1.* 4: 33–38.

Cuénot, L. (1905). Les races pures et les combinaisons chez les souris. *Arch. Zool. Exp. Gén.* 4: 123–132.

Cui, X., Gerwin, J., Navidi, W., Li, H., Kuehn, M., and Arnheim, N. (1992). Gene-centromere linkage mapping by PCR analysis of individual oocytes. *Genomics* 13: 713–717.

D'Eustachio, P., and Clarke, V. (1993). Localization of the twitcher (twi) mutation on mouse chromosome 12. Mammal. *Genome* 4: 684–686.

Daniels, D. L., Plunkett, G., Burland, V., and Blattner, F. R. (1992). Analysis of the *Escherichia coli* genome: DNA sequence of the region from 84.5 to 86.5. *Science* 257: 771–777.

Datta, S. K., Owen, J. E., Womack, J. E., and Riblet, R. J. (1982). Analysis of recombinant inbred lines derived from "autoimmune" (NZB) and "high leukemia" (C58) strains: independent multigenic systems control B cell hyperactivity, retrovirus expression, and autoimmunity. *J. Immunol.* 129: 1539–1544.

Davisson, M. T. (1990). The Jackson Laboratory Mouse Mutant Resource. *Lab Animal* 19:

23.

Davisson, M. T. (1993). Personal communication.

Davisson, M. T., and Akeson, E. C. (1993). Recombination suppression by heterozygous Robertsonian chromosomes in the mouse. *Genetics* 133: 649–667.

Davisson, M. T., and Roderick, T. H. (1989). Linkage map. In *Genetic Variants and Strains of the Laboratory Mouse.*, Lyon, M. F. and Searle, A. G., eds. (Oxford University Press, Oxford), pp. 416–427.

Davisson, M. T., Roderick, T. H., and Doolittle, D. P. (1989). Recombination percentages and chromosomal assignments. In *Genetic Variants and Strains of the Laboratory Mouse.*, Lyon, M. F. and Searle, A. G., eds. (Oxford University Press, Oxford), pp. 432-505.

Dayhoff, M. O. (1978). Survey of new data and computer methods of analysis. In *Atlas of Protein Sequence and Structure*, 5, Suppl. 3, Dayhoff, M. O., eds. (National Biomedical Research Foundation, Silver Springs, MD), pp. 2–8.

de Boer, P., and Groen, A. (1974). Fertility and meiotic behavior of male T70H tertiary trisomics of the mouse (*Mus musculus*). A case of preferential telomeric meiotic pairing in a mammal. *Cytogenet. Cell Genet.* 13: 489–510.

DeChiara, T. M., Robertson, E. J., and Efstratiadis, A. (1991). Parental imprinting of the mouse Insulin-like growth factor II gene. *Cell* 64: 849–859.

Demant, P., and Hart, A. A. M. (1986). Recombinant congenic strains—a new tool for analyzing genetic traits determined by more than one gene. *Immunogenetics* 24: 416–422.

den Dunnen, J. T., and van Ommen, G.-J. B. (1991). Pulsed-field gel electrophoresis. In *Protocols in Human Molecular Genetics*, Methods in Molecular Biology 9, Mathew, C. G., eds. (Humana Press, Clifton, NJ), pp. 169–182.

Dietrich, W., Katz, H., Lincoln, S., Shin, H.-S., Friedman, J., Dracopoli, N. C., and Lander, E. (1992). A genetic map of the mouse suitable for typing intraspecific crosses. *Genetics* 131: 423–447.

Dobrovolskaia-Zavadskaia, N. (1927). Sur la mortification spontanée de la queue chez la souris nouveau-née et sur l'existence d'un caractère (facteur) héréditaire << non viable >> . *C. R. Séanc. Soc. Biol.* 97: 114–116.

Donehower, L. A., Harvey, M., Slagle, B. L., McArthur, M. J., Montgomery Jr., C. A., Butel, J. S., and Bradley, A. (1992). Mice deficient for p53 are developmentally normal but susceptible to spontaneous tumours. *Nature* 356: 215–221.

Donis-Keller, H., Green, P., Helms, C., Cartinhour, S., Weiffenbach, B., Stephens, K., Keith, T., Bowden, D., Smith, D., Lander, E., Botstein, D., Akots, G., Rediker, K., Gravius, T., Brown, V., Rising, M., Parker, C., Powers, J., Watt, D., Kauffman, E., Bricker, A., Phipps, P., Muller-Kahle, H., Fulton, T., Ng, S., Schumm, J., Braman, J., Knowlton, R., Barker, D., Crooks, S., Lincoln, S., Daly, M., and Abrahamson, J. (1987). A genetic linkage map of the human genome. *Cell* 51: 319–337.

Dover, G. (1982). Molecular drive: A cohesive mode of species evolution. *Nature* 299: 111–117.

Drouet, B., and Simon-Chazottes, D. (1993). The microsatellite found in the DNA sequence with the code name MMMYOGG1 (GenBank) does not correspond to the myogenin gene (Myog) but to myoglobin (Mb) and maps to mouse Chromosome 15. *Mammal. Genome* 4: 348.

Dunn, L. C. (1965). *A Short History of Genetics.* (McGraw-Hill, New York).

Eddy, E. M., O'Brien, D. A., and Welch, J. E. (1991). Mammalian sperm development in vivo and in vitro. In *Elements of Mammalian Fertilization*, Vol. 1, Wassarman, P. M., ed. (CRC Press, Boston), pp. 1–28.

Eicher, E. M. (1971). The identification of the chromosome bearing linkage group XII in the mouse. *Genetics* 69: 267–271.

Eicher, E. M. (1978). Murine ovarian teratomas and parthenotes as cytogenetic tools. *Cytogenet. Cell Genet.* 20: 232–239.

Eicher, E. M., and Shown, E. P. (1993). Molecular markers that define the distal ends of mouse autosomes 4, 13, and 19 and the sex chromosomes. *Mammal. Genome* 4: 226–229.

Elliott, R. (1979). Mouse variants studied by two-dimensional electrophoresis. *Mouse News Lett.* 61: 59.

Elliott, R. W., and Yen, C.-H. (1991). DNA variants with telomere probe enable genetic mapping of ends of mouse chromosomes. *Mammal. Genome* 1: 118–122.

Elston, R. C., and Stewart, J. (1971). A general model for the genetic analysis of pedigree data. *Hum. Hered.* 21: 523–542.

Ephrussi, B., and Weiss, M. C. (1969). Hybrid somatic cells. *Sci. Amer.* 220 (April): 26–34.

Eppig, J. T., and Eicher, E. M. (1983). Application of the ovarian teratoma mapping method in the mouse. *Cytogenet. Cell Genet.* 20: 232–239.

Eppig, J. T., and Eicher, E. M. (1988). Analysis of recombination in the centromere region of mouse chromosome 7 using ovarian teratoma and backcross methods. *J. Hered.* 79: 425–429.

Erickson, R. P. (1989). Why isn't a mouse more like a man. Trend. *Genet.* 5: 1–3.

Erlich, H. A., ed. (1989). *PCR Technology* (Stockton Press, New York).

Evans, E. P. (1989). Standard Normal Chromosomes. In *Genetic Variants and Strains of the laboratory mouse.*, Lyon, M. F., and Searle, A. G., eds. (Oxford University Press, Oxford), pp. 576–581.

Farr, C. J. (1991). The analysis of point mutations using synthetic oligonucleotide probes. In *Protocols in Human Molecular Genetics*, Methods in Molecular Biology 9, Mathew, C., ed. (Humana Press, Clifton, NJ), pp. 69–84.

Farr, C. J., and Goodfellow, P. N. (1992). Hidden messages in genetic maps. *Science* 258: 49.

Feingold, M. (1980). Preface. In *Josephine: The Mouse Singer (A Play)*, McClure, M., ed. (New Directions Books, New York), pp. i–viii.

Ferguson-Smith, A. C., Reik, W., and Surani, M. A. (1990). Genomic imprinting and cancer. *Cancer Surv.* 9: 487–503.

Ferris, S. D., Sage, R. D., and Wilson, A. C. (1982). Evidence from mtDNA sequences that common laboratory strains of inbred mice are descended from a single female. *Nature* 295: 163–165.

Ferris, S. D., Sage, R. D., Huang, C.-M., Nielson, J. T., Ritte, U., and Wilson, A. C. (1983a). Flow of mitochondrial DNA across a species boundary. *Proc. Nat. Acad. Sci. USA* 80: 2290–2294.

Ferris, S. D., Sage, R. D., Prager, E. M., Titte, U., and Wilson, A. C. (1983b). Mitochondrial DNA evolution in mice. *Genetics* 105: 681-721.

Field, K. G., Olsen, G. J., Lane, D. J., Giovannoni, S. J., Ghiselin, M. T., Raff, E. C., Pace, N. R., and Raff, R. A. (1988). Molecular phylogeny of the animal kingdom. *Science* 239: 748–753.

Fiering, S., Kim, C. G., Epner, E. M., and Groudine, M. (1993). An "in-out" strategy using gene targeting and FLP recombinase for the functional dissection of complex DNA regulatory elements: analysis of the β-globin locus control region. *Proc. Natl. Acad. Sci. USA* 90: 8469–8473.

Fischer, S. G., and Lerman, L. S. (1983). DNA fragments differing by single-base pair substitutions are separated in denaturing gradient gels: Correspondence with melting theory. *Proc. Nat. Acad. Sci. USA* 80: 1579–1583.

Fisher, R. A. (1936). Has Mendel's work been rediscovered? *Ann. of Sci.* 1: 115–137.

Fitzgerald, J., Wilcox, S. A., Graves, J. A. M., and Dahl, H.-H. M. (1993). A eutherian X-linked gene, PDHA1, is autosomal in marsupials: A model for the evolution of a second, testis-specific variant in eutherian mammals. *Genomics* 18: 636–642.

Flaherty, L. (1981). Congenic strains. In *The Mouse in Biomedical Research*, Vol. 1, Foster, H. L., Small, J. D., and Fox, J. G., eds. (Academic Press, N.Y.), pp. 215–222.

Foote, S., Vollrath, D., Hilton, A., and Page, D. C. (1992). The human Y chromosome: overlapping DNA clones spanning the euchromatic region. *Science* 258: 60–66.

Forrest, S. (1993). Genetic algorithims: principles of natural selection applied to computation. *Science* 261: 872–878.

Frankel, W. N., Stoye, J. P., Taylor, B. A., and Coffin, J. M. (1990). A genetic linkage map of endogenous murine leukemia proviruses. *Genetics* 124: 221–236.

Frankel, W. N., Lee, B. K., Sotye, J. P., Coffin, J. M., and Eicher, E. M. (1992). Characterization of the endogenous nonecotropic murine leukemia viruses of NZB/BINJ and SM/J inbred strains. *Mammal. Genome* 2: 110–122.

Friedrich, G., and Soriano, P. (1991). Promoter traps in embryonic stem cells: A genetic screen to identify and mutate developmental genes in mice. *Genes Dev.* 5: 1513–23.

Gall, J. G., and Pardue, M. L. (1969). Formation and detection of RNA-DNA hybrid molecules in cytological preparations. *Proc. Nat. Acad. Sci.* 63: 378–382.

Gardner, M. B., Kozak, C. A., and O'Brien, S. J. (1991). The Lake Casitas wild mouse: Evolving genetic resistance to retroviral disease. *Trends Genet.* 7: 22–27.

Garrels, J. I. (1983). Quantitative two-dimensional gel electrophoresis of proteins. *Meth. Enzymol.* 100: 411–423.

Gasser, D. L., Sternberg, N. L., Pierce, J. C., Goldner, S. A., Feng, H., Haq, A. K., Spies, T., Hunt, C., Buetow, K. H., and Chaplin, D. D. (1994). P1 and cosmid clones define the organization of 280 kb of the mouse H-2 complex containing the Cps-1 and Hsp70 loci. *Immunogenetics* 39: 48–55.

Geliebter, J., and Nathenson, S. G. (1987). Recombination and the concerted evolution of the murine MHC. *Trends Genet* 3: 107–112.

Gendron-Maguire, M., and Gridley, T. (1993). Identification of transgenic mice. In *Guides to Techniques in Mouse Development*, Methods in Enzymology 225, Wassarman, P. M. and DePamphilis, M. L., eds. (Academic Press, San Diego), pp. 794–799.

Giacalone, J., Friedes, J., and Francke, U. (1992). A novel GC-rich human macrosatellite VNTR in Xq24 is differentially methylated on active and inactive X chromosomes. *Nature Genetics* 1: 137-143.

Goradia, T. M., Stanton, V. P., Cui, X., Aburatani, H., Li, H., Lange, K., Housman, D. E., and Arnheim, N. (1991). Ordering three DNA polymorphisms on human chromosome 3 by sperm typing. *Genomics* 10: 748–755.

Gordon, J. W., and Ruddle, F. H. (1981). Integration and stable germ line transmission of genes injected into mouse pronuclei. *Science* 214: 1244–1246.

Gossler, A., Joyner, A. L., Rossant, J., and Skarnes, W. C. (1989). Mouse embryonic stem cells and reporter constructs to detect developmentally regulated genes. *Science* 244: 463–5.

Graff, R. J. (1978). Minor histocompatibility genes and their antigens. In *Origins of Inbred Mice*, Morse, H. C., eds. (Academic Press, New York), pp. 371–389.

Graur, D., Hide, W. A., and Li, W.-H. (1991). Is the guinea-pig a rodent? *Nature* 351: 649–652.

Gray, J. W., Dean, P. N., Fuscoe, J. C., Peters, D. C., and Trask, B. J. (1990). High-speed chromosome sorting. *Science* 238: 323–329.

Green, E. D., and Olson, M. V. (1990). Systematic screening of yeast artificial-chromosome libraries by use of the polymerase chain reaction. *Proc. Nat. Acad. Sci. USA* 87: 1213–1217.

Green, E. L. (1981). *Genetics and Probability in Animal Breeding Experiments.* (Oxford University Press, New York).

Green, E. L., and Roderick, T. H. (1966). Radiation Genetics. In *Biology of the Laboratory Mouse*, Green, E. L., eds. (McGraw-Hill, New York), pp. 165–185.

Green, M. C. (1966). Mutant genes and linkages. In *Biology of the Laboratory Mouse*, Green, E. L., ed. (McGraw-Hill, New York), pp. 87–150.

Green, M. C. (1989). Catalog of mutant genes and polymorphic loci. In *Genetic Variants and Strains of the Laboratory Mouse*, Lyon, M. F. and Searle, A. G., eds. (Oxford University Press, Oxford), pp. 12-403.

Green, M. C., and Witham, B. A., eds. (1991). *Handbook on Genetically Standardized JAX Mice* 4th edition (The Jackson Laboratory, Bar Harbor).

Gropp, A., Winking, H., Zech, L., and Muller, H. (1972). Robertsonian chromosomal variation and identification of metacentric chromosomes in feral mice. *Chromosoma* 39: 265–288.

Grosveld, F., Blom van Assendelft, G., Greaves, D. R., and Kollias, G. (1987). Position-independent, high-level expression of the human β-globin gene in transgenic mice. *Cell* 51: 975–985.

Grüneberg, H. (1943). *The Genetics of the Mouse.* (Cambridge University Press, Cambridge).

Guenet, J.-L., Nagamine, C., Simon-Chazottes, D., Montagutelli, X., and Bonhomme, F. (1990). Hst3: an X-linked hybrid sterility gene. *Genet. Res. Camb.* 56: 163.

Gyllensten, U., and Wilson, A. C. (1987). Interspecific mitochondrial DNA transfer and the colonization of Scandinavia by mice. *Genet. Res. Camb.* 49: 25–29.

Hagag, N. G., and Viola, M. V. (1993). *Chromosome Microdissection and Cloning.* (Academic Press, San Diego).

Haig, D., and Graham, C. (1991). Genomic imprinting and the strange case of the insulin-like growth factor II receptor. *Cell* 64: 1045–1046.

Haldane, J. B. S. (1919). The mapping function. *J. Genet.* 8: 299-309.

Haldane, J. B. S. (1922). Sex ratio and unisexual sterility in hybrid animals. *J. Genet.* 12: 101–109.

Haldane, J. B. S., and Waddington, C. H. (1931). Inbreeding and linkage. *Genetics* 16: 357–374.

Haldane, J. B. S., Sprunt, A. D., and Haldane, N. M. (1915). Reduplication in mice. *J. Genet.* 5: 133–135.

Hamada, H., and Kakunaga, T. (1982a). Potential Z-DNA forming sequences are highly dispersed in the human genome. *Nature* 298: 396–398.

Hamada, H., Petrino, M. G., and Kakunaga, T. (1982b). A novel repeated element with Z-DNA-forming potential is widely found in evolutionarily diverse eukaryotic genomes. *Proc. Natl. Acad. Sci. USA* 79: 6465–6469.

Hammer, M. F., and Silver, L. M. (1993). Phylogenetic analysis of the alpha-globin pseudogene-4 (Hba-ps4) locus in the house mouse species complex reveals a stepwise evolution of *t* haplotypes. *Mol. Biol. Evol.* 10: 971–1001.

Hammer, M. F., Schimenti, J., and Silver, L. M. (1989). Evolution of mouse chromosome 17 and the origin of inversions associated with *t* haplotypes. *Proc. Natl. Acad. Sci. USA* 86: 3261–3265.

Hammer, M. F., Bliss, S., and Silver, L. M. (1991). Genetic exchange across a paracentric inversion of the mouse *t* complex. *Genetics* 128: 799–812.

Hanscombe, O., Whyatt, D., Fraser, P., Yannoutsos, N., Greaves, D., Dillon, N., and

Grosveld, F. (1991). Importance of globin gene order for correct developmental expression. *Genes Dev.* 5: 1387-1394.

Harbers, K., Jahner, D., and Jaenisch, R. (1981). Microinjection of cloned retroviral genomes into mouse zygotes: Integration and expression in the animal. *Nature* 293: 540–542.

Harding, R. M., Boyce, A. J., and Clegg, J. B. (1992). The evolution of tandemly repetitive DNA: recombination rules. *Genetics* 132: 847-859.

Harper, M. E., Ullrich, A., and Saunders, G. F. (1981). Localization of the human insulin gene to the distal end of the short arm of chromosome 11. *Proc. Natl. Acad. Sci. USA* 78: 4458–4460.

Hasties, N. D. (1989). Highly repeated DNA families in the genome of *Mus musculus*. In *Genetic Variants and Strains of the Laboratory Mouse.*, Lyon, M. F. and Searle, A. G., eds. (Oxford University Press, Oxford), pp. 559–573.

Hasties, N. D., and Bishop, J. O. (1976). The expression of three abundance classes of messenger RNA in mouse tissues. *Cell* 9: 761–774.

Hasty, P., Ramirez-Solis, R., Krumlauf, R., and Bradley, A. (1991). Introduction of a subtle mutation into the Hox 2.6 locus in embryonic stem cells. *Nature* 351: 234–246.

Hatada, I., Hayashizaki, Y., Hirotsune, S., Komatsubara, H., and Mukai, T. (1991). A genomic scanning method for higher organisms using restriction sites as landmarks. *Proc. Natl. Acad. Sci. USA* 88: 9523–7.

Hearne, C. M., Ghosh, S., and Todd, J. A. (1992). Microsatellites for linkage analysis of genetic traits. *Trends Genet.* 8: 288–294.

Helmuth, R. (1990). Nonisotopic detection of PCR products. In *PCR Protocols*, Innis, M. A., Gelfand, D. H., Sninsky, J. J., and White, T. J., eds. (Academic Press, San Diego), pp. 119–128.

Henry, I., Bonaiti-Pellie, C., Chehensse, V., Beldjord, C., Schwartz, C., Utermann, G., and Junien, C. (1991). Uniparental paternal disomy in a genetic cancer-predisposing syndrome. *Nature* 351: 665–667.

Henson, V., Palmer, L., Banks, S., Nadeau, J. H., and Carlson, G. A. (1991). Loss of heterozygosity and mitotic linkage maps in the mouse. *Proc. Nat. Acad. Sci. USA.* 88: 6486–6490.

Herman, G. E., Berry, M., Munro, E., Craig, I. W., and Levy, E. R. (1991). The construction of human somatic cell hybrids containing portions of the mouse X chromosome and their use to generate DNA probes via interspersed repetitive sequence polymerase chain reaction. *Genomics* 10: 961–970.

Herman, G. E., Nadeau, J. H., and Hardies, S. C. (1992). Dispersed repetitive elements in mouse genome analysis. Mammal. *Genome* 2: 207–214.

Herrmann, B. G., Barlow, D. P., and Lehrach, H. (1987). A large inverted duplication allows homologous recombination between chromosomes heterozygous for the proximal *t* complex inversion. *Cell* 48: 813–825.

Hilgers, J., and Arends, J. W. A. (1985). A series of recombinant inbred strains between the BALB/cHeA and STS/A strains. *Curr. Top. Microbiol. Immunol.* 122: 31–37.

Hilgers, J., and Poort-Keeson, R. (1986). Strain distribution pattern of genetic polymorphisms between BALB/cHeA and STS/A strains. *Mouse News Lett.* 76: 14–20.

Hillyard, A. L., Doolittle, D. P., Davisson, M. T., and Roderick, T. H. (1992). Locus map of the mouse. *Mouse Genome* 90: 8–21.

Himmelbauer, H., and Silver, L. M. (1993). High resolution comparative mapping of mouse chromosome 17. *Genomics* 17: 110–120.

Himmelbauer, H., Artzt, K., Barlow, D., Fischer-Lindhal, K., Lyon, M., Klein, J., and Silver, L. M. (1993). Mouse chromosome 17. *Mammal. Genome* 4: S230–S252.

Hochgeschwender, U. (1992). Toward a transcriptional map of the human genome. *Trends Genet* 8: 41–44.

Hogan, B., Beddington, R., Costantini, F., and Lacy, E. (1994). *Manipulating the Mouse Embryo: A Laboratory Manual*, 2nd ed. (Cold Spring Harbor Laboratory Press, Cold Spring Harbor).

Hood, L., Kronenberg, M., and Hunkapiller, T. (1985). T cell antigen receptors and the immunoglobulin supergene family. *Cell* 40: 225-229.

Horiuchi, Y., Agulnik, A., Figueroa, F., Tichy, H., and Klein, J. (1992). Polymorphisms distinguishing different mouse species and t haplotypes. *Genet. Res.* 60: 43–52.

Hörz, W., and Altenburger, W. (1981). Nucleotide sequence of mouse satellite DNA. *Nucl. Acids Res.* 9: 683–696.

Huntington's Disease Collaborative Research Group (1993). A novel gene containing a trinucleotide repeat that is expanded and unstable on Huntington's disease chromosomes. *Cell* 72: 971–983.

Huxley, A. (1932). *Brave New World*. (Doubleday, Doran & Co., Garden City, NY).

Innis, M. A., Gelfand, D. H., Sninsky, J. J., and White, T. J., eds. (1990). *PCR Protocols* (Academic Press, San Diego).

Jaeger, J.-J., Tong, H., and Denys, C. (1986). The age of *Mus-Rattus* divergence: paleontological data compared with the molecular clock. *C. R. Acad. Sci. Paris 302, Ser. II*: 917–922.

Jaenisch, R. (1976). Germ line integration and Mendelian transmission of the exogenous Molony leukemia virus. *Proc. Nat. Acad. Sci.* 73: 1260–1264.

Jahn, C. L., Hutchison, C. A., Phillips, S. J., Weaver, S., Haigwood, N. L., Voliva, C. F., and Edgell, M. H. (1980). DNA sequence organization of the beta-globin complex in the BALB/c mouse. *Cell* 21: 159–168.

Jakobovits, A., Moore, A. L., Green, L. L., Vergara, G. J., Maynard-Currie, C. E., Austin, H. A., and Klapholz, S. (1993). Germ-line transmission and expression of a human-derived yeast artificial chromosome. *Nature* 362: 255–258.

Jan, Y. N., and Jan, L. Y. (1993). Functional gene cassettes in development. *Proc. Natl. Acad. Sci. USA* 90: 8305–8307.

Jarman, A. P., and Wells, R. A. (1989). Hypervariable minisatellites: recombinations or innocent bystanders? *Trends Genet.* 5: 367–371.

Jeang, K. T., and Hayward, G. S. (1983). A cytomegalovirus DNA sequence containing tracts of tandemly repeated CA dinucleotides hybridizes to highly repetitive dispersed elements in mammalian cell genomes. *Mol. Cell. Biol.* 3: 1389.

Jeffreys, A. J., Wilson, V., and Thein, S. L. (1985). Individual-specific "fingerprints" of human DNA. *Nature* 316: 76–79.

Jeffreys, A. J., Wilson, V., Kelly, R., Tayor, B. A., and Bulfield, G. (1987). Mouse DNA "fingerprints:" Analysis of chromosome localization and germ-line stability of hypervariable loci in recombinant inbred strains. *Nucl. Acids. Res.* 15: 2823–2836.

Jeffreys, A. J., Royle, N. J., Wilson, V., and Wong, Z. (1988). Spontaneous mutation rates to new length alleles at tandem-repetitive hypervariable loci in human DNA. *Nature* 332: 278–281.

Jenkins, N. A., and Copeland, N. G. (1985). High frequency germline acquisition of ecotropic MuLV proviruses in SWR/J-RF/J hybrid mice. *Cell* 43: 811–819.

Jenkins, N. A., Copeland, N. G., Taylor, B. A., and Lee, B. K. (1982). Organization, distribution, and stability of endogenous ecotropic murine leukemia virus DNA sequences in chromosomes of *Mus musculus*. *J. Virol.* 43: 26–36.

John, H., Birnstiel, M. L., and Jones, K. N. (1969). RNA-DNA hybrids at the cytological level. *Nature* 223: 582–585.

Johnson, D. R. (1974). Hairpin-Tail: a case of post-reductional gene action in the mouse egg? *Genetics* 76: 795–805.

Joyner, A., ed. (1993). *Gene Targeting: A Practical Approach* (Oxford University Press, New York).

Joyner, A. L., Skarnes, W. C., and Rossant, J. (1989). Production of a mutation in the mouse En-2 gene by homologous recombination in embryonic stem cells. *Nature* 338: 153–156.

Joyner, A. L., Auerbach, A., and Skarnes, W. C. (1992). The gene trap approach in embryonic stem cells: the potential for genetic screens in mice. *Ciba Found. Symp.* 165: 277–88.

Julier, C., DeGouyon, B., Georges, M., Guenet, J.-L., Nakamura, Y., and Avner, P. (1990). Minisatellite linkage maps in the mouse by cross-hybridization with human probes containing tandem repeats. *Proc. Nat. Acad. Sci. USA* 87: 4585–4589.

Kasahara, M., Figueroa, F., and Klein, J. (1987). Random cloning of genes from mouse chromosome 17. *Proc. Natl. Acad. Sci. USA* 84: 3325–3328.

Keeler, C. E. (1931). *The Laboratory Mouse. Its Origin, Heredity, and Culture.* (Harvard University Press, Cambridge).

Keller, E. F. (1983). *A Feeling for the Organism: The Life and Work of Barbara McClintock.* (W. H. Freeman, New York).

Kerr, R. (1991). Extinction potpourri: Killers and victims. *Science* 254: 942–943.

Kerr, R. A. (1992). Extinction by a one–two comet punch? *Science* 255: 160–161.

Kerr, R. A. (1993). Second crater points to killer comets. *Science* 259: 1543.

Keshet, E., and Itin, A. (1982). Patterns of genomic distribution and sequence heterogeneity of a murine "retrovirus-like" multigene family. *J. Virol.* 43: 50–58.

Kevles, D. J., and Hood, L., eds. (1992). *The Code of Codes: Scientific and Social Issues in the Human Genome Project* (Harvard University Press, Boston).

Kidd, S., Lockett, T. J., and Young, M. W. (1983). The notch locus of *Drosophila melanogaster. Cell* 34: 421–433.

King, T. R., Dove, W. F., Herrmann, B., Moser, A. R., and Shedlovsky, A. (1989). Mapping to molecular resolution in the T to H-2 region of the mouse genome with a nested set of meiotic recombinants. *Proc. Nat. Acad. Sci. USA* 86: 222–226.

Kingsley, D. M., Bland, A. E., Grubber, J. M., Marker, P. C., Russell, L. B., Copeland, N. G., and Jenkins, N. A. (1992). The mouse short ear skeletal morphogenesis locus is associated with defects in a bone morphogenetic member of the TGFbeta superfamily. *Cell* 71: 399-410.

Kit, S. (1961). Equilibrium sedimentation in density gradients of DNA preparations from animal tissues. *J. Mol. Biol.* 3: 711–716.

Klein, J. (1986). *Natural History of the Major Histocompatibility Complex.* (John Wiley & Sons, New York).

Knight, A. M., and Dyson, P. J. (1990). Detection of DNA polymorphisms between two inbred mouse strains—limitations of restriction fragment length polymorphisms (RFLPs). *Mol Cell Probes* 4: 497–504.

Koop, B. F., and Hood, L. (1994). Striking sequence similarity over almost 100 kilobases of human and mouse T-cell receptor DNA. *Nature Cenet.* 7: 48–53.

Korenberg, J. R., and Rykowski, M. C. (1988). Human genome organization: Alu, Lines, and the molecular structure of metaphase chromosome bands. *Cell* 53: 391–400.

Kosambi, D. D. (1944). The estimation of map distances from recombination values. *Ann.*

*Eugenics* 12: 172–175.

Kozak, C., Peters, G., Pauley, R., Morris, V., Michalides, R., Dudley, J., Green, M., Davisson, M., Prakash, D., and Vaidya, A. (1987). A standardized nomenclature for endogenous mouse mammary tumor viruses. *J. Virol.* 61: 1651–1654.

Kramer, J. M., and Erickson, R. P. (1981). Developmental program of PGK-1 and PGK-2 isozymes in spermatogenic cells of the mouse: Specific activities and rates of synthesis. *Dev. Biol.* 87: 37–45.

Kunkel, L. M., Monaco, A. P., Middlesworth, W., Ochs, H. D., and Latt, S. A. (1985). Specific cloning of DNA fragments absent from the DNA of a male patient with an X-chromosome deletion. *Proc. Nat. Acad. Sci. USA* 82: 4778–4782.

Kusumi, K., Smith, J. S., Segre, J. A., Koos, D. S., and Lander, E. S. (1993). Construction of a large-insert yeast artificial chromosome library of the mouse genome. *Mammal. Genome* 4: 391–392.

Kwiatkowski, D. J., Dib, C., Slaugenhaupt, S. A., Povey, S., Gusella, J. F., and Haines, J. L. (1993). An index marker map of chromosome 9 provides strong evidence for positive interference. *Am. J. Hum. Genet.* 53: 1279–1288.

Laird, C. D. (1971). Chromatid structure: relationship between DNA content and nucleotide sequence diversity. *Chromosoma* 32: 378-406.

Landegren, U., Kaiser, R., Sanders, J., and Hood, L. (1988). A ligase-mediated gene detection technique. *Science* 241: 1077–1080.

Landegren, U., Kaiser, R., and Hood, L. (1990). Oligonucleotide ligation assay. In *PCR Protocols*, Innis, M. A., Gelfand, D. H., Sninsky, J. J., and White, T. J., eds. (Academic Press, San Diego), pp. 92–98.

Lander, E. S., Green, P., Abrahamson, J., Barlow, A., Daly, M. J., Lincoln, S. E., and Newburg, L. (1987). MAPMAKER: An interactive computer package for constructing primary genetic linkage maps of experimental and natural populations. *Genomics* 1: 174–81.

Larin, Z., Monaco, A. P., and Lehrach, H. (1991). Yeast artificial chromosome libraries containing large inserts from mouse and human DNA. *Proc. Nat. Acad. Sci. USA* 88: 4123–4127.

Larin, Z., Monaco, A. P., Meier-Ewert, S., and Leharch, H. (1993). Construction and characterization of yeast artificial chromosome libraries from the mouse genome. In *Guides to Techniques in Mouse Development*, Methods in Enzymology 225, Wassarman, P. M. and DePamphilis, M. L., eds. (Academic Press, San Diego), pp. 623-637.

Laurie, D. A., and Hulten, M. A. (1985). Further studies on chiasma distribution and interference in the human male. *Ann. Hum. Genet.* 49: 203–214.

Lawrence, J. B. (1990). A fluorescence *in situ* hybridization approach for gene mapping and the study of nuclear organization. In *Genetic and Physical Mapping*, Genome Analysis 1, Davies, K. E. and Tilghman, S., eds. (Cold Spring Harbor Laboratory Press, Gold Spring Harbor, NY), pp. 1–328.

Leder, A., Swan, D., Ruddle, F., D'Eustachio, P., and Leder, P. (1981). Dispersion of alpha-like globin genes of the mouse to three different chromosomes. *Nature* 293: 196–200.

LeRoy, H., Simon-Chazottes, D., Montagutelli, X., and Guénet, J.-L. (1992). A set of anonymous DNA clones as markers for mouse gene mapping. *Mammal. Genome* 3: 244–246.

Levenson, C., and Chang, C.-a. (1990). Nonisotopically labelled probes and primers. In *PCR Protocols*, Innis, M. A., Gelfand, D. H., Sninsky, J. J., and White, T. J., eds. (Academic Press, San Diego), pp. 99–112.

Li, H., Gyllensten, U. B., Cui, X., Saiki, R. K., Erlich, H. A., and Arnheim, N. (1988).

Amplification and analysis of DNA sequences in single human sperm and diploid cells. *Nature* 335: 414–417.

Lindsay, S., and Bird, A. P. (1987). Use of restriction enzymes to detect potential gene sequences in mammalian DNA. *Nature* 327: 336–338.

Lisitsyn, N., Lisitsyn, N., and Wigler, M. (1993). Cloning the differences between two complex genomes. *Science* 259: 946-951.

Lister, C., and Dean, C. (1993). Recombinant inbred lines for mapping RFLP and phenotypic markers in *Arabidopsis thaliana*. *Plant J.* 4: 745–750.

Little, C. C., and Bagg, H. J. (1924). The occurrence of four inheritable morphological variations in mice and their possible relation to treatment with X-rays. *J. Exp. Zool.* 41: 45–92.

Little, P. (1993). The end of the beginning. *Nature* 362: 408–409.

Love, J. M., Knight, A. M., McAleer, M. A., and Todd, J. A. (1990). Towards construction of a high resolution map of the mouse genome using PCR-analyzed micro-satellites. *Nucl. Acids Res.* 18: 4123–4130.

Lovett, M., Kere, J., and Hinton, L. M. (1991). Direct selection: A method for the isolation of cDNAs encoded by large genomic regions. *Proc. Nat. Acad. Sci. USA* 88: 9628–9632.

Lowe, T., Sharefkin, J., Yang, S.-Q., and Dieffenbach, C. W. (1990). A computer program for selection of oligonucleotide primers for polymerase chain reactions. *Nucl. Acids Res.* 18: 1757–1761.

Ludecke, H.-J., Senger, G., Claussen, U., and Horsthemke, B. (1989). Cloning defined regions of the human genome by microdissection of banded chromosomes and enzymatic amplification. *Nature* 338: 348–350.

Lueders, K. K., and Kuff, E. L. (1977). Sequences associated with intracisternal A-particles are reiterated in the mouse genome. *Cell* 12: 963–972.

Lyon, M. F., and Kirby, M. C. (1992). Mouse chromosome atlas. *Mouse Genome* 90: 22–44.

Lyon, M. F., and Searle, A. G., eds. (1989). *Genetic Variants and Strains of the Laboratory Mouse*. 2nd edition (Oxford University Press, Oxford).

Maniatis, T., Fritsch, E. F., and Sambrook, J. (1982). *Molecular Cloning: A Laboratory Manual*. (Cold Spring Harbor Laboratory, Cold Spring Harbor, NY).

Manly, K. F. (1993). A Macintosh program for storage and analysis of experimental genetic mapping data. *Mammal. Genome* 4: 303-313.

Marchuk, D. A., and Collins, F. S. (1994). The use of YACs to identify expressed sequences: cDNA screening using total YAC insert. In *YAC Libraries, A User's Guide*, Nelson, D. L. and Brownstein, B. H., eds. (W.H. Freeman and Company, New York), pp. 113–126.

Mariat, D., and Vergnaud, G. (1992). Detection of polymorphic loci in complex genomes with synthetic tandem repeats. *Genomics* 12: 454–458.

Marshall, C. J. (1991). Tumor suppressor genes. *Cell* 64: 313.

Marshall, J. D., Mu, J.-L., Nesbitt, M. N., Frankel, W. N., and Paigen, B. (1992). The AXB and BXA set of recombinant inbred mouse strains. *Mammal. Genome* 3: 669–680.

Marshall, J. T. (1981). Taxonomy. In *The Mouse in Biomedical Research*, Vol. 1, Foster, H. L., Small, J. D., and Fox, J. G., eds. (Academic Press, New York), pp. 17–26.

Martin, G. B., Brommonschenkel, S. H., Chunwongse, J., Frary, A., Ganal, M. W., Spivey, R., Wu, T., Earle, E. D., and Tanksley, S. D. (1993). Map-based cloning of a protein kinase gene conferring disease resistance in tomato. *Science* 262: 1432–1436.

Martin, S. L. (1991). LINEs. *Curr. Opin. Genet. Dev.* 1: 505–508.

Matsuda, Y., and Chapman, V. M. (1991). In situ analysis of centromeric satellite DNA segregating in *Mus* species crosses. *Mammal. Genome* 1: 71–77.

Matsuda, Y., Manly, K. F., and Chapman, V. M. (1993). In situ analysis of centromere segregation in C57BL/6 × *Mus spretus* interspecific backcrosses. *Mammal. Genome* 4: 475–480.

McClelland, M., and Ivarie, R. (1982). Asymmetrical distribution of CpG in an "average" mammalian gene. *Nucl. Acids Res.* 10: 7865-7877.

McDonald, J. D., Shedlovsky, A., and Dove, W. F. (1990). Investigating inborn errors of phenylalanine metabolism by efficient mutagenesis of the mouse germ line. In *Biology of Mammalian Germ Cell Mutagenesis*, Allen, J. W., Bridges, B. A., Lyon, M. F., Moses, M. J., and Russell, L. B., eds. (Cold Spring Harbor Press, Cold Spring Harbor, NY), pp. 259–270.

McGinnis, W., and Krumlauf, R. (1992). Homeobox genes and axial patterning. *Cell* 68: 283–302.

McGrath, J., and Solter, D. (1983). Nuclear Transplantation in the mouse embryo by microsurgery and cell fusion. *Science* 220: 1300–1302.

McKusick, V. (1988). *Mendelian Inheritance in Man: Catalogs of Autosomal Dominant, Autosomal Recessive, and X-Linked Phenotypes.* 8th edition (The Johns Hopkins University Press, Baltimore).

Meisler, M. H. (1992). Insertional mutation of 'classical' and novel genes in transgenic mice. *Trends Genet.* 8: 341–344.

Meitz, J. A., and Kuff, E. L. (1992). Intracisternal A-particle-specific oligonucleotides provide multilocus probes for genetic linkage studies in the mouse. *Mammal. Genome* 3: 447–451.

Michaud, J., Brody, L. C., Steel, G., Fontaine, G., Martin, L. S., Valle, D., and Mitchell, G. (1992). Strand-separating conformational polymorphism analysis: Efficacy of detection of point mutations in the human ornithine d-aminotransferase gene. *Genomics* 13: 389–394.

Michelmore, R. W., Paran, I., and Kesseli, R. V. (1991). Identification of markers linked to disease-resistance genes by bulked segregant analysis: A rapid method to detect markers in specific genomic regions by using segregating populations. *Proc. Nat. Acad. Sci. USA* 88: 9828–32.

Michiels, F., Burmeister, M., and Lehrach, H. (1987). Derivation of clones close to met by preparative field inversion gel electrophoresis. *Science* 236: 1305–1308.

Miesfeld, R., Krystal, M., and Arnheim, N. (1981). A member of new repeated sequence family which is conserved throughout eucaryotic evolution is found between the human delta- and beta-globin genes. *Nucl. Acids Res.* 9: 5931.

Miller, O. J., and Miller, D. A. (1975). Cytogenetics of the mouse. *Ann. Rev. Genet.* 9: 285–303.

Milner, C. M., and Campbell, R. D. (1992). Genes, genes and more genes in the human major histocompatibility complex. *BioEssays* 14: 565–571.

Montagutelli, X. (1990). GENE-LINK: A program in PASCAL for backcross genetic analysis. *J. Hered.* 81: 490–491.

Moore, D. S., and McCabe, G. P. (1989). *Introduction to the Practice of Statistics.* (W.H. Freeman & Co., New York).

Moore, T., and Haig, D. (1991). Genomic imprinting in mammalian development: A parental tug-of-war. *Trends Genet* 7: 45–49.

Morgan, T. H., and Cattell, E. (1912). Data for the study of sex-linked inheritance in *Drosophila. J. Exp. Zool.* 13: 79–101.

Morse, H. C. (1978). Introduction. In *Origins of Inbred Mice*, Morse, H. C., ed. (Academic

Press, New York), pp. 1–31.

Morse, H. C. (1981). The laboratory mouse—A historical perspective. In *The Mouse in Biomedical Research*, Vol. 1, Foster, H. L., Small, J. D., and Fox, J. G., eds. (Academic Press, New York), pp. 1–16.

Morse, H. C. (1985). The Bussey Institute and the early days of mammalian genetics. *Immunogenet.* 21: 109–116.

Morton, N. (1955). Sequential tests for the detection of linkage. *Am. J. Hum. Genet.* 7: 277–318.

Moseley, W. S., and Seldin, M. F. (1989). Definition of mouse chromosome 1 and 3 gene linkage groups that are conserved on human chromosome 1: Evidence that a conserved linkage group spans the centromere of human chromosome 1. *Genomics* 5: 899-905.

Moulia, C., Aussel, J. P., Bonhomme, F., Boursot, P., Nielsen, J. T., and Renaud, F. (1991). Wormy mice in a hybrid zone: a genetic control of susceptibility to parasite infection. *J. Evol. Biol.* 4: 679–687.

Moyzis, R. K., Buckingham, J. M., Cram, L. S., Dani, M., Deaven, L. L., Jones, M. D., Meyne, J., Ratliffe, R. L., and Wu, J.-R. (1988). A highly conserved repetitive DNA sequence (TTAGGG) present at the telomeres of human chromosomes. *Proc. Nat. Acad. Sci. USA* 85: 6622–6626.

Mu, J. L., Naggert, J. K., Nishina, P. M., Cheach, Y. C., and Paigen, B. (1993). Strain distribution pattern in AXB and BXA recombinant inbred strains for loci on murine chromosomes 10, 13, 17, and 18. *Mammal. Genome* 4: 148–152.

Muller, H. J. (1916). The mechanism of crossing-over. *Am. Nat.* 50: 193–221.

Muller, H. J. (1927). Artificial transmutation of the gene. *Science* 66: 84–87.

Myers, R. M., Fischer, S. G., and Maniatist, T. (1985). Nearly all single base substitutions in DNA fragments joined to a GC-clamp can be detected by denaturing gradient gel electrophoresis. *Nucl. Acids Res.* 13: 3131–3145.

Nadeau, J. H. (1984). Lengths of chromosomal segments conserved since divergence of man and mouse. *Proc. Nat. Acad. Sci.* 81: 814-818.

Nadeau, J. H., Herrmann, B., Bucan, M., Burkart, D., Crosby, J. L., Erhart, M. A., Kosowsky, M., Kraus, J. P., Michiels, F., Schnattinger, A., Tchetgen, M.-B., Varnum, D., Willison, K., Lehrach, H., and Barlow, D. (1991). Genetic maps of mouse chromosome 17 including 12 new anonymous DNA loci and 25 anchor loci. *Genomics* 9: 78–89.

Nadeau, J. H., Bedigian, H. G., Bouchard, G., Denial, T., Kosowksy, M., Norberg, R., Pugh, S., Sargeant, E., Turner, R., and Paigen, B. (1992). Multilocus markers for mouse genome analysis: PCR amplification based on single primers of arbitrary nucleotide sequence. *Mammal. Genome* 3: 55–64.

Nakamura, Y., Leppert, M., O'Connell, P., Wolff, R., Holm, T., Culver, M., Martin, C., Fujimoto, E., Hoff, M., Kumlin, E., and White, R. (1987). Variable number of tandem repeat (VNTR) markers for human gene mapping. *Science* 235: 1616–1622.

Nebel, B. R., Amarose, A. P., and Hackett, E. M. (1961). Calender of gametogenic development in the prepuberal male mouse. *Science* 134: 832–833.

Nei, M. (1987). *Molecular Evolutionary Genetics*. (Columbia University Press, New York).

Nelson, D. L., and Brownstein, B. H., eds. (1994). *YAC Libraries, A User's Guide*. (W.H. Freeman and Company, New York).

Nelson, D. L., Ledbetter, S. A., Corbo, L., Victoria, M. F., Ramirez-Solis, R., Webster, T. D., Ledbetter, D. H., and Caskey, T. (1989). Alu polymerase chain reaction: A method for rapid isolation of human-specific sequences from complex DNA sources. *Proc. Nat. Acad. Sci. USA* 86: 6686–6690.

Neumann, P. E. (1990). Two-locus linkage analysis using recombinant inbred strains and Bayes' Theorem. *Genetics* 126: 277–284.

Neumann, P. E. (1991). Three-locus linkage analysis using recombinant inbred strains and Bayes' theorem. *Genetics* 128: 631–638.

Nichols, R. D., Knoll, J. H. M., Butler, M. G., Karam, S., and Lalande, M. (1989). Genetic imprinting suggested by maternal heterodisomy in non-deletion Prader-Willi syndrome. *Nature* 342: 281–285.

North, M. A., Sanseau, P., and Buckler, A. J. (1993). Efficiency and specificity of gene isolation by exon amplification. *Mammal. Genome* 4: 466–474.

O'Brien, S., Womack, J. E., Lyons, L. A., Moore, K. J., Jenkins, N. A., and Copeland, N. G. (1993). Anchored reference loci for comparative genome mapping in mammals. *Nature Genet.* 3: 103–112.

O'Farrell, P. H. (1975). High resolution two-dimensional electrophoresis of proteins. *J. Biol. Chem.* 250: 4007–4021.

Oakberg, E. F. (1956a). A description of spermiogenesis in the mouse and its use in the analysis of the cycle of the seminiferous epithelium and germ cell renewal. *Amer. J. Anat.* 99: 391–413.

Oakberg, E. F. (1956b). Spermatogenesis in the mouse and timing of stages of the cycle of the seminiferous epithelium. *Amer. J. Anat.* 99: 507–516.

Ohno, S. (1967). *Sex Chromosomes and Sex-Linked Genes.* (Springer Verlag, Berlin).

Okada, N. (1991). SINEs. *Curr. Opin. Genet. Develop.* 1: 498–504.

Olds-Clarke, P., and Peitz, B. (1985). Fertility of sperm from *t/+* mice: evidence that +-bearing sperm are dysfunctional. *Genet. Res.* 47: 49–52.

Ollmann, M. M., Winkes, B. M., and Barsh, G. S. (1992). Construction, analysis, and application of a radiation hybrid mapping panel surrounding the mouse agouti locus. *Genomics* 13: 731–740.

Orita, M., Iwahana, H., Kanazawa, H., Hayashi, K., and Sekiya, T. (1989a). Detection of polymorphisms of human DNA by gel electrophoresis as single-strand conformation polymorphisms. *Proc. Nat. Acad. Sci. USA* 86: 2766–2770.

Orita, M., Suzuki, Y., Sekiya, T., and Hayashi, K. (1989b). Rapid and sensitive detection of point mutations and DNA polymorphims using the polymerase chain reaction. *Genomics* 5: 874–879.

Painter, J. S. (1928). Chromosome number in the house mouse. *Genetics* 13: 180–189.

Palmiter, R. D., and Brinster, R. L. (1986). Germ-line transformation of mice. *Ann. Rev. Genet.* 20: 465–499.

Papaioannou, V. E., and Festing, M. F. W. (1980). Genetic drift in a stock of laboratory mice. *Lab. Animals* 14: 11–13.

Pardue, M. L., and Gall, J. G. (1970). Chromosomal localization of mouse satellite DNA. *Science* 168: 1356–1358.

Parimoo, S., Patanjali, S. R., Shukla, H., Chaplin, D. D., and Weissman, S. M. (1991). cDNA selection: Efficient PCR approach for the selection of cDNAs encoded in large chromosomal DNA fragments. *Proc. Nat. Acad. Sci.* 88: 9623–9627.

Parrish, J. E., and Nelson, D. L. (1993). Methods for finding genes: a major rate-limiting step in positional cloning. *Gen. Anal. Tech. Appl.* 10: 29–41.

Patanjali, S. R., Parimoo, S., and Weismann, S. M. (1991). Construction of a uniform abundance (normalized) cDNA library. *Proc. Nat. Acad. Sci. USA* 88: 1943–1947.

Pedersen, R. A., Papaioannou, V., Joyner, A., and Rossant, J. (1993). *Targeted Mutagenesis in Mice.* Audiovisual material from Cold Spring Harbor Laboratory Press, Cold Spring Harbor, NY.

Pickford, I. (1989) Ph.D. thesis, Imperial Cancer Research Foundation, London, UK.

Pierce, J. C., and Sternberg, N. L. (1992). Using bacteriophage P1 system to clone high molecular weight genomic DNA. *Meth. Enzymol.* 216: 549–74.

Pierce, J. C., Sternberg, N., and Sauer, B. (1992). A mouse genomic library in the bacteriophage P1 cloning system: Organization and characterization. *Mammal Genome* 3: 550–8.

Popp, R. A., Bailiff, E. G., Skow, L. C., Johnson, F. M., and Lewis, S. E. (1983). Analysis of a mouse alpha-globin gene mutation induced by ethylnitrosourea. *Genetics* 105: 157–167.

Potter, M., Nadeau, J. H., and Cancro, M. P., eds. (1986). *The Wild Mouse in Immunology.* (Springer-Verlag, New York).

Povey, S., Smith, M., Haines, J., Kwiatkowski, D., Fountain, J., Bale, A., Abbott, C., Jackson, I., Lawrie, M., and Hultén, M. (1992). Report on the First International Workshop on Chromosome 9. *Ann. Hum. Genet.* 56: 167–221.

Punnett, R. C. (1911). *Mendelism.* (Macmillan, New York).

Rajan, T. V., Halay, E. D., Potter, T. A., Evans, G. A., Seidman, J. G., and Margulies, D. H. (1983). H-2 hemizygous mutants from a heterozygous cell line: Role of mitotic recombination. *EMBO. J* 2: 1537–1542.

Rattner, J. B. (1991). The structure of the mammalian centromere. *Bioessays* 13: 51–56.

Reeves, R. H., Crowley, M. R., Moseley, W. S., and Seldin, M. F. (1991). Comparison of interspecific to intersubspecific backcrosses demonstrates species and sex differences in recombination frequency on mouse chromosome 16. *Mammal. Genome* 1: 158–164.

Ridley, R. M., Frith, C. D., Farrer, L. A., and Conneally, P. M. (1991). Patterns of inheritance of the symptoms of Huntington disease suggestive of an effect of genomic imprinting. *J. Med. Genet.* 28: 224–231.

Rikke, B. A., and Hardies, S. C. (1991). LINE-1 repetitive DNA probes for species-specific cloning from *Mus spretus* and *Mus domesticus* genomes. *Genomics* 11: 895–904.

Rikke, B. A., Garvin, L. D., and Hardies, S. C. (1991). Systematic identification of LINE-1 repetitive DNA sequence differences having species specificity between *Mus spretus* and *Mus domesticus*. *J. Mol. Biol.* 219: 635–643.

Riley, J., Butler, R., Ogilvie, D., Finniear, R., Jenner, D., Powell, S., Anand, R., Smith, J. C., and Markham, A. F. (1990). A novel rapid method for the isolation of terminal sequences from yeast artificial chromosome (YAC) clones. *Nucl. Acid Res.* 18: 2887-2890.

Rinchik, E. M., and Russell, L. B. (1990). Germ-line deletion mutations in the mouse: Tools for intensive functional and physical mapping of regions of the mammalian genome. In *Genetic and Physical Mapping*, Genome Analysis 1, Davies, K. E. and Tilghman, S. M., eds. (Cold Spring Harbor Laboratory Press, Cold Spring Harbor, NY), pp. 121–158.

Rinchik, E. M., Carpenter, D. A., and Selby, P. B. (1990b). A strategy for fine-structure functional analysis of a 6- to 11-centimorgan region of mouse chromosome 7 by high-efficiency mutagenesis. *Proc. Nat. Acad. Sci. USA* 87: 896–900.

Rinchik, E. M. (1991). Chemical mutagenesis and fine-structure functional analysis of the mouse genome. *Trends. Genet.* 7: 15–21.

Rinchik, E. M., Bangham, J. W., Hunsicker, P. R., Cacheiro, N. L. A., Kwon, B. S., Jackson, I. J., and Russell, L. B. (1990a). Genetic and molecular analysis of chlorambucil-induced germ-line mutations in the mouse. *Proc. Nat. Acad. Sci. USA* 87: 1416–1420.

Robertson, E. J. (1991). Using embryonic stem cells to introduce mutations into the mouse germ line. *Biol. Reprod.* 44: 238–45.

Röhme, D., Fox, H., Herrmann, B., Frischauf, A.-M., Edström, J.-E., Mains, P., Silver, L. M., and Lehrach, H. (1984). Molecular clones of the mouse *t* complex derived from microdissected metaphase chromosomes. *Cell* 36: 783–788.

Rossant, J., Vijh, M., Siracusa, L. D., and Chapman, V. M. (1983). Identification of embryonic cell lineages in histological sections of *M. musculus* < — > *M. caroli* chimaeras. *J. Embryol. Exp. Morph.* 73: 179–191.

Rossi, J. M., Burke, D. T., Leung, J. C., Koos, D. S., Chen, H., and Tilghman, S. M. (1992). Genomic analysis using a yeast artificial chromosome library with mouse DNA inserts. *Proc. Nat. Acad. Sci. USA.* 89: 2456–60.

Rowe, F. P. (1981). Wild house mouse biology and control. *Symp. Zool. Soc. Lond.* 47: 575–589.

Rugh, R. (1968). *The Mouse: Its Reproduction and Development.* (Oxford University Press, New York).

Russell, E. S. (1978). Origins and history of mouse inbred strains: contributions of Clarence Cook Little. In *Origins of Inbred Mice*, Morse, H. C., ed. (Academic Press, New York), pp. 33–43.

Russell, E. S. (1985). A history of mouse genetics. *Ann. Rev. Genet.* 19: 1–28.

Russell, L. B. (1990). Patterns of mutational sensitivity to chemicals in poststem-cell stages of mouse spermatogenesis. *Prog. Clin. Biol. Res.* 340: 101.

Russell, L. B., Hunsicker, P. R., Cacheiro, N. L. A., Bangham, J. W., Russell, W. L., and Shelby, M. D. (1989). Chlorambucil effectively induces deletion mutations in mouse germ cells. *Proc. Natl. Acad. Sci. USA.* 86: 3704–3708.

Russell, L. B., Russell, W. L., Rinchik, E. M., and Hunsicker, P. R. (1990). Factors affecting the nature of induced mutations. In *Biology of Mammalian Germ Cell Mutagenesis*, Allen, J. W., Bridges, B. A., Lyon, M. F., Moses, M. J., and Russell, L. B., eds. (Cold Spring Harbor Press, Cold Spring Harbor, NY), pp. 271–289.

Russell, W. L., and Hurst, J. G. (1945). Pure strain mice born to hybrid mothers following ovarian transplantation. *Proc. Nat. Acad. Sci. USA* 31: 267–273.

Russell, W. L., Kelly, P. R., Hunsicker, P. R., Bangham, J. W., Maddux, S. C., and Phipps, E. L. (1979). Specific-locus test shows ethylnitrosourea to be the most potent mutagen in the mouse. *Proc. Nat. Acad. Sci. USA* 76: 5918–5922.

Russell, W. L., Hunsicker, P. R., Raymer, G. D., Steele, M. H., Stelzner, K. F., and Thompson, H. M. (1982). Dose response curve for ethylnitrosourea specific-locus mutations in mouse spermatogonia. *Proc. Nat. Acad. Sci. USA.* 79: 3589–3591.

Ruvinsky, A., Agulnik, A., Agulnik, S., and Rogachova, M. (1991). Functional analysis of mutations of murine chromosome 17 with the use of tertiary trisomy. *Genetics* 127: 781–788.

Sage, R. D. (1981). Wild mice. In *The Mouse in Biomedical Research*, Vol. 1, Foster, H. L., Small, J. D., and Fox, J. G., eds. (Academic Press, New York), pp. 40–90.

Sage, R. D. (1986). Wormy mice in a hybrid zone. *Nature* 324: 60-63.

Sage, R. D., Whitney, J. B., and Wilson, A. C. (1986). Genetic analysis of a hybrid zone between domesticus and musculus mice (*Mus musculus* complex): Hemoglobin polymorphisms. *Curr. Topics Micro. Immunol.* 127: 75–85.

Saiki, R. K., Bugawan, T. L., Horn, G. T., Mullis, K. B., and Erlich, H. A. (1986). Analysis of enzymatically amplified beta-globin and HLA-DQ alpha DNA with allele-specific oligonucleotide probes. *Nature* 324: 163–166.

Sambrook, J., Fritsch, E. F., and Maniatis, T. (1989). *Molecular Cloning: A Laboratory Manual.* 2nd edition (Cold Spring Harbor Laboratory, Cold Spring Harbor, NY).

Sapienza, C. (1989). Genome imprinting and dominance modification. *Ann. N.Y. Acad. Sci.* 534: 24–38.

Sapienza, C. (1991). Genome imprinting and carcinogenesis. *Biochem. Biophys. Acta* 1072: 51–61.

Scalenghe, F., Turco, E., Edstrom, J. E., Pirotta, V., and Melli, M. (1981). Microdissection and cloning of DNA from a specific region of *Drosophila melanogaster* polytene chromosomes. *Chromosoma* 82: 205–216.

Schedl, A., Montoliu, L., Kelsey, G., and Schütz, G. (1993). A yeast artificial chromosome covering the tyrosinase gene confers copy number-dependent expression in transgenic mice. *Nature* 362: 258–261.

Schwartz, D. C., and Cantor, C. R. (1984). Separation of yeast chromosome-sized DNAs by pulsed field gel electrophoresis. *Cell* 37: 67–75.

Searle, A. G. (1982). The genetics of sterility in the mouse. In *Genetic Control of Gamete Production and Function*, Crosignani, P. G. and Rubin, B. L., eds. (Academic Press, London), pp. 93–114.

Searle, A. G. (1989). Chromosomal variants: numerical variants and structural rearrangements. In *Genetic Variants and Strains of the Laboratory Mouse.*, Lyon, M. F. and Searle, A. G., eds. (Oxford University Press, Oxford), pp. 582–616.

Sedivy, J. M., and Joyner, A. L. (1992). *Gene Targeting*. (W.H. Freeman, New York).

Seldin, M. F., Howard, T. A., and D'Eustachio, P. (1989). Comparison of linkage maps of mouse chromosome 12 derived from laboratory strain intraspecific and *Mus spretus* interspecific backcrosses. *Genomics* 5: 24–28.

Serikawa, T., Montagutellie, X., Simon-Chazottes, D., and Guénet, J.-L. (1992). Polymorphisms revealed by PCR with single, short-sized, arbitrary primers are reliable markers for mouse and rat gene mapping. *Mammal. Genome* 3: 65–72.

Shedlovsky, A., King, T. R., and Dove, W. F. (1988). Saturation germ line mutagenesis of the murine t region including a lethal allele at the quaking locus. *Proc. Nat. Acad. Sci. USA.* 85: 180–4.

Sheehan, P. M., Fastovsky, D. E., Hoffmann, R. G., Berghaus, C. B., and Gabriel, D. L. (1991). Sudden extinction of the dinosaurs: Latest Cretaceous, Upper Great Plains, USA. *Science* 254: 835-839.

Sheffield, V. C., Cox, D. R., Lerman, L. S., and Myers, R. M. (1989). Attachment of a 40-base pair G + C-rich sequence (GC-clamp) to genomic DNA fragments by the polymerase chain reaction results in improved detection of single-base changes. *Proc. Nat. Acad. Sci. USA.* 86: 232–236.

Sheffield, V. C., Beck, J. S., Kwitek, A. E., Sandstrom, D. W., and Stone, E. M. (1993). The sensitivity of single-strand conformation polymorphism analysis for the detection of single base substitutions. *Genomics* 16: 325–332.

Sheppard, R., Montagutelli, X., Jean, W., Tsai, J.-Y., Guenet, J.-L., Cole, M. D., and Silver, L. M. (1991). Two-dimensional gel analysis of complex DNA families: Methodology and apparatus. *Mammal. Genome* 1: 104–111.

Sheppard, R. D., and Silver, L. M. (1993). Methods for two-dimensional analysis of repetitive DNA families. In *Guides to Techniques in Mouse Development*, Methods in Enzymology 225, Wassarman, P. M. and DePamphilis, M. L., eds. (Academic Press, San Diego), pp. 701–715.

Shizuya, H., Birren, B., Kim, U.-J., Mancino, V., Slepak, T., Tachiiru, Y., and Simon, M. (1992). Cloning and stable maintenance of 300-kilobase-pair fragments of human DNA in *Escherichia coli* using an F-factor-based vector. *Proc. Nat. Acad. USA* 89: 8794–8797.

Silver, J. (1985). Confidence limits for estimates of gene linkage based on analysis using recombinant inbred strains and backcrosses. *J. Hered.* 76: 436–440.

Silver, J., and Buckler, C. E. (1986). Statistical considerations for linkage analysis using recombinant inbred strains and backcrosses. *Proc. Nat. Acad. Sci. USA* 83: 1423–1427.

Silver, L. M. (1985). Mouse t haplotypes. *Ann. Rev. Genet.* 19: 179-208.

Silver, L. M. (1986). A software package for record-keeping and analysis of a breeding mouse colony. *J. Hered.* 77: 479.

Silver, L. M. (1988). Mouse t haplotypes: a tale of tails and a misunderstood selfish chromosome. *Curr. Top. Microbiol. Immunol.* 137: 64–69.

Silver, L. M. (1990). At the crossroads of developmental genetics: the cloning of the classical mouse t locus. *Bioessays* 12: 377-380.

Silver, L. M. (1993a). The peculiar journey of a selfish chromosome: mouse t haplotypes and meiotic drive. *Trends Genet.* 9: 250–254.

Silver, L. M. (1993b). Recordkeeping and database analysis of breeding colonies. In *Guides to Techniques in Mouse Development*, Methods in Enzymology Vol. 225, Wassarman, P. M. and DePamphilis, M. L., eds. (Academic Press, San Diego), pp. 3–15.

Silver, L. M., Uman, J., Danska, J., and Garrels, J. I. (1983). A diversified set of testicular cell proteins specified by genes within the mouse *t* complex. *Cell* 35: 35–45.

Silver, L. M., Hammer, M., Fox, H., Garrels, J., Búcan, M., Herrmann, B., Frischauf, A. M., Lehrach, H., Winking, H., Figueroa, F., and Klein, J. (1987). Molecular evidence for the rapid propagation of mouse t haplotypes from a single, recent, ancestral chromosome. *Mol. Biol. Evol.* 4: 473–482.

Silver, L. M., Nadeau, J. H., and Goodfellow, P. N. (1993). Encyclopedia of the Mouse Genome III. Mammal. *Genome* 4 (Special issue): S1-S283.

Silvers, W. K. (1979). *The Coat Colors of the Mouse.* (Springer-Verlag, New York).

Simmler, M.-C., Cox, R. D., and Avner, P. (1991). Adaptation of the interspersed repetitive sequence polymerase chain reaction to the isolation of mouse DNA probes from somatic cell hybrids on a hamster background. *Genomics* 10: 770–778.

Singer, M. F. (1982). SINEs and LINEs: Highly repeated short and long interspersed sequences in mammalian genomes. *Cell* 28: 433–434.

Siracusa, L. D., Jenkins, N. A., and Copeland, N. G. (1991). Identification and applications of repetitive probes for gene mapping in the mouse. *Genetics* 127: 169–179.

Slizynski, B. M. (1954). Chiasmata in the male mouse. *J. Genet.* 53: 597–605.

Smithies, O. (1993). Animal models of human genetic diseases. *Trends Genet.* 9: 112–116.

Snell, G. D., ed. (1941). *Biology of the Laboratory Mouse.* (Blakiston Co., New York).

Snell, G. D. (1978). Congenic resistant strains of mice. In *Origins of Inbred Mice*, Morse, H. C., ed. (Academic Press, New York), pp. 1–31.

Snell, G. D., and Reed, S. (1993). William Ernest Castle, pioneer mammalian geneticist. *Genetics* 133: 751–753.

Snouwaert, J. N., Brigman, K. K., Latour, A. M., Malouf, N. N., Boucher, R. C., Smithies, O., and Koller, B. H. (1992). An animal model for cystic fibrosis made by gene targeting. *Science* 257: 1083–1088.

Sokal, R. R., Oden, N. L., and Wilson, C. (1991). Genetic evidence for the spread of agriculture in Europe by demic diffusion. *Nature* 351: 143–145.

Soller, M. (1991). Mapping quantitative trait loci affecting traits of economic importance in animal populations using molecular markers. In *Gene-mapping Techniques and Applications*, Schook, L. B., Lewin, H. A., and McLaren, D. G., eds. (Marcel Dekker Inc., New York), pp. 21–49.

Stallings, R. L., Ford, A. F., Nelson, D., Torney, D. C., Hildebrand, C. E., and Moyzis, R.

K. (1991). Evolution and distribution of (GT)n repetitive sequences in mammalian genomes. *Genomics* 10: 807-815.

Steinmetz, M., Uematsu, Y., and Fischer-Lindhal, K. (1987). Hotspots of homologous recombination in mammalian genomes. *Trends Genet.* 3: 7–10.

Stevens, L. C. (1957). A modification of Robertson's technique of homoiotopic ovarian transplantation in mice. *Transplant. Bull.* 4: 106–107.

Strong, L. C. (1978). Inbred mice in science. In *Origins of Inbred Mice*, Morse, H. C., ed. (Academic Press, New York), pp. 45–59.

Sturtevant, A. H. (1913). The linear arrangement of six sex-linked factors in Drosophila, as shown by their mode of association. *J. Exp. Zool.* 14: 43–59.

Takahashi, N., and Ko, M. S. H. (1993). The short 3′-end of complementary DNAs as PCR-based polymorphic markers for an expression map of the mouse genome. *Genomics* 16: 161–168.

Taketo, M., Schroeder, A. C., Mobraaten, L. E., Gunning, K. B., Hanten, G., Fox, R. R., Roderick, T. H., Stewart, C. L., Lilly, F., Hansen, C. T., and Overbeek, P. A. (1991). FVB/N: An inbred mouse strain preferable for transgenic analyses. *Proc. Nat. Acad. Sci. USA* 88: 2065–2069.

Talbot, D., Collis, P., Antoniou, M., Vidal, M., Grosveld, F., and Greaves, D. R. (1989). A dominant control region from the human b-globin locus conferring integration site-independent gene expression. *Nature* 338: 352–355.

Taylor, B. A. (1978). Recombinant inbred strains: use in gene mapping. In *Origins of Inbred Mice*, Morse, H. C., ed. (Academic Press, New York), pp. 423–438.

ten Berg, R. (1989). Recombinant inbred series, OXA. *Mouse News Lett.* 84: 102–104.

Townes, T. M., and Behringer, R. R. (1990). Human globin locus activation region (LAR): Role in temporal control. *Trends Genet.* 6: 219–223.

Trask, F. (1991). Fluorescence in situ hybridization. *Trends. Genet.* 7: 149–155.

Tucker, P. K., Lee, B. K., Lundrigan, B. L., and Eicher, E. M. (1992). Geographic origin of the Y chromosome in "old" inbred strains of mice. *Mammal. Genome* 3: 254–261.

Updike, J. (1991). Introduction. In *The Art of Mickey Mouse*, Yoe, C. and Morra-Yoe, J., eds. (Hyperion, New York), pp. 1–9.

Valancius, V., and Smithies, O. (1991). Testing an "in–out" targeting procedure for making subtle genomic modifications in mouse embryonic stem cells. *Mol. Cell. Bio.* 11: 1402–1408.

van Deursen, J., and Wieringa, B. (1992). Targeting of the creatine kinase M gene in embryonic stem cells using isogenic and nonisogenic vectors. *Nucl. Acids Res.* 20: 3815–3820.

Vanlerberghe, F., Boursot, P., Catalan, J., Gerasimov, S., Bonhomme, F., Botev, B. A., and Thaler, L. (1988). Analyse génétique de la zone d'hybridation entre les deux sous-espèces de souris *Mus musculus domesticus* et *Mus musculus musculus* en Bulgarie. *Genome* 30: 427–437.

Wagner, E. F., Stewart, T. A., and Mintz, B. (1981a). The human beta-globin gene and a functional thymidine kinase gene in developing mice. *Proc. Nat. Acad. Sci.* 78: 5016–5020.

Wagner, T. E., Hoppe, P. C., Jollick, J. D., Scholl, D. R., Hodinka, R. L., and Gault, J. B. (1981b). Microinjection of a rabbit beta-globin gene into zygotes and its subsequent expression in adult mice and their offspring. *Proc. Nat. Acad. Sci.* 78: 6376–6380.

Wallace, M. E. (1981). Wild mice heterozygous for a single Robertsonian translocation. *Mouse News Lett.* 64: 49.

Wassarman, P. (1993). Mammalian eggs, sperm and fertilisation: dissimilar cells with a common goal. *Dev. Biol.* 4: 189–197.

Wassarman, P. M. (1990). Profile of a mammalian sperm receptor. *Development* 108: 1–17.

Wassarman, P. M., and DePamphilis, M. L., eds. (1993). *Guide to Techniques in Mouse Development* (Academic Press, San Diego).

Watson, J. D., and Tooze, J. (1981). *The DNA Story: a Documentary History of Gene Cloning.* (W. H. Freeman, San Francisco).

Watson, M. L., D'Eustachio, P., Mock, B. A., Steinberg, A. D., Morse, H. C., Oakey, R. J., Howard, T. A., Rochelle, J. M., and Seldin, M. F. (1992). A linkage map of mouse chromosome 1 using an interspecific cross segregating for the gld autoimmunity mutation. *Mammal. Genome* 2: 158–171.

Weber, J. L. (1990). Informativeness of human (dC-dA)n·(dG-dT)n polymorphisms. *Genomics* 7: 524–530.

Weber, J. L., and May, P. E. (1989). Abundant class of human DNA polymorphisms which can be typed using the polymerase chain reaction. *Am. J. Hum. Genet.* 44: 388–396.

Weber, J. L., Wang, Z., Hansen, K., Stephenson, M., Kappel, C., Salzman, S., Wilkie, P. J., Keats, B., Dracopoli, N. C., Brandriff, B. F., and Olsen, A. S. (1993). Evidence for human meiotic recombination interference obtained through construction of a short tandem repeat-polymorphism linkage map of chromosome 19. *Am. J. Hum. Genet.* 53: 1079–1095.

Weinberg, R. A. (1991). Tumor suppressor genes. *Science* 254: 1138–1146.

Weiss, R. (1991). Hot prospect for new gene amplifier: How LCR works. *Science* 254: 1292–1293.

Welsh, J., and McClelland, M. (1990). Fingerprinting genomes using PCR with arbitrary primers. *Nucl. Acids Res.* 18: 7213–7218.

Welsh, J., Peterson, C., and McClelland, M. (1991). Polymorphisms generated by arbitrarily primed PCR in the mouse: applications to strain identification and genetic mapping. *Nucl. Acids Res.* 19: 303–306.

West, J. D., Frels, W. I., Papaioannou, V. E., Karr, J. P., and Chapman, V. M. (1977). Development of interspecific hybrids of *Mus. J. Embryol. exp. Morph.* 41: 233–243.

Whittingham, D. G., and Wood, M. J. (1983). Reproductive physiology. In *The Mouse in Biomedical Research,* Vol. 3, Foster, H. L., Small, J. D., and Fox, J. G., eds. (Academic Press, NY), pp. 137-164.

Williams, J. G. K., Kubelik, A. R., Livak, K. J., Rafalski, J. A., and Tingey, S. V. (1990). DNA polymorphisms amplified by arbitrary primers are useful as genetic markes. *Nucl. Acids Res.* 18: 6531-6535.

Wills, C. (1992). *Exons, Introns, and Talking Genes: The Science Behind the Human Genome Project.* (Basic Books, New York).

Wilson, A. C., Ochman, H., and Prager, E. M. (1987). Molecular scale for evolution. *Trends Genet.* 3: 241–247.

Wimer, R. E., and Fuller, J. L. (1966). Patterns of behavior. In *Biology of the Laboratory Mouse,* Green, E. L., ed. (McGraw-Hill, New York), pp. 629–653.

Winking, H., and Silver, L. M. (1984). Characterization of a recombinant mouse t haplotype that expresses a dominant maternal effect. *Genetics* 108: 1013–1020.

Womack, J. E. (1979). Single gene differences controlling enzyme properties in the mouse. *Genetics* 92: s5-s12.

Wood, S. A., Pascoe, W. S., Schmidt, C., Kemler, R., Evans, M. J., and Allen, N. D. (1993). Simple and efficient production of embryonic stem cell—embryo chimeras by coculture. *Proc. Nat. Acad. Sci. USA* 90: 4582–4585.

Woodward, S. R., Sudweeks, J., and Teuscher, C. (1992). Random sequence oligonucleotide primers detect polymorphic DNA products which segregate in inbred strains of mice. *Mammal. Genome* 3: 73–78.

Woychik, R. P., Wassom, J. S., and Kingsbury, D. (1993). TBASE: A computerized database for transgenic animals and targeted mutations. *Nature* 363: 375–376.

Wright, S. (1952). The genetics of quantitative variability. In *Quantitative Genetics*, Reeve, E. C. R. and Waddington, C. H., eds. (HMSO, London), pp. 5–41.

Wright, S. (1966). Mendel's ratios. In *The Origin of Genetics: A Mendel Source Book*, Stern, C., and Sherwood, E. R., eds. (W.H. Freeman, San Francisco), pp. 92–99.

Yamazaki, K., Beauchamp, G. K., Matsuzaki, O., Kupniewski, D., Bard, J., Thomas, L., and Boyse, E. A. (1986). Influence of a genetic difference confined to mutation of H-2K on the incidence of pregnancy block in mice. *Proc. Nat. Acad. Sci.* 83: 740–741.

Yoe, C., and Morra-Yoe, J. (1991). The Art of Mickey Mouse. (Hyperion, New York).

Yonekawa, H., Moriwaki, K., Gotoh, O., Watanabe, J., Hayashi, J.-I., Miyashita, N., Petras, M. L., and Tagashira, Y. (1980). Relationship between laboratory mice and subspecies *mus musculus domesticus* based on restriction endonuclease cleavage patterns of mitochondrial DNA. *Japan. J. Genetics* 55: 289–296.

Yonekawa, H., Moriwaki, K., Gotoh, O., Hayashi, J.-I., Watanabe, J., Miyashita, N., Petras, M. L., and Tagashira, Y. (1981). Evolutionary relationships among five subspecies of *Mus musculus* based on restriction enzyme cleavage patterns of mitochondrial DNA. *Genetics* 98: 801–816.

Yonekawa, H., Moriwaki, K., Gotoh, O., Miyashita, N., Matsushima, Y., Shi, L., Cho, W. S., Zhen, X.-L., and Tagashira, Y. (1988). Hybrid origin of Japanese mice "Mus musculus molossinus": evidence from restriction analysis of mitochondrial DNA. *Mol. Biol. Evol.* 5: 63–78.

Zhang, L., Cui, X., Schmitt, K., Hubert, R., Navidi, W., and Arnheim, N. (1992). Whole genome amplification from a single cell: implications for genetic analysis. *Proc. Nat. Acad. Sci. USA* 89: 5847–5851.

Zhang, Y., and Tycko, B. (1992). Monoallelic expression of the human H19 gene. *Nature Genet.* 1: 40–44.

Zimmerer, E. J., and Passmore, H. C. (1991). Structural and genetic properties of the Eb recombinational hotspot in the mouse. *Immunogenetics* 33: 132–140.

Zoghbi, H. Y., and Chinault, A. C. (1994). Generation of YAC contigs by walking. In *YAC Libraries, A User's Guide*, Nelson, D. L. and Brownstein, B. H., eds. (W.H. Freeman and Company, New York), pp. 93–112.

Zurcher, C., van Zwieten, M. J., Solleveld, H. A., and Hollander, C. F. (1982). Aging Research. In *The Mouse in Biomedical Research*, Vol. 4, Foster, H. L., Small, J. D., and Fox, J. G., eds. (Academic Press, New York), pp. 11–35

# Index